THE
MANAGEMENT AND DISEASES
OF
SHEEP

Papers presented at a
British Council Special Course
Edinburgh
5 - 17 March 1978

Published by
The British Council
10 Spring Gardens
London SW1A 2BN

and

The Commonwealth Agricultural Bureaux
Farnham House
Farnham Royal
Slough SL2 3BN

ISBN 0 85198 451 7

JM1979

CONTENTS

Acknowledgements

Introduction

BREEDING

Breeding for improvement	Professor J.C. Bowman	1
Breed substitution and cross-breeding as a means of increasing productivity	Dr G. Wiener	13
Genetics and disease	Dr A.G. Dickinson	32

PHYSIOLOGY AND REPRODUCTION

The immune response in sheep	Dr P.W. Wells	58
The physiology of reproduction	Dr W. Haresign	66
Embryo transplantation in sheep	Dr S.M. Willadsen	69
Disease and reproduction in sheep	Dr W.A. Watson	86
Response of sheep to cold exposure in relation to selection for survival ability	Dr J. Slee	100
Genetic and physiological variation in reproductive performance	Dr R.B. Land	114
Factors affecting puberty and mating behaviour in sheep	J.L. Lees	124
Nutrition and the reproductive function in female sheep	Dr J.M. Doney	152
The physiology of pregnancy	Dr D.J. Mellor	161

NUTRITION

Energy metabolism	Professor A.J.F. Webster	166
Mineral metabolism	Dr A.C. Field	175
Protein metabolism	Dr J.C. MacRae	183
Utilisation of roughages	Dr J.F.D. Greenhalgh	201

Utilisation of cereals for sheep and implications of processing	Dr E.R. Ørskov	213
The nutrition of the pregnant ewe	Dr A.J.F. Russel	221
The nutrition of the lactating ewe	Dr T.T. Treacher	242
DISEASES OF EWES		
Management and disease control in lambing ewes	B. Mitchell	257
Scrapie disease of sheep	Dr J.T. Stamp	265
Non-parasitic respiratory disease of sheep	Dr W.B. Martin	277
NUTRITION, GROWTH AND DISEASES OF LAMBS		
Artificial rearing	Dr P.D. Penning	282
Diseases of lambs	Dr J.A. Watt	297
Diarrhoea in neonatal lambs	Dr D.R. Snodgrass	304
GRASSLAND UTILISATION		
Utilisation of grassland for sheep production	Dr J. Hodgson	307
Conserved forages in sheep production	Dr R.J. Wilkins	324
PARASITISM		
Recent advances in the epidemiology of sheep endoparasites	Professor J. Armour	339
Effects of parasitism on host metabolism	Dr A.R. Sykes and Dr R.L. Coop	345
GROWTH AND CARCASS QUALITY		
Growth and development of lambs	Dr J.H.D. Prescott	358
Sheep carcass and eating quality	A. Cuthbertson and A.J. Kempster	377
SYSTEMS DEVELOPMENT AND ANALYSES		
Economic techniques for the assessment of sheep systems	Dr T.J. Maxwell	400

Intensive systems on lowland grass	J.E. Newton	421
Intensive systems	Dr J.J. Robinson	431
INVITED PAPERS		
The question of purpose in sheep production	Professor H.P. Donald	447
Wool production in semi-arid areas with special reference to African conditions	Dr F. de K. Kotze	450
The breeding season of Iceland sheep	Dr O.R. Dýrmundsson	458
Gestational failure: a major problem in reproductive efficiency of the angora goat breed in South Africa	P.S. Pretorius	463

ACKNOWLEDGMENTS

Permission to reprint material on the pages indicated was kindly granted by the following.

Academic Press Inc (London) Ltd: pages 24, 25, 26.

Agricultural Research Council: pages 183 to 200. Dr MacRae's paper also appears in the ARC Seminar on 'Ruminant Digestion and Feed Evaluation', edited by D.E. Osbourn, D.E. Beever and D.J. Thomson; ARC, London, 1978.

American Dairy Science Association: page 9.

CSIRO Editorial and Publications Service, Australia: pages 107, 288, 312, 314.

Cambridge University Press: pages 3,4,27,131,133,135,139,207,216,217,243, 317, 318, 321, 382, 383, 433.

Churchill Livingstone: page 104.

Institut National de la Recherche Agronomique, France: pages 436,437.

Longman Group Ltd: pages 2,16,17,18,20,22,202,209,215,291,359,381,384, 387,389,441.

Ministry of Agriculture, Fisheries and Food: pages 86 to 99.

New Zealand Journal of Agricultural Research: pages 361,362,364.

The Veterinary Record: page 14.

INTRODUCTION

The range and amount of research undertaken on sheep in the United Kingdom is very comprehensive. Much of this work is centred in the Edinburgh area which is a natural focus for a course on sheep husbandry and diseases since three research institutes and the University of Edinburgh all have a considerable component of work on sheep in their research programmes. In constructing this course we received a ready response from scientists in many disciplines and so were able to devise a wide range of lectures. The papers presented, because of the nature of the material and the rapid advance in science and change in scientific knowledge, should not be considered as providing a detailed analysis of each subject. None the less, they represent a most valuable summary of much of what is known. The papers have received only minimal editorial change and should be regarded as lecture notes rather than as a comprehensive review of particular topics. We believe that these papers will be a valuable reference for those who attended the course and of interest to students, teachers and researchers on sheep.

We wish to express our gratitude to all those who participated for the excellence of their contribution.

J.M.M. CUNNINGHAM

J.T. STAMP

W.B. MARTIN

BREEDING FOR IMPROVEMENT

PROFESSOR J.C. BOWMAN

Department of Agriculture and Horticulture, University of Reading,
Earley Gate, Reading

INTRODUCTION

The sheep is farmed for the production of three commodities; meat, wool and milk, the latter being the least important. (Table 1) With the development of artificial fibres in the past 30 years, wool has declined in importance but the future relative costs of petroleum as the production base of artificial fibres will probably ensure a revival of its economic importance. Sheep meat and wool are likely to be the main products in future.

Table 1 World production of sheep products

Area	No. slaughtered ('000) 1976	Milk ('000 MT) 1976	Wool (greasy) ('000 MT) 1976
Africa	47,383	532	193
N.C. America	8,628	-	63
S. America	20,054	31	194
Asia	91,853	3,523	285
Europe	70,132	1,855	261
Oceania	66,330	-	1,066
USSR	60,000	100	431
World	364,380	7,503	2,590

Source: FAO Production Yearbook, 1976

OBJECTIVES

Having determined the main commodities to be produced, it is necessary to determine the essential combination of biological characteristics of the sheep to be used in the production systems. The specific circumstances which determine these characteristics are:

(i) The market requirements - variation in demand for different carcass weights and qualities, wool qualities and sheep milk for processing.
(ii) The production location - particularly land, topography, climate and latitude.
(iii) The nutritional regimen - of extensive grazing versus intensive concentrate and intermediate systems - byproduct utilisation.
(iv) The relative economic costs of production inputs - labour, feed,

housing and equipment, veterinary services and supplies.

(v) Government financial support programmes based on social and land use needs.

The genetically determined biological characteristics of the sheep which may be combined in different ways include:

(i) Reproduction - ram libido and fertility, ewe fertility and litter size, seasonality of reproduction and age of sexual maturity.
(ii) Fleece weight and quality - this character is particularly related to characteristic (vi) and may in part alter nutritional requirements.
(iii) Growth rate and mature size - probably related to stocking rate.
(iv) Carcass quality - proportions of bone, lean and fat, proportions of different butchers' joints and area of longissimus dorsi muscle.
(v) Feed conversion - includes stocking rate probably related to mature size.
(vi) Ability to survive and produce in different, sometimes extreme, climatic conditions.
(vii) Disease resistance - affects replacement rate and longevity.
(viii) Milk production - for processing and for lamb growth.

It is difficult to be explicit in economic terms about the relative importance of these traits without considering specific situations but, in terms of two measures of biological efficiency and with regard to sheep meat production, the work of Large (1970, 1973) is relevant. The biological efficiency of meat production (E) is defined as the weight of carcass produced by 100 units of digestible organic matter (DOM) consumed or as the ratio of N output to N input. The effects of frequency of lambing, litter size, size of ewe and replacement and mortality rates are shown in Table 2 and Figures 1 and 2.

Table 2 The effect of litter size and frequency of lambing on E for Suffolk x Scottish halfbreds

Frequency of lambing/year	Litter Size 1	2	3	4	5
0.5	2.2	3.9	5.3	6.5	7.4
1.0	3.9	6.4	8.2	9.5	10.5
1.5	5.3	8.2	10.0	11.3	12.3
2.0	6.5	9.5	11.2	12.5	13.3

$$E = \text{mean } \frac{\text{carcass wt. of each lamb x no. of lambs}}{\text{DOM eaten by ewes + DOM eaten by lambs}} \times 100$$

Source: Table 1 of Large (1970)

Figure 1

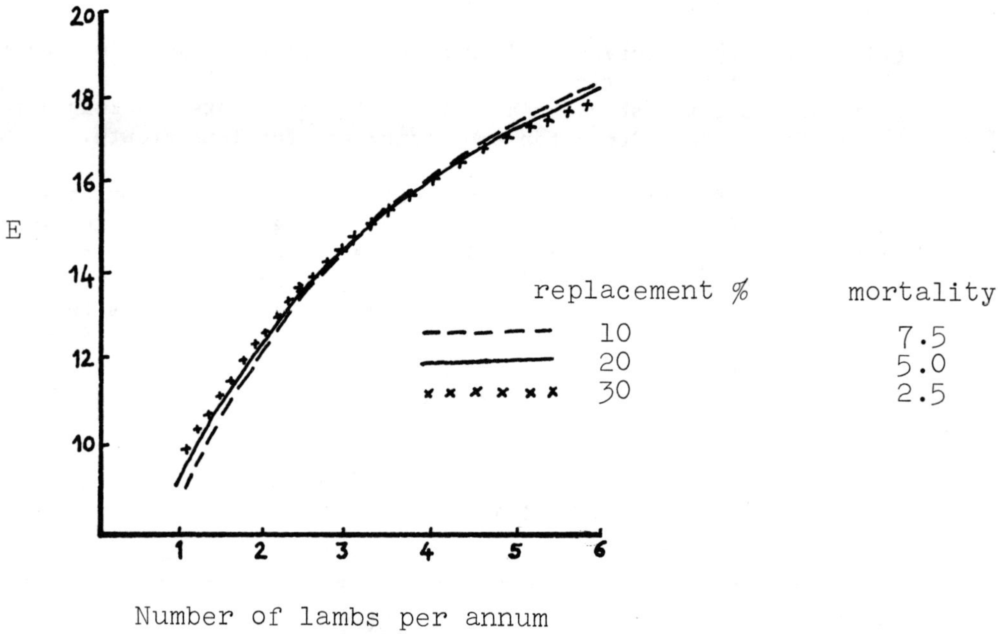

$$E = \frac{N \ \text{output}}{N \ \text{input}} \times 100/\text{annum}$$

Fig. 1. Population efficiency (replacement rate) (from Large 1973).

Figure 2

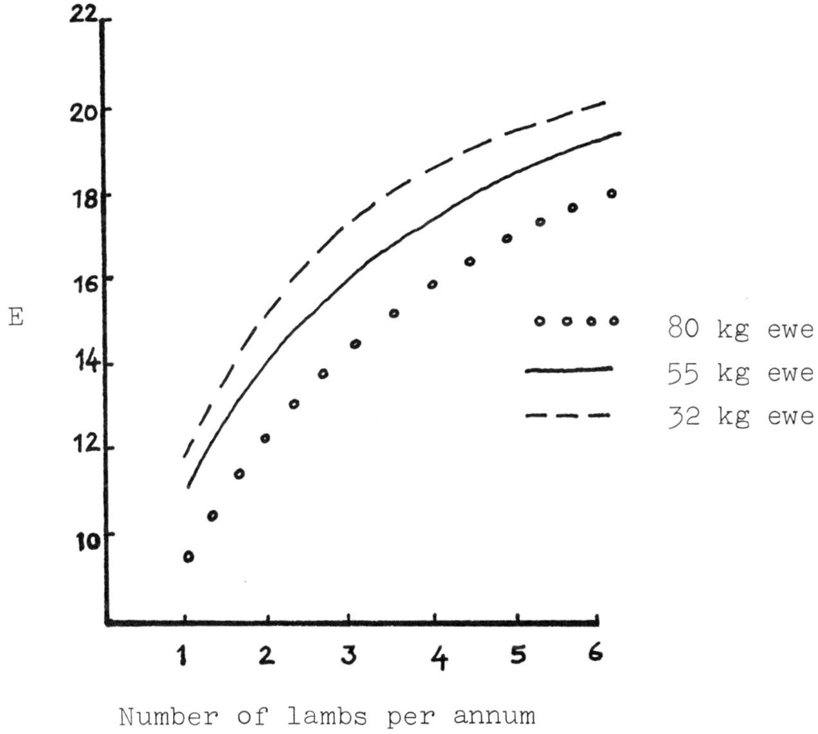

$$E = \frac{N\ output}{N\ input} \times 100/annum$$

Fig.2. Population efficiency (size of ewe) (from Large 1973).

The conclusions from these calculations by Large are that (I) efficiency is hardly affected by variation in mortality/replacement rates (II) efficiency is markedly affected by litter size and frequency of lambing and (III) the highest efficiency is obtained by mating large rams (100 kg) to small ewes with a high litter size. Similar conclusions, based on less calculation, were reached independently by Dickerson (1970). It is appreciated that the economically optimum combination of traits may differ from the optimum combination for biological efficiency. A major consideration facing the sheep producer, and therefore determining the sheep breeder's choice of objectives, is the type and level of nutrition to be used. The question is whether the sheep should be used principally as a ruminant, making use of fibrous feeds of relatively low digestibility and nutritive value or become a mixed ruminant/non-ruminant using a combination of roughage feeds, cereals and high protein feeds (essential to the non-ruminant). If the sheep is to stay as a ruminant throughout its productive life, the level of reproduction attainable for this system of feeding has been reached. If the sheep can justify economically the feeding of cereals, high protein feeds and milk replacers, the breeder has a substantial task ahead to develop strains which can produce litters of 6 to 8 viable lambs at regular intervals (every 8 months) from one year of age. Almost certainly both types of sheep will be required.

AVAILABLE BIOLOGICAL RESOURCES

The world contains many breeds which represent considerable variation in the traits referred to above, not only between breeds but between individuals within breeds. To date individuals with extreme performance have not received the attention they merit, probably because methods suitable for multiplying these individuals in large numbers have not been available. Initially, a marked improvement in genetic potential could be obtained by breed substitution both in production from purebreeds and for crossbreeds. This requires more information on breed performance throughout the world and a more flexible approach to the movement of animals across national boundaries. A summary of breeds available in different parts of the world, together with the performance they achieve in their own ecological niche, i.e. on a non-comparative basis, would be a useful addition to our literature.

GENETIC RESOURCES FOR PRODUCTION

There are several alternative breeding schemes which could be adopted for the development of sheep for commercial production purposes. These include:

(i) Purebreeding
(ii) Crossbreeding - 2 way cross, 3 way cross, 4 way cross
(iii) Inbreeding and crossbreeding
(iv) Crisscrossing and rotational crossbreedig
(v) New breed development

The relative merits of these schemes in general terms have been discussed by Dickerson (1969). Specifically with reference to sheep, inbreeding has been found to have deleterious effects (Bowman 1966, Turner 1969), and to date has been considered to be too expensive to be pursued. Reproductive rates which are low may also have been a major factor in deterring the development of inbred lines of sheep. The choice at present lies essentially between the use of purebreds or crossbreds.

The advantages of purebreds over crossbreds are:

(i) The design of selection programme can be relatively simple. Animals at all levels of the population can be chosen as parents of improved stock, thus increasing the possible selection intensity.
(ii) All animals are of a type suitable for commercial production; thus 'byproduct' animals which could spoil the market for the prime product are not produced.
(iii) Each producer can breed his own replacements or need only buy males, thus reducing disease risks and avoiding possible 'mongrelisation'.

The advantages of crossbreds over purebreds (see Figures 3 and 4) are:

(i) Increased performance from hybrid vigour.
(ii) The performance of the commercial animal can be changed quite quickly by substituting males of other breeds for male lines in the breeding scheme.
(iii) The separate breeds and crosses may be capable of using different environmental niches, thus making use of different land/farm resources.

Crisscrossing and rotational crossbreeding have not been used with sheep and appear to offer no positive advantages over more simple forms of crossbreedig.

IMPROVING THE GENETIC RESOURCES

Irrespective of whether purebreeds or crossbreeds are used for commercial production, improvement of genetic potential is dependent on selection within purebred populations. The possibilities of using different selection criteria in each of several purebred lines as a means of maximising improvement in crossbred progeny has been discussed in theoretical terms by Moav (1966a, 1966b, 1966c), Moav and Hill (1966) & Dickerson (1970 and 1973 (Table 3).

More sophisticated forms of selection for improving crossbred performance, such as recurrent or reciprocal recurrent selection, have not shown any consistent advantage over purebreed selection when tried with other species to warrant using them extensively with sheep.

The major impediments to selection in purebred populations are:

(a) A lack of records: this may be a result of farmer education, inclination, husbandry/management systems, economic disincentives and no central organisation for supervision and data analysis.
(b) Small population size and population structure (individuality of flocks).
(c) Low reproductive rate: selection in most populations is essentially a choice between males.
(d) Poor dissemination of improved stock.

Figure 3

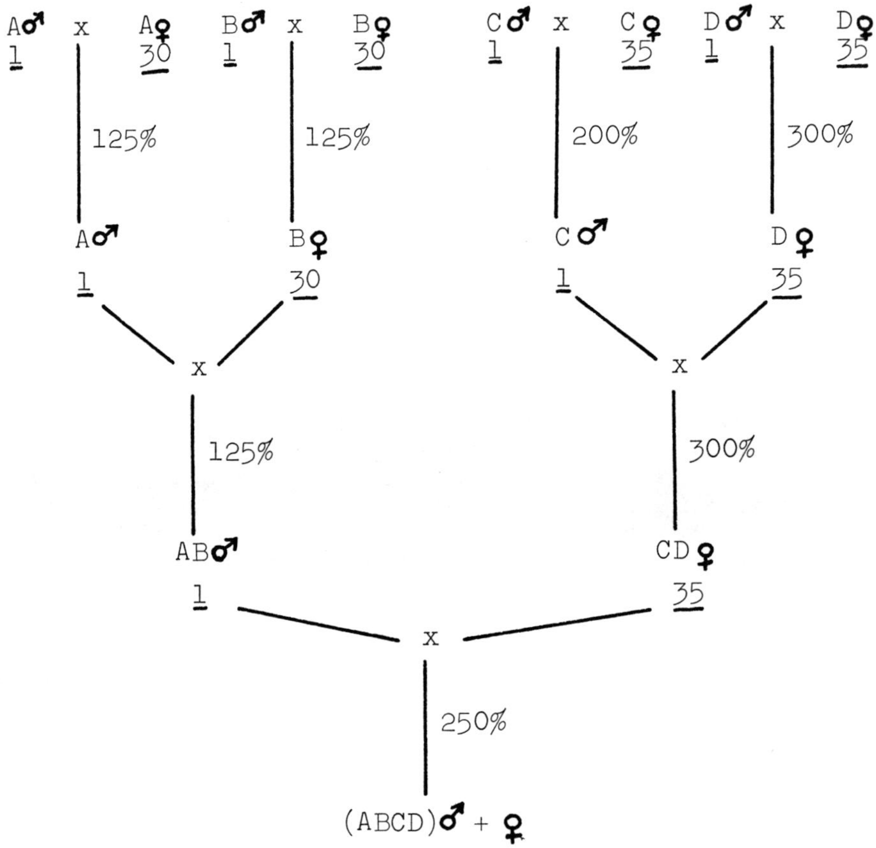

1	mating rate
%	effective lambing rate
A	genotype

Fig.3. Production of a four breed cross in sheep

Figure 4

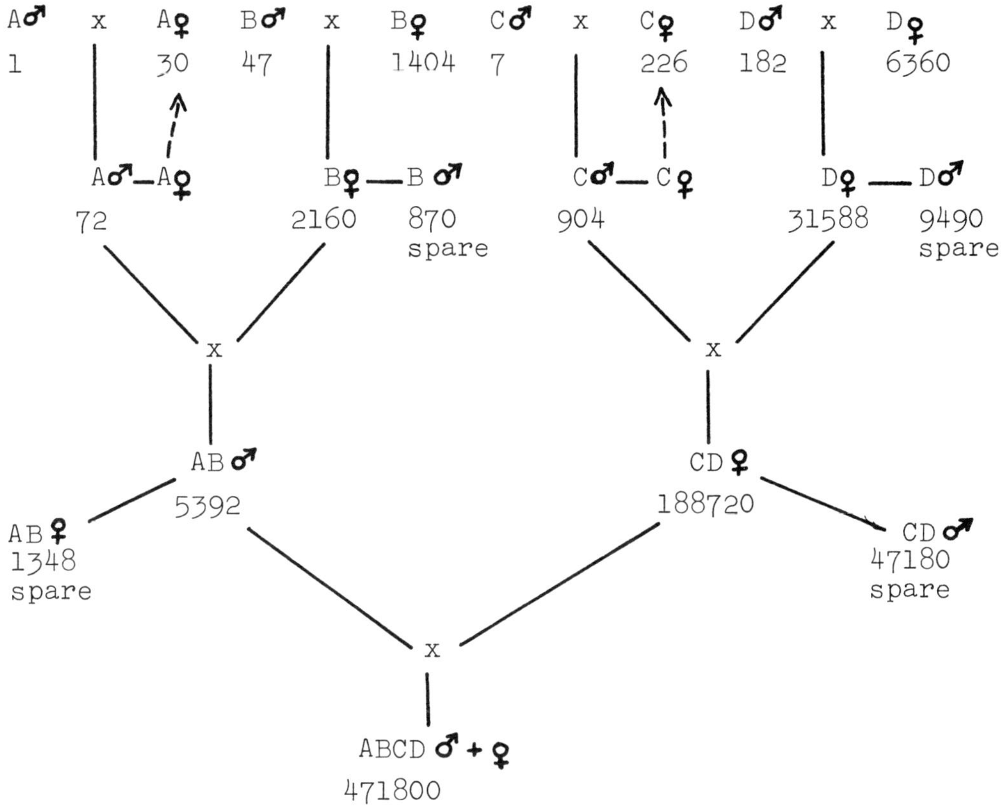

Sheep numbers - assuming sheep life of 4 years

Males	6605	Commercial	471800
Females	230488	Spares	58888

Fig. 4. Population numbers - production of a four breed cross sheep.

Table 3 Assumptions used in comparing economic efficiency for several types of breed utilisation in market lamb production

Breed	Lambs/ewe exposed %	Market growth	Lambs cutability	Relative value	Ewe costs
Sire line A	120	100	100	100	100
Sire line B	120	80	100	80	92
Dam line C	120	88	97	85	98
Dam line D	220	70	92	64	111

Relative efficiencies of alternative breeding plans

Line C ewe equivalent costs/1000 line A x C equivalent lambs

Breeding plan	Sire lines	Dam lines	Commerical
A♂ x C♀	100	100	100
A♂ (D♂ x C♀)♀	-31.3	-33.2	-21.5
A♂ x (B♂ x D♀)♀	-28	-66	-20
A♂ x (SYN B+D)♀	-22	-57	-16
A♂ x (SYN 2D, 1B, 1C)♀	-24	-59	-17
A♂ x (SYN 3D, 1B)♀	-27	-64	-19

Source: Adapted from Dickerson (1973)

These difficulties can be overcome in various ways, e.g. by recording schemes in Ireland (Timon 1971) and the UK (MLC 1972) and by cooperative schemes such as the use of ram circles in Norway (Gjedrem 1971) or nucleus flocks in Australasia (Hight & Rae 1970, Jackson & Turner 1972). Aspects of the design of nuclear breeding schemes have been studied by James (1977). I do not consider that the development of central performance or progeny testing stations is necessary, desirable or financially justified.

Reviews of the possibilities for genetically improving sheep will be found for meat in Bowman (1966, 1968), for reproduction in Turner (1969) and for meat, wool and reproduction in Turner (1977) and Turner & Young (1969).

THE FUTURE

An essential future development is to find ways of multiplying outstanding selected individuals. The increase of litter size alone would allow a marked increase in selection intensities. Special research emphasis should be placed on reliable and repeatable superovulation and simple ova transfer techniques. These developments would enable outstanding individuals to be multiplied and allow the development of inbred lines from these individuals (Figure 5).

Methods of semen dilution are probably of more importance than a method for semen storage. Artificial insemination is important for selection programmes but is much less important for commercial production than in cattle. Conception rates and litter sizes as a result of AI are not likely to be as good for flock mating.

The possibilities of using characters of general physiological significance, such as hormone levels and selection, have been suggested (Land 1973). The advantages are stated to be that these traits might be measured early in the life of the animal and in both sexes. At present and for the next few years I consider that these traits will not be used as selection criteria, because not enough is known about the causes of variation and the interrelationships with performance characters. We need to learn much more about genetic control of these physiological traits.

Figure 5

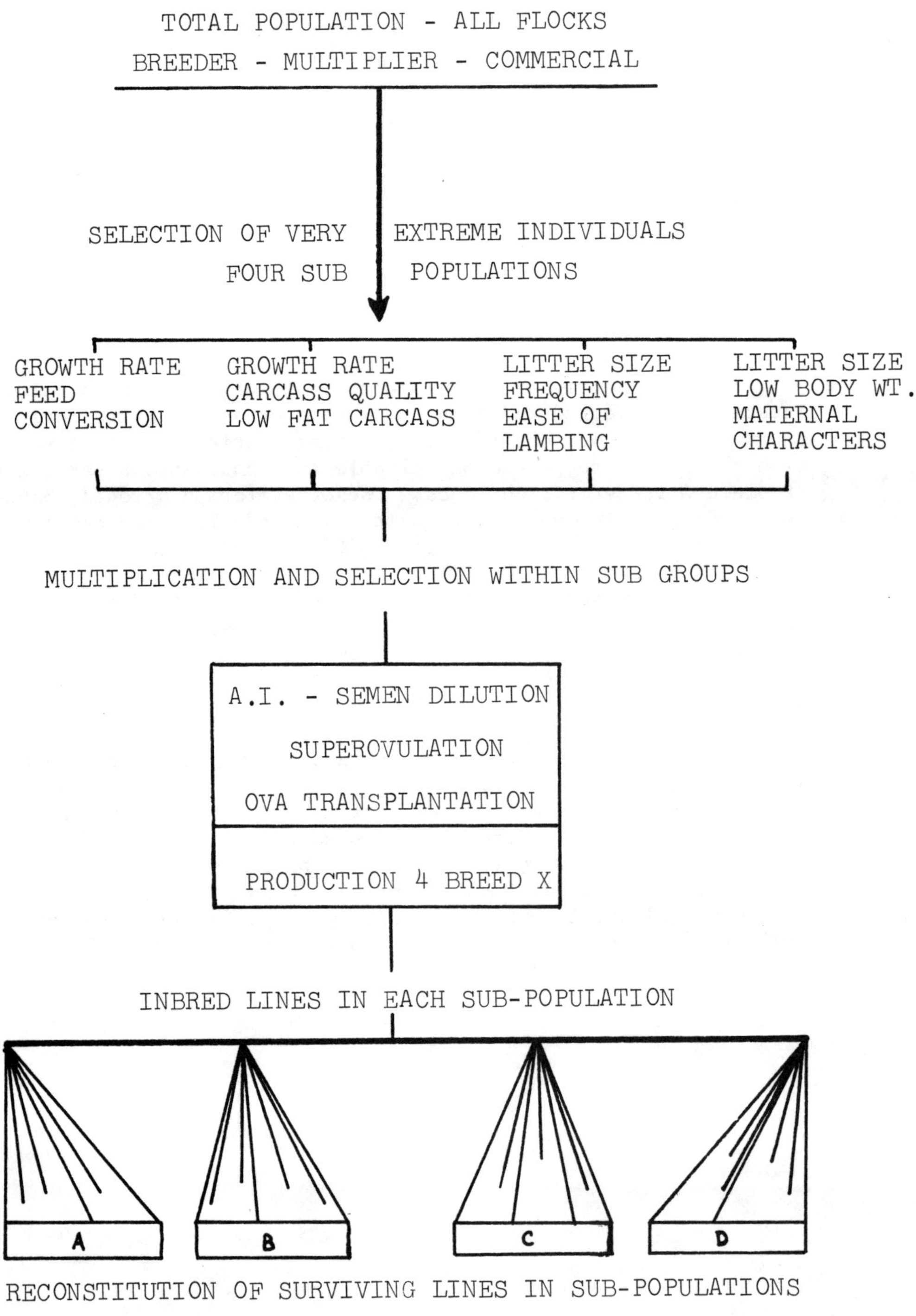

Fig. 5. Sheep breeding programme

REFERENCES

Bowman, J.C. (1966) Meat from sheep. Animal Breeding Abstracts 34,293-319.

Bowman, J.C. (1968) Genetic variation of body weight in sheep. In: Growth and development of mammals, Proc. 14th Easter School in Agricultural Science, University of Nottingham, 1967, Butterworths, London, 291-308.

Dickerson, G.E. (1969) Experimental approaches in utilising breed resources. Animal Breeding Abstracts 37,191-202.

Dickerson, G.E. (1970) Efficiency of animal production - molding the biological components. Journal of Animal Science 30,849-859.

Dickerson, G.E. (1973) Inbreeding and heterosis in animals. Proc. Animal Breeding and Genetics Symposium in honour of Dr. Jay L. Lush, American Society Animal Science and American Dairy Science Association 54-77.

Gjedrem, T. (1971) Practical breeding methods for improving sheep prolificacy, weight gains and meat yield. In: Lamb Production, US Feed Grains Council 63-68.

Hight, G.K.;Rae, A.L. (1970) Large scale sheep breeding: its development and possibilities. Sheep Farming Annual 73-85.

Jackson, N.;Turner, H.N. (1972) Optimal structure for a co-operative nucleus breeding system. Proceedings, Australian Society for Animal Production 9,55-64.

James, J.W. (1977) Open nucleus breeding systems. Animal Production 24,287-305.

Land, R.B. (1973) Genetic improvement of fertility. Annual Report. ARC Animal Breeding Research Organisation 7-12.

Large, R.V. (1970) The biological efficiency of meat production in sheep. Animal Production 12,393-401.

Large, R.V. (1973) Factors affecting the efficiency of protein production populations of animals, Chap.12, 183-199. In: The Biological Efficiency of Protein Production, Ed. J.G.W. Jones, Cambridge University Press.

Turner, H.N. (1969) Genetic improvement of reproduction rate in sheep. Animal Breeding Abstracts 37,545-563.

Turner, H.N. (1977) Australian sheep breeding research. Animal Breeding Abstracts 45,9-31.

Turner, H.N.;Young, S.S.Y. (1969) Quantitative genetics in sheep breeding. South Melbourne, Victoria, Australia; Macmillan Co. of Australia Pty. Ltd.

Timon, V.M. (1971) Improved breeding and production methods for intensification of Irish sheep production. In: Lamb Production, US Feed Grains Council 69-78.

BREED SUBSTITUTION AND CROSSBREEDING AS A MEANS OF INCREASING PRODUCTIVITY

DR G. WIENER

A.R.C. Animal Breeding Research Organisation, West Mains Road, Edinburgh

Opportunities for improving the efficiency of sheep production are provided by the existence of different breeds. In any situation, whether it be in the context of individual flocks, of cooperative schemes or of national improvement programmes, it is worthwhile to examine whether the indigenous or currently-used breeds are likely to give the best return in relation to existing nutritional and environmental circumstances, or whether the introduction of new breeds might provide some benefits. Where improvements in nutrition, health and management are being made or contemplated, it is necessary to ask whether the responses obtained from the existing sheep population might be bettered by other breeds or combinations of breeds.

I intend (i) to illustrate the pre-requisites for making valid comparisons among breeds, (ii) to look at the evidence for the existence of genetic differences among breeds in the main components of productivity, (iii) to examine the consequences of crossbreeding, and (iv) to discuss briefly the factors which need to be considered in the choice between alternative breeding strategies involving the use of different breeds. No attempt will be made to catalogue the multitude of different breeds and their attributes.

ESSENTIALS FOR BREED COMPARISONS

The difficulty in assessing the genetic merit of different breeds arises from the fact that they are only rarely kept together under the same conditions. The point is illustrated in Table 1. The left-hand side shows the average performance of three breeds when they were run together as a single flock on a grassland farm in Peeblesshire. The right-hand side shows the same three breeds but each on a different hill farm. As expected, average body weight, fleece weight and lamb production were lower for each breed under the hill conditions, but the important point is that comparison of the three breeds when kept separately would lead to a different impression of their relative merit than would comparison of the three when kept under the same conditions. The right-hand side would lead one to suppose that the Cheviots were the heaviest, with the heaviest fleeces and that the Welsh sheep produced more lambs per ewe than either of the other two. But because each breed was kept, as is commonly the case, in a separate flock with different management and a different environment, it is impossible to decide what is due to the effects of genes and what to the effects of environment. When the three breeds were kept together in one flock the differences can validly be attributed to the capabilities of the breeds kept in these conditions.

Table 1 Average Production of Female Sheep of Three Hill Breeds[1]

Breed	Grassland Farm[2]			Hill Farm			Source of data
	Mature live-weight (kg)	Fleece weight (kg)[3]	Lambs weaned 100 ewes mated[4]	Mature live-weight (kg)	Fleece weight (kg)[3]	Lambs weaned/ 100 ewes mated[4]	
Scottish Blackface	67	2.6	151	41	1.7	81	ABRO
Cheviot (SC)	67	2.3	137	53	2.1	71	HFRO
Welsh Mountain	47	1.8	134	34	1.1	105	ABRO

[1] Wiener (1973)

[2] Adapted from Wiener (1967)

[3] Average of first four shearings

[4] Average of first three lamb crops

GENOTYPE-ENVIRONMENT INTERACTIONS

It is sometimes claimed that breeds are adapted to particular environments, often in quite limited localities. This must be taken to imply that breed A, for example, does better in a particular locality than does breed B whereas somewhere else breed B might be at an advantage over breed A. If such interactions were common it would make it difficult to generalise from breed comparisons in only one environment. For the relatively limited range of conditions and 'simple' traits for which this matter has been examined experimentally, there is little evidence for such interactions. Given more extreme test conditions such as artificially induced stresses or the large, if ill-defined, differences between temperate northern regions and hot arid or tropical regions, a different story might emerge. Under those circumstances breed comparisons are likely to be most informative if they are made under the conditions under which the animals are to be kept. It should be said however that, while statistically significant interactions might be found, what is often of greater practical importance are changes in ranking. Considerations confined to single sets of attributes such as growth or reproduction may provide a different answer from considerations concerned with combinations of traits or with overall profitability. In the most general terms one might ignore interactions in the first instance for traits of high heritability such as mature body size or fleece quality, but probably not for traits of low heritability like reproductive rate, survival and disease resistance. And what, one might ask, is the heritability of profit?

BREED DIFFERENCES AND CROSSBREEDING

Provided that breeds are compared under the same sets of conditions and the breed groups are large enough and representative, we can assume that most of the differences between them are heritable. It is this presumption wich makes us expect the crossbred to be intermediate in its performance between the levels of the two parental breeds. And it is on this basis that one can consider the bringing together of different desirable attributes from two parent lines. A practical example in this country is the combination of the good maternal performance of hill breeds with the extra size derived from rams of for example "Longwool breeds". Whilst the first expectation is that the crossbred will be intermediate in performance between the levels of the parental breeds, hybrid vigour, a positive deviation from the parental mean, can be expected as a bonus for some traits, particularly those associated with reproduction and survival, and in some breed combinations. These points will now be illustrated with examples from an experiment in which three breeds, the Scottish Blackface, the Cheviot and the Welsh Mountain, were bred pure and also crossed with each other, and additionally with the Lincoln.

Body Weight

Table 2 gives weights at birth, weaning (15 weeks), 72 weeks and 4½ years old for four breeds. The breed values presented were adjusted for differences in maternal environment and in twinning rate. The figures show significant breed variation in weight at all four ages. The lower part of Table 2 shows weights of crosses as deviations from the mean of the parental pure breeds expressed as a percentage of the parental breed mean. Again, the figures were adjusted for variations in maternal environment and for litter size. It can be seen that crossbreds showed no heterosis in birth or weaning weight, though by 72 weeks the crosses with Lincoln deviated significantly and among the hill breeds only the Blackface Cheviot cross showed significant heterosis.

Conformation

Table 3 indicates that breeds can also differ in conformation when at or close to mature size. For each body part the average size of each breed (at 54-72 weeks old) was expressed as a deviation from the mean of all the breeds. On this basis some, like the Lincoln, were clearly above average in all linear dimensions while a breed like the Welsh Mountain was consistently below. To standardise the comparisons of shape among the breeds the deviation of each body part was expressed as a difference from the percentage deviation averaged over all six linear body parts measured for each breed. The breeds can be seen to differ mainly in the relationship of body width to leg length. Thus, Southdowns in particular,

Table 2 Comparative body weights of four breeds of sheep and crosses (Adapted from Wiener & Hayter (1974))

a. Live-weights of Breeds (kg)
(fitted mean values for breeds)

	Birth	15 weeks (weaning)	72 weeks	4¼ years
Blackface	4.6	34.4	57.2	73.1
Cheviot	4.7	33.6	55.7	70.7
Welsh	4.0	27.8	41.5	53.9
Lincoln	5.4	35.9	62.6	(88 approx.)[1]

b. Live-weight of crossbred sheep (reciprocal crosses pooled)
(as deviation (%) from mean of parental pure breeds)

	Birth	15 weeks (weaning)	72 weeks	4¼ years
Blackface x Cheviot	0.9	0.7	3.9	9.4
Blackface x Welsh	-2.4	0.9	1.2	1.5
Cheviot x Welsh	0.4	0.9	-1.5	-3.4
Blackface x Lincoln	-2.1	5.3	9.5	-
Cheviot x Licoln	-6.5	0.7	6.8	-
Welsh x Lincoln	0.2	1.5	8.0	-

[1]from Wiener (1967)
- an unadjusted average weight based on the previous generation of these sheep on the same farm.

but to a lesser extent Cheviot, are seen to be wide and short-legged, while Lincoln, Merino and to some extent Blackface are narrow and long in the leg. In other words, Table 3 shows breed differences in conformation, differences which exist independently of weight; a Southdown is not just a smaller-scale version of a Lincoln.

Table 3 Relative Body Conformation of 6 Breeds Near Maturity (Adapted from Wiener & Hayter (1974)) (expressed for each breed and body part as a deviation (%) from the average size difference of each breed from the mean of all six breeds)

	Blackface	Cheviot	Welsh	Lincoln	Southdown	Merino
Body length	1.1	0.2	0.4	-1.9	0.4	-0.2
Shoulder width	-4.4	1.3	-0.4	-4.3	10.6	-2.9
Hook width	-1.6	4.9	2.0	-3.2	3.9	-6.0
Head width	1.8	1.0	-2.5	-2.3	3.2	-1.1
Tibia length	2.9	-1.2	1.6	1.7	-8.4	3.3
Cannon length	0.2	-6.2	-1.1	9.9	-9.7	6.8

Reproductive performance

It is convenient to look at some of its components and to see what evidence there is for breed variation and crossbred advantage. Table 4 shows three of these components; the proportion of ewes lambing per ewe mated, the number of lambs born per ewe lambing and the lamb survival per lamb born. For the first of these, the differences between the breeds and crosses was not significant in this particular experiment, nor is there much evidence that it is so for a whole range of breeds. Although barrenness can be influenced by a number of factors, it is not generally a trait which breeding or breed substitution can do a great deal to change. The number of lambs born per ewe lambing on the other hand shows significant variation even for the breeds involved here. When we also consider breeds such as the Finnish Landrace, the significance of breed differences for this characteristic is quite marked. There is moreover an indication for this trait of some positive advantage from crossbreeding, although, as shown in Table 4, only one of the crosses is significantly better than mid-parental level. For the third component, lamb survival, there is gain some significant variation among the breeds and crosses and in this case there is clear indication that cross-bred lambs born to crossbred mothers have a 5-7% better survival than expected from the average of the parental breeds involved; put another way, the mortality of the crossbred lambs was about one third lower than that of the purebreds.

Table 4 Maternal performance of breeds and crosses
(Adapted from Wiener & Hayter (1975))

a. Maternal Performance of Breeds
(fitted mean values for breed)

	% ewes lambing per ewe mated	No. lambs born per ewe lambing	% lamb surival per lamb born
Blackface	97	1.70	82
Cheviot	94	1.45	76
Welsh	96	1.37	90

b. Maternal Performance of Crossbred sheep (reciprocal crosses pooled)
(as deviation (%) from mean of parental pure breeds)

	% ewes lambing per ewe mated		No. lambs born per ewe lambing		% lamb survival per lamb born	
	F_1 mate	F_2 mate	F_1 mate	F_2 mate	F_1 mate	F_2 mate
Blackface x Cheviot	-0.8	4.3	9.6	3.1	7.5	15.3
Blackface x Welsh	-0.7	6.4	-0.1	17.0	4.1	40.8
Cheviot x Welsh	-1.1	8.3	2.3	15.8	5.7	7.9

Table 4b also shows, in respect of the crosses, the crossbred advantage when instead of ewes being mated to their F_1 rams, they were mated to their F_2 sons, to produce inbred progeny. Under these circumstances the advantage of the crossbreds was very substantially greater although in this case the level of performance was lower than the figures shown in Table 4a. This information on crossbred advantage, when inbred rather than non-inbred progeny are produced, shows that the crossbred advantage, the heterosis, depends quite markedly on the genetic background of the population against which it is measured. Moreover the comparison shows the crossbred advantage to be greater when, in one sense at least, the pure breeds are more homozygous or, in other words, more "pure".

Table 5 shows total litter weights at weaning per ewe mated and per ewe lambing. For both these traits there was both breed variation and crossbred advantage, but as Table 5 shows, some crosses, in this case the Blackface-Cheviot crosses, showed much greater advantage relative to the pure breeds than did the other two crosses. In other words, while one can claim a general advantage from crossbreeding for litter weight at weaning, there are also sufficient differences between breeds in the way in which they combine to make it worthwhile to find the most favourable combinations.

In the experiment from which these data are drawn females were kept up to a maximum of four lamb crops. No sheep were deliberately culled but some died. The total costs of production of the flock are affected by the number of ewes which have to be replaced, and the total lifetime production of a group of sheep depends not only on what the sheep produce when they are living but also on what they will not contribute if they die. The evidence however suggests that although there were differences among the breeds and crosses in ewe mortality, these were not distinguishable from random effects. Like barrenness, it does not look as though breeding would, at least on an individual flock basis, affect this part of total reproductive performance.

Total productivity and economic considerations

Tables 6 and 7 deal with the performance of a different series of breeds. In this case rams of the five breeds shown in the Tables were mated to Scottish Blackface hill ewes. The crossbred female offspring as adults were mated in turn to the same breed or breeds of ram. Table 6 shows the lamb production of these crossbred ewes both in terms of lambs weaned per ewe mated and total litter weight at weaning. In their first year the performance of the crossbred ewes of the first three types shown did not differ from each other very much; the Finnish Landrace was slightly better but the Merino very much worse, largely, in this case, because so few of them had any lambs at all. In the second and third crop all the groups show a marked improvement as would be expected and the Finnish Landrace crossbred ewes gave about half a lamb more than the average and the Merinos half a lamb less than average. The last column shows however that though the numbers born are an important component, the advantage of the Finnish Landrace in the crosses is no longer as great in terms of total litter weight, although the ranking of the five types is precisely the same as that for litter size.

Table 5 Litter weights at weaning of different breeds (Adapted from Wiener & Hayter (1975))

a. Maternal Performance of Breeds

	Litter weight at weaning (kg)	
	per ewe mated	Per ewe lambing
Blackface	46.6	48.0
Cheviot	34.3	36.5
Welsh	31.1	32.4

b. Maternal Performance of Crossbred sheep (reciprocal crosses pooled)
(as deviation (%) from mean of parental pure breeds)

	Litter weight at weaning (kg)	
	per ewe mated	per ewe lambing
Blackface x Cheviot	16.4	18.6
Blackface x Welsh	6.6	7.2
Cheviot x Welsh	10.4	11.5

Table 7 shows the carcass weight produced per ewe varying from 32 kilograms for crosses of Blackface with Finnish Landrace to half that from the Merino cross Blackface ewes. The next column in the table shows the relative receipts per ewe and the final column the age at slaughter of the lambs. It was interesting that the price per kilogram of carcass meat did not differ markedly over the period that the lambs were slaughtered; neither was there a marked difference in the weight of the individual lambs when they were sold nor in the receipts per lamb. The differences in the receipts per ewe were almost completely related to the carcass weight produced per ewe which in turn related to the number of lambs per ewe.

Receipts do not however equal profits. The components of profit will certainly include the time that the lambs have to be kept, particularly if the lambs have to be given additional food to get them ready for slaughter. If therefore we were to divide the receipts per ewe by the number of days that the lambs took to reach slaughter the story would change again. The receipts per day would show the Finnish Landrace crosses, the Border Leciester crosses and the Dorset Horn crosses all fairly similar, with the Clun Forest a little way behind and the Merino crosses by an even greater margin at the tail end. It is not suggested that receipts per day are any clearer guide to profits than the total receipts. Both are over-simplifications of the economics involved. But the point to note is that the balance of advantage as to which breed and which cross to use will depend on the circumstances for which they are kept. If food is cheap, or grass is available which would merely go to waste, and if the problems of small lambs do not put a strain on management, clearly the total number of lambs and the total carcass weight will be of paramount importance. If however the additional costs of keeping lambs for a longer time are important, or if the prices received get less as the season advances, perhaps it is more important to have a fast growing, large lamb than many smaller, slower-growing ones.

Attention should be drawn again to that aspect of Table 6 which suggests that there is breed variation in the age when sheep first start producing lambs. A majority of Finnish Landrace crosses successfully produced lambs at 1 year old, but only a very small proportion of the Merino crosses did so. Evidence of this type from different breeds shows that this is an important aspect of reproductive performance for which breed differences exist and particular attention is drawn in this respect to findings of Land et al (1974).

Carcass composition

Table 8 illustrates evidence of breed differences in carcass attributes based in this case on the use of different sire breeds each mated to the same ewe breeds in Ireland. The differences are based on the performance of crossbred lambs only and, as in the case of the ABRO experiment summarised in Tables 6 and 7, could be affected by differences in the manifestation of heterosis in the different crosses. Notwithstanding this complication it appears that Dorset Down and Texel crossbred lambs had the best lean to bone ratio, due in the former case to a low proportion of bone in the carcass and in the latter to an above-average development of lean. Recent experiment in ABRO (Smith,

Table 6 Lamb Production of Crossbred Ewes (from Donald et al (1968)

Sire of ewe	Lambs weaned/ewe mated		Total litter weight at weaning (kg)[1]
	1st year	Av. of 2nd & 3rd	
Border Leicester	0.78	1.40	127
Clun Forest	0.71	1.35	117
Dorset Horn	0.70	1.37	122
Finnish Landrace	0.82	1.91	139
Merino	0.38	0.99	89

[1]Total for three lamb crops, years 1-3.

Table 7 Lamb Production Per Ewe (Adapted from Table 5, Donald et al (1970)

Sire of ewe	Carcass wt. produced per ewe (kg)	Receipts per ewe as % of average	Age of lamb at slaughter (days)
Border Leicester	27.1	105	181
Clun Forest	24.2	96	187
Dorset Horn	26.2	101	172
Finnish Landrace	32.3	132	219
Tasm. Merino	15.8	66	223

unpublished) support the evidence for high lean, low fat development in Texel crosses. One problem in the comparison of breeds and crosses in respect of carcass attributes arises from the choice of criteria for slaughter, for example whether this is on the basis of weight, or of 'finish', on proportion of mature size reached or on age. Market considerations could greatly influence the relative advantages of different biological attributes of breeds.

Table 8 Least Square Estimates of Sire Breed Means for Carcass Composition Characteristics (From Timon & Ireland 1974)

Breed of Sire	Lean %	Fat %	Bone %	Lean to Bone	L. dorsi area (sq.cm)
Suffolk	55.9	27.4	16.7	3.39	11.24
Texel	59.9	22.7	17.0	3.58	12.01
Dorset Horn	57.4	25.8	16.6	3.50	11.59
Hampshire	55.7	27.6	16.4	3.44	11.20
Oxford Down	56.1	26.6	17.1	3.32	11.07
Lincoln	56.1	26.2	17.5	3.22	10.67
Ile de France	55.1	26.8	17.2	3.34	11.30
Dorset Down	54.7	29.8	15.2	3.59	11.92
Mean	56.4	26.6	16.7	3.42	11.38
Significance of breed difference	****	****	****	****	****

Wool

For fleece weight and for components of the fleece, there are large differences among breeds particularly if the Merino and Merino-derived breeds on the one hand are compared with non-Merino breeds on the other. But even among British breeds there are appreciable differences both in quantity and quality of wool. In no other aspect of sheep production have breed differences been so well documented or in commercial terms so well categorised. Hybrid vigour however does not appear to be an important expectation from crossbreeding for wool attributes but breed substitution and crossbreeding are easy ways of changing the average type. If wool is an important part of the income derived from a group of sheep, and if the particular type of wool produced is important, the choice of breeds used for crossing cannot ignore fleece type. This however will limit the opportunities for change in other directions. It should be borne in mind that most wools find a ready market and the costs of a change of fleece type must be assessed alongside the returns from improvements in other directions.

MORTALITY, DISEASE AND BIOCHEMICAL VARIATION

It was indicated above that in terms of general lamb survival crossbreds have an advantage. I now wish to turn to a more specific point relating to metabolic disorders. For a number of these there is evidence of genetic variation and the best recorded to date in sheep relates to work by Dr Field and myself, involving disorders associated with copper metabolism from swayback at one end of the scale to copper poisoning at the other. Table 9 shows that in the particular flock from which several examples have already been drawn there was a strikingly different incidence of swayback among the breeds and crosses, varying from 40% for the Scottish Blackface to zero for the Welsh Mountain.

Table 9 Incidence of Swayback up to 18 weeks (from Wiener (1966))

Breed of cross[1] of lamb	No. lambs born alive	% of Swayback
Scottish Blackface	48	39.6
Cheviot	35	11.4
Welsh Mountain	56	0.0
Blackface x Cheviot	92	20.6
Blackface x Welsh	106	15.1
Cheviot x Welsh	81	4.9
All breeds	418	14.8

[1]The crossbred lambs are from matings of crossbred rams of the same type e.g. (B x C) ewes x (C x B) rams

In the experimental flock under consideration, close inbreeding ws superimposed on both the pure breeds and the crossbred groups. It is possible therefore to classify the sheep into four groups as shown in Table 10. The first part of the table shows that with respect to losses from swayback the four classes did not differ significantly from each other. Inbreeding did not increase the losses from swayback and crossbreeding was no advantage; this is quite different from the comparable figures for general mortality. Here, by far the lowest mortality was amongst crossbred lambs which were not inbred and by far the highest mortality among the purebred sheep which were also inbred.

Table 10 The relationship of lamb losses (%) to breeding system (from Wiener (1966))

a. Swayback losses

	Outbred	Inbred	Average
Purebred	15.4	16.8	16.1
Crossbred	17.3	12.7	15.0
Average	16.4	14.8	15.6

b. General mortality (excluding Swayback and Stillbirths)

	Outbred	Inbred	Average
Purebred	11.5	28.3	19.9
Crossbred	4.0	19.1	11.6
Average	7.8	23.7	15.7

Table 11 gives some information on the underlying variation in copper concentration in these breeds and crosses. Both for the concentration of copper in liver and in blood there was very significant breed variation. In the case of blood copper however the crossbreds were at the level of the higher parent contributing to the cross; this is not so obvious for liver copper where the particular results for these crosses (taking into account the relatively small number of animals involved) are not significantly different from the mid-parental values. Later studies (Wiener et al 1976) suggest that crossbreds have liver copper concentrations slightly above the mid-parental level but nothing like the very marked heterosis shown for blood copper level. It is interesting to note however that the heterosis for blood copper level is not related to a corresponding advantage in terms of lower incidence of swayback (as seen in Table 9). It appears therefore that at the breed level the incidence of swayback and the level of copper in the blood are to some extent independent.

Table 11 Copper Concentrations of Liver (PPM of DM) and of Whole Blood (ug/100ml) (from Wiener & Field (1969))

Class[1]	No. of animals	All 68 sheep: Copper in Liver: Fitted Value	SE	Blood: Fitted Value	SE
Blackface	12	14.3	13.8	59.8	7.6
Cheviot	7	44.7	17.0	66.8	9.4
Welsh	13	49.7	13.1	75.7	7.2
Blackface x Cheviot	12	26.2	14.3	80.8	7.9
Blackface x Welsh	12	23.2	13.9	78.6	7.7
Cheviot x Welsh	12	54.9	13.6	84.4	7.5

[1]The breed values shown refer to old females of the outbred class with no history of swayback among their lambs in 1964.

Preliminary work suggests that there are breed differences in the rate of absorption of dietary copper and that this may account for the differences in copper levels now shown by us to exist for a number of different breeds and situations. Figure 1 shows results of changes in plasma copper levels during repletion with 3 ppm Cu in the diet following depletion. It should be noted that the rate of repletion of the crossbred sheep resembles that of the parental breed, the 'North Ronaldsay', with the higher rate (Wiener et al 1978).

There is evidence suggesting that susceptibility to copper poisoning may also show breed differences.

While these results are specific to copper, evidence has been reported of breed differences in the concentrations of other minerals and metabolites in the blood of both sheep and cattle (for reviews of the earlier evidence, see Wiener & Field 1971, Wiener 1979). The relevance

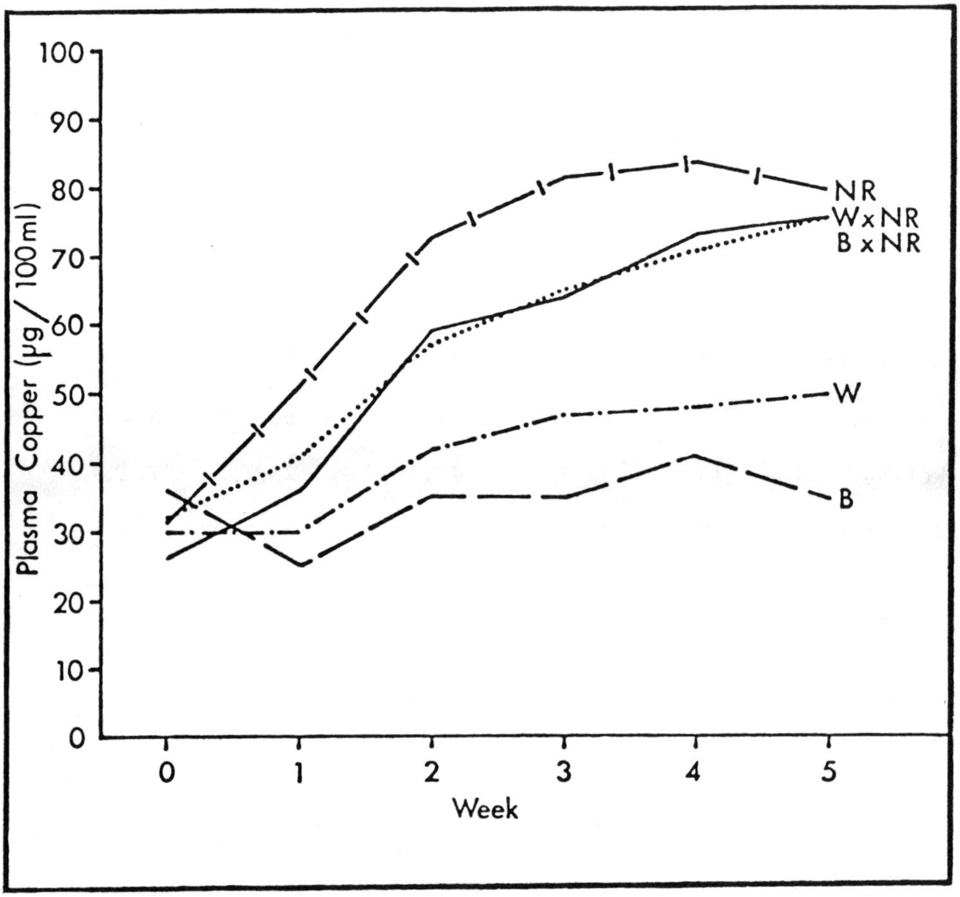

Fig.1. Changes in plasma Cu concentration (µg/100ml) of 3 breeds and 2 crossbred groups of sheep during a 5-week period when the dietary Cu concentration was increased to 3 ppm following a prolonged period of depletion.

NR = North Ronaldsay , W = Welsh Mountain , B = Scottish Blackface .

(from Wiener et al 1978)

of these points in the present context is that for specific situations the existence of breed differences may be important factors in susceptibility or resistance to metabolic disorders and in determining the responses to the particular nutritional environment. To draw the appropriate inference from the example of copper it may be possible to visualise breeds or crosses which will perform adequately in copper-deficient environments and others which might be relatively resistant to conditions where copper poisoning could occur.

GENERAL CONSIDERATIONS CONCERNING BREED UTILIZATION AND CROSSBREEDING

The examples chosen demonstrate the principles. A fuller review of the literature, particularly in relation to estimates of heterosis, is given by Nitter (1978) who also goes on to consider the advantages of different crossbreeding strategies in relation to purebreeding and to each other, when assuming the levels of heterosis found in the review of the literature.

In the total flock situation the effects of crossbreeding should not be considered in isolation. Production of first crosses also requires the maintenance and replacement of purebreds to produce these crosses. The costs and revenues from the purebred section of the population have to be integrated with those of the crossbred component. Even though the individual merit of sheep of specific first crosses may exceed that of sheep of F_2 , three-way or criss-crossing systems, the overall demands of biological efficiency and economic merit of the total system should receive prime consideration. Similarly, considerations of substituting one breed for another, particularly in a purebreeding situation, will generally include factors in overall profitability for which little or no experimental evidence is available. Not least of these are problems of acclimatization of breeds to new disease, nutritional, climatic and management situations which could undermine the profitability of breed substitution for a considerable period of time. Especially when transfer of new breeds into very different environments is contemplated a process of gradual introduction and assessment through crossbreeding is to be commended rather than a massive breed substitution.

SUMMARY AND CONCLUSIONS

Breed differences in performance characteristics offer a simple and quick method of effecting change either by substituting one breed for another or by crossbreeding. The crossbred may show an advantage over the average of the parental breeds contributing to the cross (there may be hybrid vigour) for some traits but not for others and in some crosses but not in others. In general, there is a greater probability of finding heterosis in traits concerned with reproduction, survival and health than in traits related to body size, carcass attributes or fleece quality. Specific crosses however may deviate quite markedly from the average expectation in showing either more or less hybrid vigour.

Interactions between breeds or crosses and their environment are more likely to be found if the environmental differences are large and for characteristics involving health, survival and reproduction. Even then the ranking of breeds may not differ in different environments but the relative profitability could be more affected.

Crossbreeding offers the particular advantage of allowing relatively specialised sire and dam lines to be maintained which are good for different desirable attributes. These are then combined in the production of offspring to be raised for meat production. Thus the size, fast growth and carcass attributes required in the lamb can be brought in from the sire breed, while the maternal breed can be kept as small as is consistent with prolificacy and mothering ability. Small maternal size will reduce the overhead costs of production.

In deciding which breeds or crosses are most appropriate to a particular farming system, it is essential that the comparisons of breeds and crosses be made under common conditions. If genotype-environment interactions are thought to be or known to be important, the test conditions should approximate those under which the animals will be kept commercially. Since maternal effects are important it is necessary to produce reciprocal crosses between two breeds before it can be assessed whether the cross is better than the average of the two parental types. Alternatively the magnitude of heterosis can be assessed from the comparison of F_1 with F_2 crosses (F_1 x F_1) since on average F_2 crosses are expected to display only half the heterosis of F_1 crosses, when allowance is made for maternal effects.

Since some of the advantage of hybrid vigour is derived from the use of crossbred mothers, crossbreeding systems, involving criss-crossing, more than two breeds or crosses mated among themselves, should be considered in place of the continuing re-creation of first crosses from a purebred foundation. Animals derived from the more complicated crossing systems are expected in theory to display less hybrid vigour than first crosses. Hybrid vigour however is only one consideration in the costs and returns from the system; these will need to be assessed for each situation. Experiments are also required to show how closely the theoretical expectations apply in practice.

Considerations of breed substitution or crossbreeding must be related to the whole farm flock or national flock situation and not merely to the single most profitable component of the system.

REFERENCES

Bowman, J.C. (1966) Meat from sheep. Animal Breeding Abstracts 34,293-319.

Dickerson, G. (1969) Experimental approaches to utilizing breed resources. Animal Breeding Abstracts 37,191-202.

Donald, H.P.;Read, J.L.;Russell, W.S. (1968) A comparative trial of crossbred ewes by Finnish Landrace and other sires. Animal Production 10,413-421.

Donald, H.P.;Read, J.L.;Russell, W.S. (1970) Influence of litter size and breed of sire on carcass weight and quality of lambs. Animal Production 12,281-290.

Falconer, D.S. (1960) Introduction to Quantitative Genetics. Edinburgh: Oliver & Boyd.

Field, A.C.;Wiener, G.;Wood, J. (1969) The concentration of minerals in the blood of genetically diverse groups of sheep. II. Calcium, phosphorus, magnesium, potassium, sodium and chlorine concentrations for three hill-breeds and their crosses at pasture. Journal of Agricultural Science, Cambridge 73,267-274.

Land, R.B.;Russell, W.S.;Donald, H.P. (1974) The litter size and fertility of Finnish Landrace and Tasmanian Merino sheep and their reciprocal crosses. Animal Production 18,265-271.

Large, R.V. (1970) The biological efficiency of meat production in sheep. Animal Production 12,393-401.

Mason, I.L. (1957) A world dictionary of breeds, types and varieties of livestock. Tech. Com. 8 C.B.A.B.G., Commonwealth Agricultural Bureaux, Farnham Royal, England.

Rae, A.L. (1952) Crossbreeding of sheep. Animal Breeding Abstracts 20,197-207.

Nitter, C. (1978) Breed utilization for meat production in sheep - a review. Animal Breeding Abstracts 46,131-143.

Terrill, C.E. (1974) Review and application of research on crossbreeding of sheep in North America. Proceedings, 1st World Congress on Genetics applied to Livestock Production, Madrid, 1,765-777.

Timon, V.M.;Ireland, A.F.T. (1974) The evaluation of sheep breeds and breeding strategies. Proceedings, Working Symposium on Breed Evaluation and Crossing Experiments with Farm Animals, Zeist. 367-387.

Turner, H.N. (1969) Genetic improvement of reproduction rate in sheep. Animal Breeding Abstracts 37,545-563.

Wiener, G. (1966) Genetic and other factors in the occurrence of swayback in sheep. Journal of Comparative Pathology 76,435-447.

Wiener, G. (1967) A comparison of the body size, fleece weight and maternal performance of five breeds of sheep kept in one environment. Animal Production 9,177-195.

Wiener, G. (1971) Relationships between swayback incidence and concentration of copper in the blood of sheep of different breeds. Journal of Comparative Pathology 81,515-520.

Wiener, G. (1973) Breeding for increased productivity. Veterinary Record 92,609-613.

Wiener, G. (1974) Crossbreeding studies with sheep at the Animal Breeding Research Organisation, Edinburgh. Proceedings, Working Symposium on Breed Evaluation and Crossing Experiments with Farm Animals, Zeist. 493-510.

Wiener, G. (1979) Review of genetic aspects of mineral metabolism with particular reference to copper in sheep. Livestock Production Science 6,223-232.

Wiener, G.;Deeble, F.K.;Broadbent, J.S.;Talbot, M. (1973) Breed variation in lambing performance and lamb mortality in commercial sheep flocks. Animal Production 17,229-243.

Wiener, G.;Field, A.C. (1969) Copper concentrations in the liver and blood of sheep of different breeds in relation to swayback history. Journal of Comparative Pathology 79,7-14.

Wiener, G.;Field, A.C. (1971) Genetic variation in mineral metabolism of ruminants. Proceedings, Nutrition Society 30,91-101.

Wiener, G.;Field, A.C.;Wood, J. (1969) The concentration of minerals in the blood of genetically diverse groups of sheep. I Copper concentration at different seasons in Blackface, Cheviot, Welsh Mountain and crossbred sheep at pasture. Journal of Agricultural Science, Cambridge 72,93-101.

Wiener, G.;Hayter, S. (1974) Body size and conformation in sheep from birth to maturity as affected by breed, crossbreeding, maternal and other factors. Animal Production 19,47-65.

Wiener, G.;Hayter, S. (1975) Maternal performance in sheep as affected by breed, crossbreeding and other factors. Animal Production 20,19-30.

Wiener, G.;Hayter, S.;Field, A.C.;Macleod, N.S.M. (1974) Copper levels in liver and brain of dead lambs in relation to breed, age at death and cause of death. Journal of Comparative Pathology 84,27-38.

Wiener, G.;Herbert, J.G.;Field, A.C. (1976) Variation in liver and plasma copper concentrations of sheep in relation to breed and haemoglobin type. Journal of Comparative Pathology 86,101-109.

Wiener, G.;Macleod, N.S.M. (1970) Breed, body weight and age as factors in the mortality rate of sheep following copper injection. Veterinary Record 86,740-743.

Wiener, G.;Suttle,N.F.;Field, A.C.;Herbert, J.G.;Woolliams, J.A. (1978) Breed differences in copper metabolism in sheep. Journal of Agricultural Science, Cambridge 91,433-441.

GENETICS AND DISEASE

DR A G DICKINSON

Animal Breeding Research Organisation, West Mains Road, Edinburgh.

INTRODUCTION

What are the implications for livestock genetics of the existence of microorganisms which can affect animals? The theme of this paper will be diseases caused by microorganisms and the subject will be considered from a comprehensive biological viewpoint. No attention will be paid to diseases where the pathogen is a simple toxic chemical or where the disease is due to nutritional imbalance or where the cause is a mutant gene in the animal without any microorganism being involved. Of course, there are many diseases where the cause is unknown and many of the points that will be made are relevant for investigating the possibility that a microorganism is involved and in determining whether it has a primary role in causing the disease or whether its involvement is only secondary. Renal infections of cattle are an example where there is still this type of uncertainty.

Genetical studies are often complicated even when one is only dealing with the phenotypic expression, as in growth or milk yield, of the genotypes of individuals. However, "infectious disease" is the phenotypic expression of the interaction of two independent genotypes - those of the host and the pathogen. When the large variety of environmental variables which can affect a disease process are taken into account, it will readily be appreciated that detailed analysis of the situation is liable to be difficult. The differences from one disease to another are so great that it is pointless to attempt to formulate a single recommendation for breeding policy which will apply to all of them. A more fundamental decision in each case where it is known that genetical methods could be used successfully for short-term control, is whether such methods should be adopted in preference to conventional methods for prevention and treatment of disease, when long-term considerations are used.

What I will attempt, therefore, is to describe a wide variety of examples so that those who need to take action against specific diseases can decide what is the most relevant policy in their own case. To those who have no direct interest in disease but whose field is general quantitative genetics as applied to animal breeding, I hope that what follows is a useful introduction to a fascinating subject, which might be of more immediate relevance to animal breeding theory than some may suspect. There is also the practical point that genetical selection for a direct economic trait, such as growth rate, can be accompanied by an increased disease liability, which may not be recognised at the time (e.g. poultry growth and Marek's disease, Han & Smyth 1972).

DEFINITION OF THE PROBLEM

Although the details will differ from country to country, many readers will have experienced some of the adverse results from an overstatement of the case for the use of antibiotics. With the advent of antibacterial drugs we were confidently told that their widespread use and development would obviate the need to breed more resistant animals and even that improved growth rates were more reliably and economically achieved by improving nutrition with antibiotic feed additives than by long-term genetic selection. This policy leads to the selection of drug resistant bacteria which can transfer their resistance factors infectively to other types of bacteria. The consequences of this for animals and humans are potentially serious - sufficiently so that it is at last receiving official attention in various countries.

I have mentioned this at the outset so as to strike a note of caution because there are those who similarly overstate the case for selecting animals for genetically controlled resistance to disease. The unfortunate consequences of this could be very similar to those from uncontrolled use of antibiotics because the pathogenic organisms which can develop means of dealing with antibiotics can also change by natural selection into strains which evade the host's genetically controlled "resistance". In plant breeding this type of consequence has been seen in some countries where cereal rusts have mutated rapidly one step behind the introduction of a new "resistant" variety of wheat, leading to the need for a new "resistant" variety, leading to a new strain of rust - and so on, through more than 20 cycles in one recorded case (Walker 1951). The adoption in this case, of mixed cereal varieties being grown together instead of pure-stands of one variety, has apparently been a useful tactic. Single genetically homogeneous strains of poultry housed in large numbers at high density constitute similar ideal target populations where new strains of pathogenic organisms could develop even though the birds may have had all the desirable productivity and resistance characteristics at the outset. Husbandry developments and genetical techniques for other livestock are moving in similar directions of relative uniformity and large production units. With this in mind, an urgent word of warning ought to be stated lest, by default and lack of biological foresight, animals should be following the same path as the "rust-resistant" wheats. F.B. Hutt (1958) has been one of the main advocates of selecting livestock for disease resistance and, in my view, he has overstated the case by neglecting some of the long-term aspects in favour of short-term objectives. An example from his own work will illustrate this. He has been able to show with fowl 'leukosis' (Hutt 1965), that selection for twenty generations in a closed flock produced a decline in susceptibility from 15% to about 2% in one line. However, this resistant line was quite susceptible to other virus strains of the leukosis complex present in other geographical areas (Zander pers. comm.). The reason was that this genetical work was undertaken before it was recognised that several unrelated viruses were producing a number of diseases which were being grouped as 'fowl leukosis'. Some of these diseases were caused by one or more RNA viruses (leukosis-sarcoma group) and others involved DNA viruses (herpes group). Genes conferring resistance independently to several different components of the complex are known but, not surprisingly, none confers resistance to the whole complex. (Avian leukosis is well reviewed by Darcel 1973).

If genetic selection is to be used to control a disease it must involve exposure to the appropriate range of causal pathogens and their relevant substrains. Now that this point concerning policy has been stated, we can turn to more positive aspects of the problems of animal disease control. At the outset it should be made clear that I hold the belief, which most geneticists would presumably share, that one can expect to find in outbred populations of animals genetical variation affecting resistance or susceptibility to most diseases - the questions of importance are how best to locate and utilise this variation in large populations of animals with both long-term and short-term objectives in view. For any specific disease, confidence in the successful attainment of the long-term objective will depend upon a thorough basic understanding of the physiological and histological interaction of the host and disease organism and the relative roles of genetical and environmental factors. Whilst I am not going so far as to suggest that each disease in each species will require a different solution, a sufficient variety of examples will be given to dispel the idea that the procedures needed in one disease will necessarily work in a different one.

One general point needs to be made here. A great deal can be learnt about a disease, and the steps at which genetic controls of the disease operate, by using various types of artificial exposure to the infection - injection into different organs for example. However, in a breeding programme with economic objectives, the method of exposure to the pathogen(s) must not bypass any of the defence mechanisms of the host, so that the test gives a true indication of response to natural exposure. This is not easy to achieve.

The concepts of susceptibility and resistance need very careful definition otherwise the use of these terms, which are so convenient to use in discussion, can be a hindrance to understanding rather than a help. The example already given of fowl leukosis "resistance" will have pointed to some of the difficulties. When I attempted to review work on genetical aspects of diseases some time ago (Dickinson & Mackay 1967) two features of the literature were outstanding. One was that relatively little work had been done and then mostly with laboratory animals; the other feature was the imprecision and lack of clear thinking often associated with use of the terms susceptibility and resistance. For example, an individual can be "susceptible" to clinical disease either because it lacks appropriate defence mechanisms or because it over-reacts to the presence of the pathogen. Or, as a later example with salmonellosis will show, a "resistant" individual can be the one which permits the more rapid tissue damage but leads to a self-limiting course to the infection. The resistant genotype may be one better able to tolerate multiplication of the pathogen or the tissue damage, either by more efficient detoxification of metabolites or more rapid tissue repair. If "resistance" is used to imply the absence of clinical signs of disease, then a genotype which permits an asymptomatic infection would be described as resistant, but this genotype is clearly not particularly desirable if it acts as a carrier and can disseminate the infection.

Resistance and susceptibility are best used as relative terms rather than absolutes, if for no other reason than the frequent possibility of infecting an otherwise "resistant" animal by using a large enough dose of the pathogen by an unconventional route or under adverse environmental conditions. Given animals of two different genotypes which respond differently to the same pathogen, it is only appropriate to describe one as more resistant if all the conditions of exposure are the same for both - dose of the microorganism, route of infection, other concurrent diseases, nutritional state, environmental factors, age at exposure, recent passage history of the disease organism and so on. The relevance of some of these factors will be dealt with later.

It is instructive to consider the range of properties which could distinguish a "resistant" animal from one which was less resistant or more susceptible. There are rare examples where a genetical defect in the immune mechanisms results in susceptibility to a range of unrelated diseases due to the inability to make an adequate immune response; a blood-complement deficient guinea-pig is one example (Rich 1923) though a similar defect in the mouse carries no such apparent disadvantage (Herzenberg et al 1963). More common than the preceding type is where there is correlated resistance to a related group of organisms, most often where these share antigenic properties. Examples of this are known for resistance to the different members of the B group of arboviruses (louping ill of sheep, yellow fever of man); a gene conferring resistance to this virus group is present in mice (Sabin 1952) but this gene confers no resistance to several other types of viruses which are antigenically unrelated to the arbovirus B group. Similarly, genetic resistance in mice to myoviruses (e.g. influenza) was not accompanied by resistance to arboviruses, Herpes simplex virus or some enteroviruses which were tested (Lindenmann et al 1963). Greater resistance to some strains of an organism may entail greater susceptibility to other strains; examples of this have been demonstrated for Rous associated virus in poultry (Vogt & Ishizaki 1965) and with scrapie in sheep and mice (Dickinson & Stamp 1969 and unpublished, Dickinson 1975).

Resistance can imply that there is no initial infection, with the pathogen effectively excluded from entry to the animal's tissues or cells (avian tumour viruses; Vogt & Ishizaki 1965), or that there is entry into the host but replication is blocked genetically in some cells (arbovirus B group) or is so slow that the host has adequate time to respond immunologically to the infection, or that the animals can produce neutralising antibodies more quickly (mousepox, Schell 1960a and b) - we will return to these aspects later. The clinical symptoms accompanying greater resistance will also vary according to the example chosen; thus resistance may indicate the absence of symptoms (Japanese encephalitis virus in man; Grossberg & Scherer 1966) or relatively trivial signs (mousepox in C57BL, Briody et al 1956) or a brief illness without complications, or survivial instead of death (salmonellosis in mice, Bohme et al 1959) or a long incubation period instead of a short one even though the outcome is fatal in both cases (scrapie in mice, Dickinson and Meikle 1971). It would also be appropriate to describe a genotype as more resistant if a larger dose is necessary to produce the same degree of disease as in the more susceptible type (foot and mouth disease in mice,

Subak-Sharpe 1961). Also, genetical differences in resistance may be readily apparent when a natural route of infection is used but the difference may be less or even absent when artificial challenge by injection is used (mousepox, Schell 1960a and b). With diseases where neutralising antibodies are important the term "more resistant" could also be applied to a genotype which may not differ from the rest at the first infection but which has more persistent antibodies.

These remarks bring us to the major problem with any attempt to reduce the economic losses due to livestock diseases by genetical methods. Namely, the uncertainty that differences between individuals or progeny groups represent inherited differences rather than differences in exposure to the infection and, even with highly infectious diseases, a large measure of uncertainty will remain in the case of individual animals. It may be possible to make confident predictions about natural resistance from experimental challenge of every animal, especially if a route close to the natural mode of entry of the pathogen can be used. This will, of course, depend upon having adequate basic information about the disease concerned, probably including information about the physiological components of host resistance. It will also depend upon being able to use, experimentally, strains of the microorganism which give results relevant to those strains which will be encountered under natural exposure in the field. Most groups of livestock will include individuals of different ages, especially where natural suckling occurs. While not forgetting that passive immunity derived from the mother will often have to be taken into account, the age of the host is one of the principal factors affecting the outcome of an infection (Siegel 1952). Susceptibility may increase or decrease with ageing; the relationship is usually not linear and the route of infection is of particular importance in age-specific effects.

Before ending this section a brief comment on nutrition should be made. Whilst it may be true in some cases that a good nutritional status will help an animal to counteract infections, other possibilities have to be borne in mind. The microorganisms have their own nutritional requirements and these can be very specific. For example, Leftwich & Mirick (1949) found that pyridoxin deficiency increased the resistance of mice to a pneumonia virus. What may be needed, therefore, is basic information about the chemical constituents of the diet and their specific modes of action in the disease. Thus, while it is of interest to know, for example, that mice died more quickly on a diet A than on diet B with any of seven tuberculosis strains (Hedgecock 1966), such information is of little predictive value but points to the usefulness of more detailed investigation.

To close this section I would like to add some general remarks about the type of control which genes can exercise. Considered historically, genes whose expression is unaffected by a wide range of environments have been easier to identify and this has helped to overemphasize the inflexibility of genetically controlled processes. Genes never act alone but only as part of a complex genotype and genes do not act in the abstract but only within specific environmental frameworks, which, in one way of looking at the problem, should be taken to include the genetic background in which each gene is operating.

There is a widespread misapprehension of what is implied by the finding that a trait is subject to a considerable degree of genetic control. Misconceptions have been fostered by the rather sterile "nature versus nurture" arguments of the past and the overstatement of the case by some who have argued for genetic determinism. Thus to find, for example, that in response to a virus infection genotype A is susceptible whereas genotype B is resistant does not necessarily imply that A will be more susceptible under other environmental circumstances. The relative status of the two genotypes can change quantitatively to various degrees even to the extreme of a reversal of relative susceptibilities (see Dickinson 1975).

THE BIOLOGICAL SPECTRUM OF DISEASE

Gregor Mendel was very fortunate in the type of problem which he decided to study. Not only did he have available pure-breeding strains of peas but he also chose characters which segregated in a clear-cut manner in the F_2 generation. We are in a less opportune position. Our livestock are normally outbred and highly variable; in the long-term view this is their greatest asset which we should strive to preserve in some form or other. But the absence, except in poultry and laboratory rodents, of pure-breeding lines in any meaningful use of the term, renders investigation difficult, slow and costly. Moreover, the complexity and diversity of economic traits, including diseases, means that characters which segregate in a clear-cut Mendelian manner are either difficult to find or are individually swamped by many genes, independently segregating, but affecting a common character of interest.

In this event there are two methods of attack; one may lump together all the important characters into a selection index of some form or one may seek individual segregating genes affecting some of these characters. I have presented these as blunt alternatives but, obviously, they are not. However, the effectiveness of a selection index is weakened for each component trait as additional characters are added or as any policy is adopted which weakens the full implementation of the index such as simultaneous selection for recognisable genes.

Given the crude alternative 'selection index' or 'individual genes' it has been fashionable in animal breeding during the last twenty or thirty years to dismiss the latter alternative as a too simple-minded approach except in infrequent special cases. I would like to attempt in this section to adjust the prevailing balance between these approaches. Incidentally, success in plant breeding has much more often come from finding and incorporating useful major genes than from employing quantitative genetic methods, but there are some good biological reasons for this difference in emphasis between plants and animals.

Quantitative genetics as implemented in animal breeding seems to have two great weaknesses. In the models which are widely used, correlations between relatives depend entirely on large numbers of chromosomal genes acting additively, while the environmental components present no quantitative or qualitative problems and, often, there is no allowance for genotype-environment interactions. Complex gene action - dominance, genic interactions etc - presents problems whether one uses a Sewell Wright or Fisher type of approach and simplifying assumptions are usually needed. Analysis of the environment is usually neglected because of its complexity and with the hope that any differences will "even out" given appropriate experimental design and, likewise, genotype-environment interactions have rarely been considered beyond the type of selection experiment which uses "high" and "low" planes of nutrition.

These commonplace types of simplifying assumptions may be fully justified in some cases but they are certainly an unsatisfactory compromise in many cases, especially with long-term experiments. As an example, with disease organisms which may present problems only erratically throughout the course of an experiment, the deficiency of considering "environment" in statistical rather than biological terms will be apparent. "Environment" is a useful term but often needs careful specification.

But the main point which I want to make now is another one. Related animals tend to be more alike than unrelated animals for many more reasons than the chromosomal genes which they have in common. In mammals, maternal sibs share similar prenatal environments, including the passive antibodies which they receive, and some of these effects may be permanent; they are also able to receive non-chromosomal inherited factors - either cytoplasmic components, prenatal infections or milk-transmitted agents.

With diseases the non-Mendelian resemblances between relatives can be of prime importance. During the last two decades the development of molecular biology with its growing understanding of the biochemical resemblances between viruses and genes has made it less easy to draw a clear demarcation between the host's hereditary properties and viruses associated with the host. Examples are known in several different host species where a virus can be integrated, in a more or less stable manner, into the host's genetic apparatus. Some viruses may be able, in rare cases, to infect early embryos and then become 'integrated' into the germ-line chromosomes (Jaenisch 1976). Thus, the same disease can on some occasions appear to be an inherited defect and on other occasions to be infectious or, at least, artificially transmissible. Certain tumour viruses are among the pathogens which have been shown to behave in this way, notably those causing leukemia in mice (Gross 1959). The difficulties of elucidating this type of complex are considerable (see Spiegelman et al 1975), especially if the work has to be undertaken in livestock. There are two sheep diseases with the potential for viral integration into the host genome due to the presence of viral reverse transcriptase. These are pulmonary adenomatosis (Jaagsiekte) where more than one type of virus is involved (Martin et al 1976) and the progressive pneumonia-encephalomyelitis complex (Maedi-Visna) (Nathanson et al 1976). Analysis of the genetics of sheep susceptibility in these diseases will be a daunting task.

A lesser degree of association between host and microorganism can also give rise to non-genetic correlations between relatives. This is when vertical transmission of the pathogen occurs, usually from mother to offspring. In mammals this can occur prenatally or through the milk or by straightforward infection of the newborn but even where it is clear that some form of vertical transmission is occurring it can be difficult to determine at which stage it takes place. Obviously, if a disease agent is still present in so-called "germ-free" caesarian-derived mammals, then the pathogen is transmitted prenatally to the embryos. This has now been shown with "germ-free" stock for a number of diseases, one being lymphocytic choriomeningitis (LCM) in mice (Pollard et al 1968).

LCM is relevant to mention here for another reason; it can be a latent virus, in the sense that no disease develops in certain cases in an infected individual. When LCM virus is transmitted prenatally a persistent tolerant infection develops without clinical disease but if the same host genotype were instead to become infected when a few weeks old, a clinical disease would have appeared (Traub 1960, Weigand & Hotchin 1961). Also strains of LCM virus differ in the dose which will produce a carrier status in a given host genotype (Camyre & Pfau 1968). Prenatal virus infection is known to be the cause of Border disease (hypomyelinogenesis congenita) of sheep, where nervous and other developmental defects are produced by infection of the embryo, although the infection is almost without symptoms in the ewe which transmits the virus (Dickinson & Barlow 1967) and there is probably persistent infection of those infected lambs which survive and the next-generation progeny were also affected from the few prenatally infected ewes which have had lambs. This type of carrier state is not confined to mammalian diseases. Congenital transmission of viruses associated with fowl leukosis has been shown to occur, in which case it is probable that this leads to a persistent immunological tolerance of the virus (Rubin et al 1961).

Even where there is simple vertical transmission of the microorganism, without such problems of persistent tolerant infection, but leading directly to the production of overt disease, problems of interpretation can arise if the interval between infection and the appearance of signs is so delayed that the infectious nature of the process can remain unnoticed. As an example scrapie in sheep combines very long incubation periods - several years - and vertical transmission of the agent (Dickinson et al 1964). The problems for analysis resulting from these peculiarities were such that one investigator was even led to conclude that scrapie was caused solely by a recessive gene in the sheep, without an infectious agent (Parry 1962).

Where vertical transmission is present it is difficult to estimate the extent to which genetical variation affects the disease unless one also has available a range of genotypes known to be free from the pathogen. Evidence that vertical transmission is present usually starts with the finding of reciprocal differences, backed up by independent information that one is dealing with a disease caused by a transmissible organism. Similarly, when dizygous twins are available there is an unexpectedly high proportion of cases where both twins are affected and a deficiency of pairs where only one member is affected.

There are other types of complications which can present difficulties for a geneticist studying diseases. Two or more microorganisms can interact to affect the course of the disease in a particular host. The following example, which is still only partially resolved, will demonstrate some of the problems with this.

In mice the well known mammary tumour virus is normally transmitted from mother to offspring in the milk. It has been found that there are many differences in response between inbred strains of mice following infection with the mammary tumour agent - in some the tumours appear in all the females, in others only in females which have been bred, while other strains are not susceptible to the development of tumours (Nandi & Bern 1960). The age at which resistance to infection with the tumour virus appears also varies between inbred strains: in BALB/c mice there is a slight increase in resistance with ageing but C3H/f mice become very highly resistant from 18 days of age. This characteristic of the C3H/f mice is not due to the genes which they possess but is the result of another virus which they carry (Nandi & DeOme 1965). This was found by injecting cell-free extracts from one strain into the other; an indicatin that some such complication was present would presumably have come from formal genetical tests involving crossing and then observing potentially segregating generations. The point which I want to make is that, in livestock-population studies, this type of effect would tend to be regarded as a property of the genotype and its true nature would be likely to remain undetected for several reasons. The lack of inbred lines would hinder appropriate formal genetical analysis in livestock, reciprocal crosses are often not considered and in livestock breeding the prevailing attitudes are against the notion that phenotypic characters may sometimes be transmissible by injection, unless one is dealing with an obvious disese organism. In the mouse example given, disease <u>resistance</u> was transmissible by injection.

Many examples of interference or synergistic action between different types of viruses have been found - the sarcoma viruses are a good example of synergism, where one virus supplies functions for which the other is defective, thereby enabling the defective one to cause a neoplastic disease. It is not my purpose to discuss these categories in detail but only to point out their repercussions for genetics in cases where there is some form of vertical transmission, latency or integration into the genetic apparatus.

In some cases the presence of a virus can have a non-specific effect in lowering the overall resistance of the host. Thus with cytomegalovirus infection, in mice, interferon and antibody production tend to be suppressed (Osborn & Medearis 1967), while the presence in mice of lactic dehydrogenase elevating (LDH) virus can enhance their capacity to produce antibodies to foreign protein (Notkins et al 1966) - without biochemical tests one is unaware of the presence of LDH virus in a mouse colony.

I hope that I have mentioned enough here to show the role which micro-organisms can sometimes play in producing correlated phenotypic effects in related animals. These remarks are intended to cast doubt on the common assumption in animal breeding practice that such correlations, especially if they persist for several generations must be due to chromosomal genes. Many other examples are known and some are given in an earlier paper (Dickinson & Mackay 1967). No mention has been made of some types of possibilities - for example, the possible mutagenic action of some viruses - but it is quite clear that an important part has often been found to be played by nonchromosomal inherited components, vertically transmitted agents and latent viruses in those organisms where it has been practicable to study the problem efficiently. In animal breeding more allowance for such possiblities ought to be made than is current practice even though the practical difficulties are great. It is a matter for speculation where the balance of importance lies between chromosomal and extrachromosomal factors and this will vary from case to case.

Let us return to where we began this second section. Problems of interpretation and choice of breeding policies are simple where Mendelian segregation involving major genes is found. Otherwise problems of biological interpretation arise and breeding predictions based on Mendelian assumptions may not hold or may not indicate the most efficient course of action. This points to the usefulness of basic biological investigation of the biochemical pathways underlying the economic characters which we measure because this is one of the best methods of identifying individual genes. With disease resistance characters I would stress the usefulness of finding the important gene-controlled steps in pathogenesis, especially when the long-term view is taken.

EXAMPLES OF THE VARIETY OF INHERITED RESISTANCE MECHANISMS

Before dealing with evidence for genetically controlled resistance where experimental challenge of the host has been used, some comments on natural disease are appropriate.

Everyone is familiar with groups of animals where only a few are affected by a disease outbreak. In such an event, age can be important and genetic differences can be expressed in the age of maturation of immune reactivity against foreign antigens (Billingham & Brent 1957). More basic, though, is the problem of interpretation, already mentioned, arising from uncertainty that there has been appropriate exposure of every animal to the disease organism. This restricts easy interpretation to disease where there is an efficient biological mechanism for exposure, being either highly infectious or vertically transmitted.

Even with the highly infectious diseases there are surprisingly few well documented cases which might indicate genetic control of resistance and, obviously, in these cases there can be no better attempt without experimental investigation, to distinguish chromosomal and non-chromosomal components. Natural selection of more resistant rabbit populations has been shown to occur in Australia in areas where there are recurrent outbreaks of myxomatosis (Marshall & Fenner 1958).

Breed differences in susceptibility often give the first hint that genetic variation is present, as in an outbreak of mousepox (extromelia) in a colony having many mouse strains (Briody et al 1956). Three of the strains were resistant to the disease but the C57BL strain showed only slight clinical signs, became a carrier of the virus and served to spread the disease.

Subsequent work on experimentally produced mousepox by Schell (1960a and b) emphasised the role of immune responses as the basis for the strain differences in susceptibility. When the virus is administered as a small dose by a relatively natural route - into the skin of the footpad - there is slow multiplication for several days in both susceptible and resistant mice at the site of injection; when virus reaches local lymph nodes the titre then rises higher in the susceptible ones, whereas neutralising antibodies start appearing relatively earlier in resistant mice; the disease progresses in the susceptible mice by large amounts of virus reaching target organs in the viscera. If, instead, very large doses or highly artificial routes of injection are used then there is less difference in resistance between strains; the age at exposure also affects the response. Resistant and susceptible strains could be infected by the same small dose given into the footpad but the former required 100,000 times as much virus as the latter before the result was equally lethal. In some cases change of route of injection also altered the type of histopathological damage and hence the immediate cause of death. Thus a strain which was the more resistant to footpad injection was the one which died more quickly when given large doses of the virus intranasally.

Although we do not know the detailed way in which genes may control the immune response to mousepox virus, the examples given illustrate how an inherited resistance mechanism which can efficiently control natural infection can be obscured by experimental procedures if care is not taken in the choice of methods. Alternatively, there are many examples in which genetic differences revealed by experimental methods may not be directly relevant to the disease as it occurs naturally - frequently this is due to strain differences between the microorganism used experimentally and those present in the field.

Several lines of White Leghorn poultry were selectively inbred for natural resistance or susceptibility to lymphomatosis. In a report on the results after 17 years of selection (Burmester et al 1960) it was shown that there was 24% difference in natural incidence of the disease between the extreme selection lines but when tested by injection of the RPL12 strain of visceral lymphomatosis virus the maximum difference was almost 90%. Further work with these inbreds involving diallel crossing of three lines and testing their response to injection with various doses of either RPL virus, erythroblastosis virus, myeloblastosis virus or Rous sarcoma virus showed that there were significant genetic effects with each virus but the ranking of the different genotypes showed large genotype-virus interactions. The leukosis mortality in crosses of inbred lines could not be predicted because it was often higher in the crosses than in either parent. (Crittenden et al 1964a and b, 1972.) These authors concluded that "no artificial exposure method was found which would increase specific mortality and provide efficient criteria of resistance under all circumstances".

Cole (1966) was able to select poultry for increased or decreased resistance to Marek's disease, one of the major subdivisions of avian leukosis. However, there is no genetic correlation between resistance to Marek's disease and resistance to the lymphoid leukosis group of diseases.

Several groups of workers have shown that the susceptibility of chickens to Rous sarcoma virus can be genetically determined by a single gene with susceptibility being dominant (Waters & Brumester 1961, Crittenden et al 1964b, Payne & Biggs 1964, Bower et al 1965). Indeed, this conclusion has been substantiated by segregation, population allele-frequency and variance fall-ratio types of studies. However, the control of resistance and susceptibility to RSV is far more complex than this. There are both helper viruses and interfering viruses, the presence of which can obscure the underlying genic variation. Thus in screening programmes aimed at leukosis eradication the possible presence of resistance due to Rous interfering virus (RIF) must be considered. Some strains of RSV are defective and for replication of infectious particles they require the presence of a helper virus - either Rous associated virus (RAV) or avian myeloblastosis virus (AMV). These helper viruses in turn have variants which differ antigenically and, because of this, in their host range; genes in the chickens determine which subtype of RAV or AMV can be present. A good account of this complex situation is given by Vogt & Ishizaki (1965).

The point which I have been emphasising about the need for care in avoiding hasty conclusions about genic control of resistance to disease will be obvious from this description and it should be remembered that this example concerns a disease-complex which is of considerable economic importance. It is a great help when individual segregating genes can be identified in these resistance complexes because it avoids the need to assume, without supporting evidence, that chromosomal genes are involved. In microbiology, Koch's postulates were formulated almost 100 years ago in order to ensure that an organism thought to cause a disease was in fact the cause and not some irrelevant secondary invader; the organism must always be present in the disease, it must be isolated, purified and then cause the disease when injected into healthy hosts, from which it must be capable of reisolation and purification. Huebner (1966) has given an amusing, up-to-date, version of Koch's postulates, thought necessary because of the unreliability of many "discoveries" in this field. I have often felt that livestock quantitative genetics - not only in disease studies - needs a comparable series of "postulates" to qualify the prevailing assumption that what is transmitted from generation to generation must necessarily follow the rules which apply to chromosomal genes. Animal breeding has strayed too far from the rigorous rules which Mendel was able to follow.

Fortunately, many examples appear to be biologically much simpler than fowl leukosis. Examples of genetical control in poultry, where reciprocal crosses between susceptible and resistant lines have been made, include Newcastle disease (Reta et al 1963), infectious bronchitis (Purchase et al 1966), pullorum disease (Roberts Card 1935) and coccidiosis (Rosenberg et al 1954).

Only in a minority of animal diseases which have been studied have single gene effects yet been found with any degree of certainty. These include the arbovirus B group dominant resistance gene in mice (Lynch & Hughes 1936), three genes involved in avian RNA tumour virus susceptibility (Pani 1976), and a gene showing dominant susceptibility to mouse hepatitis virus (Bang & Warwick 1960) and a gene in sheep giving dominant susceptibility to injection of the SSBP/1 source of scrapie agent (Dickinson et al 1968).

Only a few genes are involved in some mouse strain differences in resistance to mammary tumour agent (Heston et al 1956). This was shown by the rate of elimination of the agent in serial back-crossing to the resistant strain and it could also be shown that the contribution of the resistance genes had not been to cause the virus to change to a temporary latent form.

In the case of most diseases, many genes and, often, complex dominance relationships are involved. This has been shown, for example, in mice in the case of tuberculosis (Lynch et al 1965), foot and mouth disease (Subak-Sharpe 1961) and salmonellosis (Webster 1937).

Where a genetic disease resistance mechanism can be shown to occur in livestock it is of clear potential importance in its own right but the merit of such variation in a mouse may not seem so obvious to some. Genetics has proved a powerful tool in understanding the cause of some diseases, in understanding the pathogenesis of others - revealing the weak links in an animal's defences or the ways in which it wards off infection and disease - and, recently, some contributions have been made to our fundamental understanding of cell responses to infection. The following examples will illustrate some of these points.

The molecular nature of the agent which causes the neurological disease, scrapie, in sheep has not yet been determined. It is very small for a virus but different strains of the agent exist and these differences are copied exactly during replication; however, some workers are reluctant to accept it as a virus because of the high degree of resistance of the agent to inactivation by chemical or physical abuse. In mice an autosomal gene <u>(sinc)</u> with two alleles, s7 and p7, controls the incubation period from 4½ months to over 2 years, depending on strain of agent, route of injection and dose. Two groups of agents are known and these can only be distinguished by their incubation periods in s7s7 and p7p7 types of mice; one group has shorter incubation in the s7s7 genotype, the other group being shorter in p7p7 mice. With some agents the p7 allele is dominant; with other agents there is little or no dominance, whereas for others the s7 allele is dominant. (Dickinson & Meikle 1971, Dickinson 1975.) The evidence is that the <u>sinc</u> gene product forms a multimeric structure which is used by scrapie agents as a replication site. The function of these sites in the normal animal is unknown but there appears to be a very small number available for scrapie replication and this shortage acts as the limiting factor in this type of slow disease. The animal does not appear to have any immune defence against scrapie - indeed, some parts of the conventional immune system appear to operate to the host's disadvantage because immunologically immature or immunosuppressed mice are more difficult to infect (except by direct injection into the brain) than immunologically competent ones (Dickinson & Fraser 1976).

There has been recurrent interest in interferons as possible substances for antiviral therapy since their discover 20 years ago. However, attempts to understand the basis for mouse resistance to such arboviruses as West Nile virus have put the role of the interferon in a rather different perspective. Goodman & Koprowski (1962) have shown that the normal course of the disease is similar in both resistant and susceptible mice in the early course of the infection with the virus multiplying in the spleen and liver; their difference starts to show later in the first day after infection when a small proportion of macrophages in the susceptible mice support virus multiplication and probably serve to transport virus to the ultimate site in the brain, where the resulting inflammation leads to death. The macrophage step is effectively blocked in resistant mice. Neither neutralising antibodies nor non-specific serum inhibitors play any significant part in the difference between the types of mice. "Resistance" can be overcome by giving large doses intracerebrally and as Vainio et al (1961) have shown interferon production is simulated by the virus but the susceptible mice produce more interferon than the resistant ones; interferon production is here an indicator, not of greater resistance, but of greater virus multiplication.

Several of the genetic mechanisms which act in the intact animal also act at the cellular level, for instance in tissue cultures. Where tissue extracts can temporarily override the genetic potential of cells in terms of resistance or susceptibility, there is good prospect of understanding key steps in the disease in chemical terms. A single gene effect where such temporary susceptibility can be conferred on cultured "resistant" cells concerns mouse hepatitis virus (Kantoch & Bang 1962). Various studies in this disease have shown the key role played by macrophages and it is by way of this type of cell that the gene difference is expressed.

Other examples of control of resistance being expressed at a cellular level have been found with Rous sarcoma virus in poultry (Crittenden et al 1963) and mouse mammary tumour virus (Nandi et al 1966). On the basis of differences between six mouse strains in susceptibility to Salmonella infection, it was at one time thought possible that the degree of resistance was related to the total number of leukocytes in the blood (Gowen & Calhoun 1943). This was subsequently shown to be an incorrect interpretation by genetic selection methods. Thus Weir & Schlager (1962) were able, during the course of 25 generations of selection with inbreeding, to increase the number of leukocytes per cubic millilitre in one line from 9,000 to 17,000 and in the other line to decrease it to 4,000. When challenged with Salmonella, both "high" and "low" leukocyte numbers, per se, are a disadvantage unless the leukocytes are also able to digest the bacteria.

It may be that in salmonellosis genetical differences are expressed by way of the leukocytes other than in terms of their total numbers, because Sultzer (1968) has shown genetic control of leukocyte response to bacterial endotoxins, including toxins from Salmonella. However, of more immediate relevance to a positive understanding of the role of resistance mechanisms in Salmonella infection are the studies of Oakberg (1946) and Bohme et al (1959). In the resistant strains the rate of damage in the spleen was lower but the degree of liver damage was higher than in the more susceptible mouse strains. Surprisingly, the susceptible mice appeared to have the more efficient lymphoreticular systems. The effective genetic difference is, however, expressed in several ways - the resistant mice are better able to repair liver damage and glycogen metabolism in liver areas adjacent to damage is relatively unimpaired, whereas bacteria only multiply within macrophages in the susceptible mice. Unfortunately, the "resistant" mice can become persistent carriers of the infection.

It may be a surprise to some that the genetic differences in many cases do not involve a leading role for specific antibodies, even though these may be produced. Specific antibodies are formed in Salmonella infections (Gorer & Schutze 1938) but major genetic differences were not found in this respect. Specific antibody production can be higher in the more susceptible individuals and may merely be an indicator of the presence of abundant antigen. An example of this was with Marek's disease of chickens where indirect haemagglutination antibodies were measured (Han & Smyth 1972).

Antigenic similarities between host and microorganism are another aspect of our problem and it is possible that these are responsible for the maintenance of antigenic polymorphism in many populations. Viruses which mature by budding from host cell surfaces can incorporate host-specified cell-surface antigens. The example of the Rous complex of viruses in chickens in which this effect occurs has already been given. Here the helper virus (RAV) carries the antigenic specificity of its most recent host genotype.

The more antigenic specificities a microorganism shares with its host, the more it may be able to evade some of the host's defences because of immunological tolerance but another situation could arise where there are partial similarities; efficient antibody production could then lead to autoimmune damage if there was cross reaction with some of the host's tissues. Much of the speculation whether multiple sclerosis in humans has a viral trigger is based on this concept, even though the tissue antigens which have a somewhat higher frequency in cases of M.S. are not always the same in different populations (Compston et al 1977).

A wide variety of viruses have now been found to produce characteristic antigenic changes in host cells (Breyere & Williams 1965, Zinkernagel & Doherty 1974, Garrido et al 1976, review by Doherty et al 1977). These include tumour viruses, ones leading to acute disease and ones which can become persistent infections. Infected, antigenically changed, macrophages play a leading role in this important line of the animal's defences and this mechanism appears to provide an explanation for the occurrence and polymorphism of histocompatibility antigens. Thymus-derived lymphocytes are able to detect the infectively produced antigenic changes on cells with which these lymphocytes share some histocompatibility antigens. This results in a sequence of events leading to efficient cell-mediated lysis of infected cells. As a result of these findings there has been a great deal of theoretical speculation and reasons have even been put forward for an evolutionary advantage of heterozygosity per se at loci determining histocompatibility antigens (Doherty et al 1976).

The continued existence of the different ABO-system blood group alleles in man has been the basis for much conjecture. It has been shown that antibody on the respiratory mucesa is important in resistance to respiratory infection with some picernaviruses and myxoviruses (e.g. influenza, type C): individuals of type A are less frequently affected with influenza virus but are relatively more frequently affected with adenevirus (McDonald & Zuckerman 1962, Tyrrell et al 1968).

Two lines of mice have been developed by selection which produce either high or low amounts of antibodies in response to a variety of complex antigens (Biozzi et al 1971). This difference was not antigen-specific and related to the rate of proliferation and differentiation of cell-lineages responsible for antibody production. Infection with a microorganism likewise introduces not one antigen but many antigenic specificities and it is probable that numerous genes will be involved in the overall response, which makes genetic analysis of the details difficult.

In contrast, work using chemically simple antigens has been of great importance because it has lead to the elucidation of the genetic control of several aspects of immune responses in antibody and cellular terms and several of the genes have been located. Single genes were found a number of years ago which control reaction to specific antibodies (Cinader & Dubiski 1964) but, more recently, much attention has been concentrated on the linkage complexes between genes for cell surface antigens determining histocompatibility and a number of genes governing different aspects of immune reactions such as antibody production, cell mediated reactions and complement components.

Work with mice and guinea pigs was predominant in the early work of this type and several genes were identified where the dominant allele permitted both high antibody responses and good cellular immunity, in contrast with the absence of cellular immunity and total or partial deficiency of antibody production in the recessive homozygotes (McDevitt & Benacerraf 1969).

It soon became clear that many, but not all, the genes controlling immune response to antigens were genetically linked with histocompatibility genes. In particular, that there was a major histocompatibility complex locus (MHC) that included many genes controlling various immunological functions and that there are homologous versions of the MHC in different mammalian species (e.g. H-2 in mice, HLA in humans). Although many immune response (Ir) genes have been located, and many others are likely to be found, it is clear that there are fewer genes than immunologically different specificities to which antibodies can be made (see Benacerraf & McDevitt 1972).

The relative stability of these complex loci over a vast evolutionary time scale points to the importance of this association of genes controlling different facets of the immune system. These findings have caused a lot of excitement among research workers in this field but the events are too recent for a good perspective to have emerged. Perhaps a few points should be mentioned which could have a bearing on the assessment of the position. There is the possibility that the immune system is a specialised branch of the more basic control of tissue differentiation and morphogenesis (see Pierpauli & Serkin 1972) and it is likely that some of the cell mediated mechanisms that have been mentioned are involved in morphostatic control, that is, in monitoring the body for cells with neoantigens which could develop into tumours. So the evolutionary stability of the complex loci will only result in part, if at all, from the need to react to infections. Also, there are various types of cell-mediated cytotoxic reactions and they do not all depend on the cells sharing common major histocompatibility antigens (Zinkernagel & Oldstone 1976). Further, the genetic control of cell mediated responses is not always linked to genes for serologically detectable antigens - the mixed lymphocyte reaction in mice is one exception (Festenstein et al 1972).

The MHC linkage associations have the practical consequence that markers such as histocompatibility and Ia antigen type may in some cases give a reliable indication of the liability of an individual to develop a disease, if there is close linkage with a gene controlling an immune response, which is crucial for that disease. Marek's disease and the marker B^{21} in chickens (Briles et al 1977) and ankylosing spondylitis and the HL/a /b27 antigen in humans are good examples with high correlations. However, there are many examples where the correlations are low and the marker is of little predictive value.

One of the potentially important consequences of having markers available is that they may help in the classification of diseases of uncertain etiology. Where different but related diseases are being lumped together as a single disease it may be possible to differentiate them using markers, which would be essential if different treatments were required.

We need, however, to avoid exaggerating the usefulness of such markers for a number of reasons. Increased resistance to one disease may be accompanied by higher susceptibility to another, either due to linkage of determinants or for physiological reasons. Increased immune responses are not necessarily an advantage. For example, autoimmune conditions or some adrenavirus infections (e.g. visna in sheep) where the main damage to tissues is caused by the immune reactions to the virus. Similarly, in Aleutian disease of mink, where the ones carrying the Aleutian gene succumb more rapidly to the infection, the lethality is due to large amounts of immune complexes blocking the kidneys and producing glomerulenephritis (Henson et al 1969).

THE NEED FOR AN INTEGRATED APPROACH TO DISEASE CONTROL

Now that many of the 'simpler' diseases, particularly those caused by bacteria, are easily dealt with by standard veterinary prophylaxis and therapy, it is an unfortunate fact that the remaining problems tend to be with the more intractable diseases. Even so, there is still much to be gained by having a thorough understanding at a fundamental level, including genetic aspects, of the diseases which are at present under control.

On the question of the initial investigation of novel or perplexing diseases, what strategies should be adopted?

There have been many attempts in the short history of genetics to propose simple Mendelian explanations, on the basis of inadequate evidence, for various diseases. Three recent examples are Border disease in sheep, scrapie in sheep and kuru in humans. In the two sheep diseases the confusion arose, at least in part, because of vertical transmission of the infective agent, which was shown to be occurring (Border disease, Dickinson and Barlow 1967; scrapie, Dickinson et al 1964). With both kuru (McArthur 1964) and scrapie the inappropriateness of a simple genic hypothesis was originally shown by analysis using methods of population genetics and finding that the solely genic hypothesis would not fit the epidemiology of the disease. It was subsequently, in the case of kuru, that a transmissible agent was found to be the cause of the disease (Gajdusek et al 1966).

When a disease of unknown cause has to be dealt with, genetical methods of analysis, as defined in the broadest sense, should be considered as tools having a part to play in the primary investigation of the problem along with bacteriology, virology, histopathology, epidemiology, immunology and nutritional analysis. Unless this is done from an early stage, when the immediate cause of the disease has been determined - the microorganism, in the present context - it will not be possible to assess what immediate and long-term policies to adopt. Or, to put this more crudely, if only microbiological methods of investigation have been used there is a danger that microbiological techniques will appear to offer the only potential means of attack. Even where a concerted effort on a broad scientific front has been used there is the temptation, once a short-term expedient has been found, to discount the long-term benefit which further genetical investigation may offer.

The major objective for each disease should be to understand the sequence of steps in pathogenesis and the points at which the host could, given the appropriate genes, prevent further progress of the disease. Given this information one can much more confidently adopt short-term expedients, with the assurance that changes of policy, which may become necessary, can be decided in advance and be well founded. The ultimate objective is probably not the eradication of the infectious organism and the 'susceptibility genes' - the first may not be practicable and the second may not be desirable.

We next turn to the area of the subject which some may have assumed to be the only aspect of importance, namely, selection for increased disease resistance. As a preliminary to this one point of a very general nature should be considered.

Selection programmes often involve some degree of inbreeding and the net result is that, besides achieving some progress in the desired direction, genetical variation becomes reduced in characters of no immediate interest - this is especially liable to happen when the population size in the programme is small. This tendency to throw away genes of no current value is acceptable when there are sufficient selection lines and a diversity of outbred reserves is available. But the trend in many national schemes, particularly in some species, is to take the situation to extremes, effectively involving whole populations in selection programmes and permitting the number of established breeds to decrease. If the available gene pool is reduced, all characters are affected but the situation may be more acute for genes controlling disease resistance. Susceptibility genes when in low frequency may be unnoticed and no disadvantage because their infrequency inhibits pathogen spread. But the possibility obviously exists that some susceptibility genes could be concentrated, undetected, in a selection line and that we would only be aware of this when conditions had been created which would permit a large outbreak to occur. This possibility is quite distinct from the one mentioned previously which concerned the selection of new strains of a pathogen.

The actual procedures for selecting for resistance vary with different diseases and the more basic information we have about a particular disease, the more appropriate and efficient the overall selection can become.

In most diseases no genes are known which either act as markers or which control important steps in pathogenesis and little may be known of the physiological variables which affect the disease. In this case selection will depend on challenging the animals with the disease organism or awaiting its natural occurrence. There is no doubt that this can produce a relevant selection response in some cases but a number of the possible long-term repercussions have already been mentioned. There can also be short-term drawbacks involving slow rates of progress and hindrance of other breeding objectives. For instance, with some diseases it is not possible to challenge all the animals and then breed from more resistant ones or from survivors. In this case either progeny testing is needed or the animals must be bred before they are challenged, the progeny retained and subsequently bred according to the parental disease incidence. With a disease of long incubation period, I am unfortunately well aware from personal experience how slow this process can be in livestock. In my own case, even though an important major gene was eventually found (Dickinson et al 1968), the selection programme had to be formulated on the assumption that many genes each of small effect were operating: it was also important to have a selection programme which minimised inbreeding - but that is another question.

When sufficient is known about the disease to select on physiological or biochemical characters or on markers, instead of having to rely on challenge with the microorganism, this can be a great help. A good, early, example of this was given by the work of Hutt & Crawford (1960) with pullorum disease of chicks. This is caused by a bacterium, Salmonella pullorum, which can be transmitted from an infected hen through the eggs. More resistant stock can be produced by genetic selection and the evidence suggested that the resistant stock were better able to raise their body temperature to a "warm-blooded" status by 10 days old. The possibility existed that this ability might be largely responsible for their resistance, because the peak of mortality occurred in the susceptibles shortly after 10 days old. This causal relationship between susceptibility and delayed change to a warm-blooded condition was strongly supported by a selection experiment carried out in the absence of *Salmonella pullorum*. The selection criterion was average rectal temperature during the first six days and when first- and second-generation samples of the high- and low-temperature selection lines were challenged with *Salmonella pullorum*, the survival rates were directly related to the temperature status of the lines.

The ability to select for disease resistance characters in the absence of the disease organism is obviously a great advantage, because there is then no risk of the disease spreading to animals outside the test.

The other resort of the animal breeder seeking various genetic attributes including higher disease resistance, is to import novel breeds from other areas in the hope that new relevant genes can be introduced from the imports into local breeds. This is the point at which there can be a conflict of interest between the geneticist and those responsible for animal health regulations. So as to avoid the disaster of introducing new pathogens or more virulent strains of existing ones, the regulatory authorities normally insist on a quarantine period. However, the short quarantine periods appropriate for many acute infections are quite irrelevant for many of the diseases which have been mentioned in this paper. The trend has, therefore, been to prolong quarantine periods, not only to many months or years but in some cases to insist on the breeding of one or more generations in quarantine. For example, some countries adopt such a policy for sheep imports in an attempt to prevent the introduction of scrapie. Others ban all imports of sheep.

However, there is an accompanying trend which highlights the dilemma in such import regulations. The more isolated a country or region becomes the greater the risk of epidemic can be if certain pathogens are introduced accidentally. Foot and mouth disease in several countries is one example, another is the introduction of a new breed of sheep into Iceland in 1933 leading to various epidemics (Palsson 1976).

Genetical techniques for control should only rarely be considered for many of the simpler diseases that are adequately controlled by therapeutic measures even though selection for resistance would often be effective: the important criterion is to decide which is the most economical policy. The genetical objectives should be concentrated on the remaining, more complex diseases and some, such as respiratory diseases, are likely to increase in animal rearing. These trends, because they lead to more standard conditions of nutrition and greater uniformity of the physical environment, will make it even more important to study and understand the relationship between genetical resistance and physiological conditions.

What I have been attempting to illustrate is the need for a flexible policy of disease control or eradication. We must choose in each case what role genetics should play, based on the likely efficiency of selection. This efficiency will usually be increased by having basic information about the disease derived both from the livestock and from experimental animals. We need to know the component steps of susceptibility and resistance at the whole animal, cellular and molecular levels in order to identify the weak links which can be exploited genetically: we need to know which type of immune responses, if any, are important for the diseases in question and how they are controlled by the genotype. We should attempt to determine the range of variation of different strains of the pathogens and assess the likelihood that prolonged genetical effort might be nullified by changes in the pathogen. Such information can be used predictively for both short-term and long-term objectives, so that, when genetic selection seems appropriate, it can be efficient.

Many geneticists advocate for animal breeding a quantitative selection index approach, with the index having been formulated with little reference to the underlying biological variables: in particular, where diseases are concerned, I regard this as unsatisfactory. I hope that it has been shown that such an approach, on its own, is not necessarily as relevant, efficient and free from drawbacks as some appear to believe, nor is the quest for individual genes as short-sighted for our problem as current fashion may indicate.

REFERENCES

Bang, F.B.;Warwick, A. (1960) Proceedings, National Academy of Sciences, USA, 46, 1065.

Benacerraf, B.;McDevitt, H.O. (1972) Science 115,273.

Billingham, R.E.;Brent, L. (1957) Transplantation Bulletin 4,67.

Biozzi, G.;Stiffel, C.;Mouton, D.;Bouthillier, Y.;Decreusefond, C. (1971) In 'Progress in Immunology' (Ed) Amos. Academic Press N.Y.,529.

Bohme, D.H.;Schneider, H.A.;Lee, J.M. (1959) Journal of Experimental Medicine 110,9.

Bower, R.K.;Gyles, N.R.;Brown, C.J. (1965) Genetics 51,739.

Briles, W.E.;Stone, H.A.;Cole, R.K. (1977) Science 195,193.

Briody, B.A.;Hauschka, T.S.;Mirand, E.A. (1956) American Journal of Hygiene 63,59.

Burmester, B.R.;Fontes, A.K.;Waters, N.F.;Bryan, W.R.;Groupe, V. (1960) Poultry Science 39,199.

Camyre, K.P.;Pfau, C.J.(1968) Journal of Virology 2,161.

Cinader, B.;Dubiski, S. (1964) Nature (Lond.), 202,102.

Cole, R.K. (1966) Poultry Science 45,1077.

Compston, D.A.S.;Batchelor, J.;McDonald, W.I. (1977) Lancet, i, 1123.

Crittenden, L.B.;Okazaki, W.;Reamer, R.H. (1963) Virology 20,541.

Crittenden, L.B.;Purchase,H.G.;Okazaki, W. (1964a) Poultry Science 43,1310.

Crittenden, L.B.;Okazaki, W.;Reamer, R.H. (1964b) Proceedings, International Conference Avian Tumour Viruses. National Cancer Institute Monograph, 17,161.

Crittenden, L.B.;Purchase, H.G.;Solomon, J.J.;Okazaki, W.;Burmester, B.R. (1972) Poultry Science 51,242.

Darcel, C. le Q. (1973) Canada, Dept. Agric. Monograph 8, Ottawa.

Dickinson, A.G. (1975) Genetics 79,387.

Dickinson, A.G.;Barlow, R.M. (1967) Veterinary Record 81,114.

Dickinson, A.G.;Fraser, H. (1977) In 'Slow virus infections of the central nervous system'. (Ed) ter Meulen and Katz, Springer-Verlag, N.Y.,3.

Dickinson, A.G.;Mackay, J.M.L. (1967) In 'Methods in Virology'. Ed. Maramorosch and Koprowski, Vol.1. Acad. Press, N.Y.,19.

Dickinson, A.G.;Meikle, V.M.H. (1971) Molecular and General Genetics 29.91.

Dickinson, A.G.;Stamp, J.T.;Renwick,C.C.;Rennie, J.C. (1969) Journal of Comparative Pathology 78,313.

Dickinson, A.G.;Young, G.B.;Stamp, J.T.;Renwick, C.C. (1965) Heridity, 20,485.

Doherty, P.C.;Blanden, R.V.;Zinkernagel, R.M. (1976) Transplantation Review 29,89.

Festenstein, H.;Abbasi, K.;Sachs, J.A.;Oliver, R.T.D. (1972) Transplantation Proceedings 4,219.

Gajdusek, D.C.;Gibbs, C.J.; Alpers, M. (1966) Nature (Lond.), 209,794.

Garrido, F.;Festenstein, H.;Schirrmacher, V. (1976) Nature (Lond.), 261,705.

Goodman, G.T.;Koprowski, H. (1962) Journal of Cellular and Comparative Physiology 59,333.

Gowen, J.W.;Calhoun, M.L. (1943) Journal of Infectious Diseases 73,40.

Gross, L. (1959) Proceedings, Society for Experimental Biology and Medicine 100,102.

Grossberg, S.E.;Scherer, W.F. (1966) Proceedings, Society for Experimental Biology and Medicine 123,118.

Han, P.F.-S.;Smyth, J.R. (1972) Poultry Science 51,975.

Hedgecock, L.W. (1966) Journal of Bacteriology 92,439.

Henson, J.B.;Gorham, J.R.;Padgett, G.A.;Davis, W.C. (1969) Archives of Pathology 87,21.

Herzenberg, L.A.;Tachibana, D.K.;Herzenberg, L.A.;Rosenberg, L.T. (1963) Genetics 48,711.

Huebner, R.J. (1966) Avian Diseases 10,441.

Hutt, F.B. (1958) Genetic resistance to disease in domestic animals. Constable, London.

Hutt, F.B. (1965) In 'Genetics Today'. (Ed) Geerts, Vol.3, Macmillan, N.Y., 775.

Hutt, F.B.;Crawford, R.D. (1960) Canadian Journal of Genetics and Cytology 2,357.

Jaenisch, R. (1976) Proceedings, National Academy of Sciences, USA 73,1260.

Kantoch, M.;Bang, F.B. (1962) Proceedings, National Academy of Sciences, USA 48,1553.

Leftwich, W.B.;Mirick, G.S. (1949) Journal of Experimental Medicine 89,155.

Lilly, F. (1966) Genetics 53,529.

Lindenmann, J.;Lane, C.A.;Hobson, D. (1963) Journal of Immunology 90,942.

Lynch, C.J.;Hughes, T.P. (1936) Genetics 21,104.

Lynch, C.J.;Pierce-Chase, C.H.;Dubos, R. (1965) Journal of Experimental Medicine 121,1015.

McDevitt, H.O.;Benacerraf, B. (1969) Advances in Immunology 11,31.

McDonald, J.C.;Zuckerman, A.J. (1962) British Medical Journal 2,89.

Martin, W.B.;Scott, F.M.M.;Sharp, J.M.;Angus, K.W.;Norval, M. (1976) Nature (Lond.) 264,183.

Nandi, S.;Bern, H.A. (1960) Journal of the National Cancer Institute 24,907.

Nandi, S.;DeOme, K.B. (1965) Journal of the National Cancer Institute 35,299.

Nandi, S.; Handin, M.;Robinson, A.;Pitelka, D.R.;Webber, L.E. (1966) Journal of the National Cancer Institute 36,783.

Nathanson, N.;Panitch, H.;Petursson, G. (1976) In 'Slow virus diseases of animals and man'. (Ed) Kimberlin, North Holland, Amsterdam, 115.

Notkins, A.L.;Mahar, S.;Scheele, C.;Goffman, J. (1966) Journal of Experimental Medicine 124,81.

Osborn, J.E.;Medearis, D.N. (1967) Proceedings, Society for Experimental Biology and Medicine 124,347.

Palsson, P.A. (1976) In 'Slow virus diseases of animals and man'. (Ed) Kimberlin, North Holland, Amsterdam, 17.

Pani, P.K. (1976) Journal of General Virology 32,441.

Parry, H.B. (1962) Heredity 17,75.

Payne, L.N.;Biggs, P.M. (1964) Virology 24,610.

Pierpaoli, W.;Sorkin, E. (1972) Experientia 28,1385.

Pollard, M.;Sharon, N.;Teah, B.A. (1968) Proceedings, Society for Experimental Biology and Medicine 127,755.

Purchase, H.G.;Cunningham, C.H.;Burmester, B.R. (1966) Avian Diseases 10,162.

Reta, G.;Bohren, B.B.;Moses, H.E. (1963) Poultry Science 42,1182.

Rich, F.A. (1923) Vermont Agricultural Experimental Station Bulletin 230.

Roberts, E.;Card, L.E. (1935) Illinois Agricultural Experimental Station Bulletin No.419.

Rubin, H.;Cornelius, A.;Fanshier, L. (1961) Proceedings, National Academy of Sciences, USA 47,1058.

Sabin, A.B. (1952) Proceedings, National Academy of Sciences, USA 38,540.

Schell, K. (1960a) Australian Journal of Experimental Biology and Medical Science 38,271.

Schell, K. (1960b) Australian Journal of Experimental Biology and Medical Science 38,289.

Siegel, M.M. (1952) Annual Review of Microbiology 6,247.

Spiegelman, S.;Axel, R.;Baxt, W.;Kufe, D.;Schlom, J. (1975) Genetics 79 (Suppl.),317.

Subak-Sharpe, H. (1961) Archiv fur die gesamte Virusforschung 11,39.

Sultzer, B.M. (1968) Nature (Lond.), 219,1253.

Traub, E. (1960) Zentralblatt fur Bakteriologie und Parasitenkunde Abt I. Originale 144,472.

Tyrrell, D.A.J.;Sparrow, P.;Beare, A.S. (1968) Nature (Lond.) 220,819.

Vainio, T.;Gwatkin, R.B.L.;Koprowski, H. (1961) Virology 14,385.

Vogt, P.K.;Ishizaki, R. (1965) Virology 26,664.

Walker, J.C. (1951) In 'Genetics in the 20th Century' (Ed) Dunn, Macmillan, N.Y.,527.

Waters, N.F.;Burmester, B.R. (1961) Journal of the National Cancer Institute 27,655.

Webster, L.T. (1937) Journal of Experimental Medicine 65,261.

Weigand, H.;Hotchin, J.E. (1961) Journal of Immunology 86,401.

Weir, J.A.;Schlager, G. (1962) Genetics 47,1199.

Zinkernagel, R.M.;Doherty, P.C. (1974) Nature (Lond.), 248,701.

Zinkernagel, R.M.;Oldstone, M.B.A. (1976) Proceedings, National Academy of Sciences, USA 73,3666.

THE IMMUNE RESPONSE IN SHEEP

DR P.W. WELLS

Animal Diseases Research Association, Moredun Institute, Edinburgh

An immune response assists and augments pre-existing non-specific means by which sheep, like other animals, resist infection with the multitude of micro-organisms encountered in their environment. There are a number of non-specific factors which constitute the background of innate defence mechanisms upon which the specific immune response is superimposed.

NON-SPECIFIC FACTORS

An example of such a non-specific factor is observed in the respiratory tract secretions of sheep. This substance is of large molecular size and has certain inhibitory effects on paramyxoviruses. Among the viruses affected by this inhibitor is parainfluenza type 3 (P13) virus of sheep, which may be associated with respiratory disease. Perhaps the most important non-specific factor operating in host resistance to viruses is interferon. This protein is produced by cells in response to virus infection or stimulation by various inducers and exerts a suppressive effect on the growth of viruses in other cells by interfering with the synthesis of new virus nucleic acid. Other non-specific factors include certain fatty acids in body secretions which lower pH and directly affect bacteria. Mucus in the respiratory tract hampers virus infection of cells and aids the ciliated epithelium in the clearance of invading bacteria. Regarding defence against bacterial infection, phagocytosis is one of the most important innate mechanisms. This ability to ingest bacteria is largely associated with the macrophages of the reticulo-endothelial system and the polymorphonuclear leucocytes present in the peripheral blood. Once ingested the bacteria are exposed to the detrimental effects of enzymes and other factors contained in the lysosomes of the phagocytic cells. Although effective in many instances the contents of lysosomes cannot control the intracellular multiplication of all bacteria and it is in such instances that the immune response may exert its influence.

IMMUNE RESPONSE

An immune response is the specific response of an animal which results from introducing into the body by infection or other means materials which are recognised as foreign. This specific immune response may result in immunity and can be acquired either actively or passively. Active immunity may be stimulated by infection or vaccination and passive immunity may result from transfer of maternal immunity or from administration of specific antisera.

Actively acquired immunity is relatively long lasting and may persist throughout the life time of the sheep whereas passively acquired immunity is transient, lasting a matter of weeks. Passively acquired immunity is mediated humorally while actively acquired immunity is mediated by cellular and humoral mechanisms. In both passively and actively acquired immunity the immune processes are considered to be capable of acting at both systemic and local levels.

Humoral Immunity

Antibodies are responsible for the mediation of humoral immune responses. They constitute a heterogeneity of molecules which exhibit antibody activity but which are all glycoproteins present in the globulin fraction of serum. Known as immunoglobulin, in the ovine species three distinct classes are distinguished on the basis of their physical, chemical and antigenic characteristics; IgG, IgM and IgA.

IgG is the major immunoglobulin constituent of sheep serum, with the two subclasses IgG_1 and $2gG_2$ present in similar concentrations. Antibody molecules of the IgG type have the ability to combine with the specific antigen at two antigen combining sites. In contrast each molecule of IgM antibody appears more efficient, being able to combine with specific antigen at up to 10 sites. Except however in the early phase of an immune response, antibodies are predominantly of the IgG class in serum. Antibody activity in the secretions of the ovine respiratory and gastro-intestinal tracts is largely associated with IgA. As yet IgE has not been identified in sheep although the immediate-type hypersensitivity reactions, mediated by this immunoglobulin in other species, are recognised.

Apart from certain deleterious effects such as hypersensitivity reactions, in general the biological activity of antibodies are concerned with facilitating the existing host defences. Opsonisation is an example of the interdependence of an innate defence mechanism with the specific immune response. The combination of antibody molecules with micro-organisms bearing antigens, for which the antibody has specificity, can clump the organisms. The antibody agglutinated organisms adhere more readily to macrophages because these cells have receptors for the Fc portion of the immunoglobulin molecule and phagocytosis is promoted. Activation or fixation of the compliment system is another important result of an antigen-antibody interaction. In sheep, IgM and IgG_1 antibodies are largely responsible for triggering the series of enzyme reactions of compliment activation which result in opsonisation and cell lysis with release of chemotactic factors. Other important biological activities of antibodies include neutralisation of bacterial exotoxins and the prevention of virus infection of cells by stereochemical inhibition of virus absorption on to cells.

Cell-mediated Immunity

Cell-mediated immune responses are initiated by a particular series of lymphocytes which develop under the influence of the thymus and which have become known as T-lymphocytes. T-lymphocytes of the appropriate specificity interact with antigen and as a result of this interaction the lymphocytes release lymphokines. These non-antibody biologically active factors have a number of properties which largely involve recruitment of non-specifically committed cells to become involved in the inflammatory defence response. Lymphokines cause chemotaxis of polymorphs, macrophages and lymphocytes. They inhibit movement of the phagocytic macrophage and also activate these cells to increase their capacity for ingestion and destruction of bacteria. In addition, they exhibit interferon-like activity in preventing virus infection of cells. An important function of lymphocytes themselves in the control of virus infections may be their specific recognition and lysis of virus infected cells.

The classical manifestation of a cell-mediated immune response is the inflammatory response of delayed-type hypersensitivity observed in the skin of cattle reacting to the tuberculin test. Such intradermal testing provides valuable measurements of cell-mediated immune responses in vivo. A number of in vivo assays have been developed all of which appear to have in common difficulties as regards reproducibility and interpretation. In general they depend upon measurement of lymphokine release or lympocyte proliferation resulting from the specific interaction of lympocytes with antigen.

DEVELOPMENT OF THE IMMUNE RESPONSE

The ability of sheep to respond to antigenic stimuli is acquired during the course of foetal and neonatal development. The immunological capability of the developing lamb does not suddenly switch on at some point during gestation and become capable of responding to all antigens. Antibody to certain antigens can be produced as early as the 53rd day of gestation and the property to reject allogeneic skin-grafts, a cell-mediated immune function, is acquired around the 77th day. Unless however a foetus is exposed to infection in utero, the lamb is born in a state of immunological virginity. To overcome this immunologically compromised situation, the transfer of non-specific factors and specific antibody to the newborn lamb takes place through the ingestion of colostrum.

PASSIVELY ACQUIRED SPECIFIC IMMUNITY

The newborn lamb has no detectable immunoglobulin in its serum prior to suckling. Colostrum contains high concentrations of antibody to the organisms encountered by the ewe in the environment in which the lamb should find itself after birth. These antibodies are of the IgG_1 class of immunoglobulin and are selectively transferred from serum into the colostrum by the mammary tissue during the last two to three weeks of gestation. The newborn lamb is capable of absorbing intact IgG from the small intestine from the first day of life and it would appear that approximately 80% of lambs acquire adequate levels of maternal antibody in this way. There are a number of factors, such as the amount and quality of colostrum ingested and the time at which suckling first occurs, which may account for the wide variation in amount of maternal antibody acquired by newborn lambs. Antibody absorbed into the serum has been shown to pass into the secretions bathing the cornea and respiratory tract, offering local protection against infection at a time when local synthesis of IgA antibody is just beginning.

This passively acquired antibody is important in helping the lamb over the critical early weeks of life but its presence is transient, having a half life in serum of about two weeks. Also the amount of antibody ingested to offer some local protection in the gut, even after absorption has ceased, declines rapidly during the first week with the transitiion from colostrum to milk.

One possible disadvantage of the passively acquired antibody may be its immunosuppressive effect on vaccines administered early in life. It seems however that priming is still possible under the protection of maternal antibody if certain types of vaccine are used. Louping ill virus vaccine given at 10 days of age may not result in an obvious primary immune response superimposed on a declining serum titre of maternally derived antibody. Nevertheless, on subsequent injection of antigen at 6 months of age a typical accelerated and enhanced secondary immune response results.

IMMUNITY TO VIRUS INFECTIONS

Both cell-mediated and humoral immune responses may be involved in the resistance of sheep to infection with viral agents. The relative importance of these mechanisms probably depends upon such factors as the route of infection and the mode of spread within the body. Let us consider P13 virus, affecting the respiratory tract of sheep. Despite the naturally occurring inhibitors to this virus present in the respiratory tract secretions, P13 is capable of establishing an infection when virus is instilled into the nasal cavity. Passively acquired maternal antibody present in the serum and respiratory secretions of lambs can prevent this infection, indicating the significance of specific, acquired humoral immunity in protecting against this virus. Cell-mediated immunity would appear to be unimportant as an immune mechanism in this infection.

With regard to active immunity to viruses infecting the respiratory tract, it has become evident that titres of serum antibody are a poor correlate with resistance to disease and that the presence of antibody in the secretions bathing the mucosa of the respiratory tract is of importance. In sheep it has been shown that live P13 virus given intranasally will stimulate comparable serum IgG but higher nasal secretion IgA antibody titres than the same dose of live virus given intramuscularly. Virus inactivated by treatment with formalin and given by either route fails to stimulate nasal secretion antibody and results in low or undetectable serum titres. Immunity to aerosol exposure to P13 virus is conferred by inoculation with live virus given either intranasally or intramuscularly but not by inactivated virus given by either route. If however the inactivated P13 virus is incorporated in an oil adjuvant, high titres of serum antibody are produced accompanied by appreciable levels of nasal secretion antibody. This immune response protects against aerosol challenge and examination of the nasal secretions show that, whereas normally IgA is the predominant immunoglobulin, in lambs vaccinated in this way IgG was the major immunoglobulin component. The antibody activity in the secretions was associated with this immunoglobulin which appears to have been transferred from serum. In other experimental systems where clinical disease can be induced by infection with P13 virus, this oil adjuvant vaccine of the formalin inactivated virus has proved successful and may be of practical value. It seems to depend upon elevation of serum antibody titres to such an extent that serum antibody is transferred to the respiratory tract secretions in sufficient amount so as to provide protection.

Another virus infection of sheep in which the humoral immune response would appear to be important is louping ill, an infectious, tick-borne, viral encephalomyelitis. Once more, passively acquired serum antibody is effective in providing protection against infection in the newborn lamb, probably by inhibiting the viraemic stage in the development of the disease. An inactivated louping ill virus vaccine incorporating oil adjuvant is protective against infection and is commercially available.

A virus infection which seems to be more important in the causation of clinical disease in newborn lambs than in adults is that due to rotavirus. Unlike louping ill virus, protection against rotavirus does not appear to be related to the levels of maternal antibody in the serum of the neonate. Resistance to rotavirus infection has however been shown to depend on humoral immunity. The ingestion of antibody to rotavirus by the newborn lamb will prevent virus infection and the associated clinical signs of dullness and diarrhoea. Such protection can be achieved in the absence of circulating serum antibody and appears to depend upon the continued local presence of antibody to rotavirus in the gut.

In contrast to these examples of paramyxovirus, togavirus and rotavirus infection where humoral immunity appears important, resistance to infection with poxviruses appears to be dependent largely upon cell-mediated immunity. Two important diseases of sheep caused by poxviruses are contagious pustular dermatitis (orf) and sheep pox. In both of these conditions, resistance to infection is not well related to levels of circulating serum antibody. It is significant that resistance to disease caused by these viruses is conferred with live rather than inactivated vaccines. This may be a further indication of the significance of cell-mediated immunity in resistance to orf and sheep pox since cell-mediated immunity would appear to be more readily stimulated by exposure to the live agent.

IMMUNITY TO BACTERIAL INFECTIONS

As with the virus infections, the relative importance of humoral and cellular immunity to a species of bacteria depends on its mode of infection and the pathogenesis of the disease it causes.

Phagocytosis is an important defence mechanism involved in the clearance and destruction of bacteria but certain organisms can remain viable even after ingestion into macrophages. Such bacteria which can affect sheep are *Listeria monocytogenes* and *Brucella abortus.* In these bacterial infections, cell-mediated immune responses which augment the pre-existing non-specific phagocytic mechanisms are of significance. Macrophages activated under the influence of specifically released lymphokines are more capable of controlling infection by these organisms.

In pasteurellosis of sheep it would appear that immunity is not well correlated with the presence of serum antibody. An experimental model of pneumonic pasteurellosis has been developed in which the disease can be reproduced by infection with P13 virus followed by superinfection with Pasteurella haemolytica. If lambs are given antiserum to P. haemolytica by intraperitoneal injection before challenge, they are not protected against the effects of the bacterial superinfection. This would suggest that cell-mediated immune responses are important and, moreover, lambs which show no serum antibody response following immunization can be shown to be immune to experimental infection. Cell-mediated immune responses have been demonstrated, following immunization against P. haemolytica, by examining the proliferative response of peripheral blood lymphocytes cultured in vitro in the presence of P. haemolytica antigen. Clearly monitoring of cell-mediated immune responses to these bacteria may be of significance in the development of more effective vaccines against pasteurellosis.

In contrast to these bacterial diseases, against which sheep appear dependent upon cell-mediated immune responses, humoral immunity is important in protecting against the clostridial diseases. These diseases include black disease, struck, lamb dysentery and pulpy kidney and in all an exotoxin produced by the bacteria is responsible for the pathogenic effects. Antisera to these bacteria and their toxins appear to provide transient protection against the disease, demonstrating the significance of antibody in immunity. Also lambs are protected against infection by passive transfer of colostral antibody. Vaccination programmes against clostridial disease take account of this and include a 'boosting' injection during pregnancy to ensure adequate passive protection of the young lamb. On the whole vaccination has proved highly successful in controlling these diseases.

CONCLUSION

While non-specific factors provide a valuable first line of defence aginst infectious disease, specific acquired immunity constitutes an essential part of the overall host defence mechanism. Humoral immunity, mediated by antibodies, and cellular immunity, mediated largely by lymphocytes and macrophages, both contribute in varying degrees to the specific immune response. Humoral immunity is important in the initial protection of the newborn lamb and in the adult is readily used as a serological indicator of prior infection in diagnosis. Artificial induction of specific acquired immunity through vaccination offers a valuable prophylaxis in a number of diseases and hopefully with further research the range of effective vaccines available may be increased.

REFERENCES

General Immunology

Roitt, I. (1974) Essential Immunology, 2nd edn. Blackwell Scientific Publications, London

Veterinary Immunology

Buxton, A.;Fraser, G. (1977) Animal Microbiology. Volume 1. Blackwell Scientific Publications, London

Ovine Immunoglobulins

Jonas, W.E. (1968) The immunoelectrophoretic patterns of some body fluids of sheep. Research in Veterinary Science 9,324-350.

Pahud, J.J.;Mach, J.D. (1970) Identification of secretory IgA, free secretory piece and serum IgA in the ovine and caprine species. Immunochemistry 7,679.

Smith, W.D.;Dawson, A.McL.;Wells, P.W.;Burrells, C. (1975) Immunoglobulin concentrations in ovine body fluids. Research in Veterinary Science 19,189-194.

Watson, D.L.;Lascelles, A.K. (1971) IgA in the body fluids of sheep and cattle. Research in Veterinary Science 12,503-507.

Watson, D.L.;Lascelles, A.K. (1973) Comparisons of Immunoglobulin secretion in the salivary and mammary glands of sheep. Australian Journal of Experimental Biology and Medical Science 51,255.

Maternally-Derived Passive Humoral Immunity in Lambs

Halliday, R. (1975) In 'Maternofoetal Transmission of Immunoglobulins'. (Ed) W.A. Hemmings. Cambridge University Press, 1975.

Reid, H.W.;Boyce, J.B. (1976) The effect of colostrum-derived antibody on louping-ill virus infections in lambs. Journal of Comparative Pathology 81,291.

Reid, J.F.S. (1972) Serum immune globulin concentrations of newborn hill lambs. Veterinary Record 90,371-372.

Smith, W.D.;Wells, P.W.;Burrells, C.;Dawson, A.McL. (1976) Immunity to parainfluenza 3 virus in the ovine respiratory tract. Proceedings 20th World Veterinary Congress Thessaloniki, 1975.2,1371-1372.

Snodgrass, D.R.;Wells, P.W. (1976) Rotavirus infection in lambs studies on passive protection. Archives of Virology 52,201-205.

Wells, P.W.;Reid, H.W. (1978) Antibody responses to vaccination against louping-ill virus in newborn lambs. Journal of Comparative Pathology 88,425-431.

Development of the Immune Response in the Foetal Lamb

Silverstein, A.M.;Uhr, J.W.;Kraner, K.L.;Lukes, R.J. (1963) Foetal response to antigenic stimulus. Journal of Experimental Medicine 117,799.

Cell-Mediated Immunity to Infections

World Health Organisation Technical Report Series No.519 (1973) Cell-mediated immunity and resistance to infection. World Health Organisation, Geneva.

Immunity to Virus Infections in Sheep

Brotherston, J.G.;Boyce, J.B. (1969) A new vaccine against louping ill. Veterinary Record 84,514-515.

Martin, W.B.;Ergin, H.;Koylu, A. (1973) Tests in sheep of attenuated sheep pox vaccines. Research in Veterinary Science 14,53-61.

Snodgrass, D.R.;Wells, P.W. (1978) The immunoprophylaxis of rotavirus infections in lambs. Veterinary Record. In press.

Wells, P.W.;Smith, W.D.;Burrells, C.;Dawson, A.McL.;Sharp, J.M. (1977) Immunity in the respiratory tract of sheep. Biochemical Society Transactions 5,1574.

Wells, P.W.;Sharp, J.M.;Rushton, B.;Gilmour, N.J.L.;Thompson, D.A. (1978) The effect of vaccination with a parainfluenza type 3 virus on pneumonia resulting from infection with parainfluenza type 3 virus and Pasteurella haemolytica. Journal of Comparative Pathology 88, in press.

Immunity to Bacterial Infections in Sheep

Kerry, J.B. (1975) Experience in the United Kingdom of the vaccination of sheep against foot rot. Proceedings of 20th World Veterinary Congress, Thessaloniki, Greece.

Sterne, M.;Batty, I.;Thomson, A.;Robertson, J.M. (1962) Immunisation of sheep with multicomponent clostridial vaccines. Veterinary Record 74,909-913.

PHYSIOLOGY OF REPRODUCTION

DR W. HERESIGN

University of Nottingham School of Agriculture, Sutton Bonington, Loughborough, Leics.

In animals such as the sheep, which have an oestrous cycle, ovulation usually occurs in close relation to a period of sexual receptivity known as "oestrus" or "heat". This ensures that deposition of spermatozoa in the female tract is synchronised with ovulation, thus increasing the chances of conception. The natural control of the timing of the breeding season and the events of the oestrous cycle in the sheep has been the subject of much interest since the pioneering studies of Marshall in 1903. The accumulation of information was very slow at first but has accelerated since about 1965, largely because of the effort devoted to the development of hormone measurement techniques, together with a realisation that the central nervous system plays a key role in controlling these events and is capable itself of secreting hormones. This revolutionary idea owes a great deal to the work of the late Geoffrey Harris whose studies may be said to have formed the foundation of modern concepts of neuroendocrine control mechanisms and parallels in importance the pioneer work of Marshall and Hammond on reproduction and fertility of sheep.

Ovulation in the sheep can be observed by laparotomy, in which the abdomen is opened under general anaesthesia and the ovaries are observed directly. In a more recently developed technique, laparoscopy, the ovaries are observed using an optical instrument inserted through a puncture incision in the abdomen, again under general anaesthesia, a procedure which can be repeated with minimal risk of trauma and haemorrhage. There is considerable variation within a single sheep and between sheep in the length of the oestrus period and also in the timing of ovulation in relation to the onset and termination of sexual receptivity, although in most individuals ovulation occurs late in the oestrus period. This is partly due to variation in the timing relative to the onset of oestrus of a pre-ovulatory surge of release from the pituitary gland of one of the goadotrophic hormones, luteinizing hormone (LH). This surge of LH into the bloodstream is believed to be responsible alone, or in conjunction with a concurrent surge of the other pituitary gonadotrophin, follicle stimulating hormone (FSH), for acting on the ovary causing rupture of the ovarian follicle and ovulation.

Blood cncentrations of LH are measured by radioimmunoassay which is applied to frequently-taken samples of jugular vein blood, taken with an indwelling jugular vein cannula. The ideal interval between the taking of successive blood samples depends on many factors but is normally 15 minutes when samples are to be assayed for LH or FSH. If the whole cycle is being followed, the interval has to be lengthened to 2 hours to avoid excessive loss of blood. Such a sampling frequency will show basal levels of LH throughout the cycle with a surge of LH associated with each oestrus period.

The variability of timing of the pre-ovulatory LH surge in relation to the onset of oestrus causes practical problems in characterising the surge completely. In the majority of animals the surge will have begun by four hours after the onset of oestrus and will be completed by 20 hours after onset. For example, in Clun Forest ewes the pre-ovulatory LH surge lasts for approximately 8-10 hours and shows a peak of approximately 180-260 ng/ml from a baseline of 1 - 18 ng/ml. There is also a pre-ovulatory peak of FSH, coincident with that of LH, although it is less marked.

From early studies on the close anatomical relationship between them it is known that the hypothalamus plays a role in controlling the release of hormones from the pituitary gland. The nature of this control as far as the gonadotrophins are concerned was clarified by the discovery in 1961 in the neural tissue of the hypothalamus of an LH-releasing factor and four years later of an FSH-releasing factor.

Much effort has been devoted to attempts to establish whether these activities reside in the same neuro-hormone molecule or in separate but closely-related molecules. Until recently the chemical investigation of neurohormones lagged behind their physiological investigation and early studies on gonadotrophin releasing activity were based on biological assay methods using animals or pituitary gland tissue to measure concentrations of releasing activity in the tissue of the hypothalamus. Recently, radioimmunoassay methods have been developed to measure gonadotrophin releasing activity but for various reasons their use has not clarified the situation greatly.

A further important change leading up to ovulation is the presence in the blood of a surge of oestrogen occurring on the day before the onset of oestrus. This oestrogen peak disappears before the pre-ovulatory LH peak begins to rise but is believed to be responsible for triggering the LH peak. It has been shown in the sheep that high levels of oestrogen (and low levels of progesterone) make the pituitary gland more sensitive to gonadotrophin releasing hormone. This can be demonstrated in anoestrous ewes. The reason for the oestrogen surge is by no means clear but it is associated with the growth of ovarian follicles.

The other important hormone of the cycle is progesterone produced by the corpus luteum after ovulation. For "heat" to occur the ewe must be subjected to sequential progesterone and oestrogen stimulation. The length of time during which progesterone is secreted controls the length of the oestrous cycle; it has a blocking action on gonadotrophin release and ovulation and it is necessary for circulating progesterone to fall before this series of changes will occur. The decline in progesterone therefore immediately precedes the oestrogen peak on day 14-15 of the cycle, where the onset of oestrus is taken as day 1. There is evidence that the uterus directly controls the lifespan of the corpus luteum and progesterone secretion. This effect is particularly well-defined in the ewe and a "luteolytic" material, prostaglandin F_2, appears in the uterine vein blood at about day 14 of the 17-day cycle. This leads to a rapid decline in progesterone levels, oestrus and ovulation.

During the summer months the ewe fails to cycle, showing seasonal anoestrus and anovulation. The reproductive tract of the ewe at this time is in a regressed state. The ovaries are small, pale in colour and contain few follicles and the uterus shows no evidence of hormonal stimulation. That the ovaries are still under the influence of some gonadotrophin stimulation can be demonstrated by removal of the anterior pituitary gland, the source of FSH and LH. When this is done the ovaries regress completely within about 5 days. It is therefore evident that the reproductive tract of the seasonally anoestrous ewe is under a mild degree of gonadotrophic hormone stimulation, but this is much less than that found during the breeding season.

The importance of seasonal anoestrus to the sheep producer is that it limits the reproductive rate of the ewe to one lamb crop per year in most situations. Over the years, various empirical treatments have been used in an attempt to induce ovulation, oestrus and fertility in sheep out of the breeding season, but the degree of success obtained has been inconsistent and frequently poor. This is hardly surprising since from the foregoing it can be appreciated that the oestrous cycle of the ewe is governed by complex endocrine changes and inter-relationships.

Whilst the importance of day length in the control of sesonality has long been recognised the mechanisms by which it operates have remained obscure. The interval from a reduction in day length to the onset of reproductive activity in the ewe is the same no matter whether the decrease in day length is abrupt or gradual, and attempts to implicate the eyes or the pineal gland have met with little success. Nonetheless it is obvious that day length does modify the endocrinology of the ewe in some way so as to impose a seasonal effect on its reproductive activity.

For a long time it has been thought that decreasing day length in the autumn stimulated, directly or indirectly, the endocrine changes compatible with oestrous cycle activity and that seasonal anoestrus represented the removal of this stimulus. Recent experimental data however support the hypothesis that seasonal anoestrus is really a period of active inhibition, and that the breeding season reflects a loss of this inhibition. The detailed evidence to justify this hypothesis is too complex to discuss here but what it does mean is that we shall have to reconsider carefully our approach to treatments designed to overcome seasonal anoestrus in the ewe.

REFERENCES

Crighton, D.B. (1975) The natural control of ovulation in the sheep. University of Nottingham School of Agriculture Report 1974-75,53-65.

Yeates, N.T.M. (1949) The breeding season of the sheep with particular reference to its modification by artificial means using light. Journal of Agricultural Science 39(1)1-43.

Karsch, F.J.;Legan, S.J.;Ryan, K.D.;Foster, D.L. (1977) Feedback effects of ovarian steroids on gonadotrophin secretion. In 'Control of Ovulation'. Proceedings of 26th Easter School, University of Nottingham. (Eds) D.B. Crighton;G.R. Foxcroft;N.B. Haynes;G.E. Lamming. Butterworths, London, pp29-48.

EMBRYO TRANSPLANTATION IN SHEEP

S.M. WILLADSEN

ARC Institute of Animal Physiology, Animal Research Station,
307 Huntingdon Road, Cambridge

INTRODUCTION

In the ewe, as in most mammals, the ovulation rate may be increased if the ovaries are stimulated with exogenous gonadotrophins. Treatment with moderate doses of Pregnant Mare Serum Gonadotrophin (PMSG) has been used on occasion in attempts to increase the incidence of multiple births, but since the gains to be obtained are only marginal and the superovulatory response varies widely such treatment has generally been found to be unwarranted. On the other hand, superovulation used in conjunction with embryo transplantation constitutes a method of accelerating and improving the reproductive performance of selected ewes (Rowson & Adams 1957). The recent development of methods for long term storage and long distance shipment of embryos further adds to the potential value of this approach. These techniques have seen very limited application in commercial sheep breeding, mainly for economic reasons but also because they are relatively demanding in terms of planning, facilities, equipment and technical skills. They have however been of considerable interest to reproductive physiologists, first as research topics in their own right (Warwick & Berry 1949, Hunter et al 1955) and at present mainly and increasingly as experimental tools, particularly in studies concerning ovarian and early embryonic physiology. I shall describe here some of the principles and methods employed in embryo transplantatioin in the sheep and point out some of the more important results which have been obtained, with a view to practical application.

OESTRUS CONTROL AND SYNCHRONIZATION

The success of embryo transplantation is dependent upon efficient methods of detection, control and synchronization of oestrus in prospective embryo donors and recipients, so that treatments and operations may be carried out on specific days of the oestrous cycle. If naturally cycling ewes are to be used, they should be penned with vasectomised teaser rams. The rams should be crayon-harnessed or have their briskets painted daily and the ewes should be inspected twice daily for mating marks.

The use of progestagens or prostaglandins for the induction and synchronization of oestrus facilitates the planning of subsequent operations and greatly reduces the work load of oestrus detection. Progestagen treatment should be of at least ten days' duration. Daily injections of progesterone in oil, progesterone impregnated sponges inserted into the vagina and progesterone impregnated subcutaneous implants are all effective as are various other progestagen treatments (Lamond 1964, Falkenberg et al 1964, Shelton & Moore 1967). Two intramuscular injections of prostaglandin seven to twelve days apart constitute a somewhat more convenient and equally effective treatment (Haresign 1978). A single injection of prostaglandin will suffice to synchronize oestrus in the majority of cycling ewes. Examples of synchronization methods and the results obtained are listed in tables 1A and B.

Table 1A Oestrus synchronization in mature Welsh Mountain ewes following treatment with subcutaneous progesterone implants (350mg) for 10 days

Season	No. of ewes treated	No. ewes (%) detected in oestrus (h. after implant removal)			
		< 24	24 - 48	48 - 72	>72
Late anoestrus	50	18 (36)	31 (62)	1 (2)	-
Peak of breeding season	96	13 (13.5)	67 (69.8)	14 (14.6)	2 (2.1)

Table 1B Oestrus synchronization in cycling Welsh Mountain ewes following intramuscular injection of cloprostenol (100 μg)

Treatment	No. of ewes treated	No. ewes (%) detected in oestrus (h. after prostaglandin injection)			
		< 24	24 - 48	48 - 72	> 72
One injection	77	15 (19.5)	59 (76.6)	2 (2.6)	1. (1.3)
Two injections 7 days apart	43	3 (7.0)	39 (90.7)	1 (2.3)	-
One injection 9-13 days after superovulation	30	2 (6.7)	4 (13.3)	14 (46.7)	10 (33.3)

SUPEROVULATION

The approach to superovulation in the sheep has to a large extent been empirical. Age and breed of the ewe and season and time during the oestrous cycle as well as individual differences of a more obscure nature all contribute to the variability in the results obtained after a particular treatment.

PMSG is the gonadotrophin which has been used most extensively to induce superovulation. Gonadotrophins of pituitary origin have found limited use, although greater efficiency has been claimed for them (Moore & Shelton 1964a). Due to their relatively short biological half-life pituitary gonadotrophins have to be administered as several injections over a three day period whereas PMSG is usually effective when given as a single injection.

Only a relatively small proportion of the follicles present in the ovaries are responsive to PMSG, in the sense that they will mature following PMSG stimulation (Thibault 1977, Hay & Moor 1978). The dose of the hormone should only be sufficiently high to ensure that such responsive follicles actually do respond by synchronously reaching a condition in which they may be activated to ovulate by an endogenous or exogenous surge of luteinizing hormone. At the same time the dose of PMSG should be sufficiently low to minimize the risk of premature ovulation due to the LH-like activity of PMSG itself. It is questionable whether these two objectives can be met simultaneously, and therefore a high dose of PMSG should be followed by an injection of a luteinizing hormone. In most instances a dose of 20 to 50 i.u. of PMSG per kg body weight has been used.

If progestagens or prostaglandins are not to be used to induce luteal regression prospective donors should be injected with PMSG late on day 12 or early on day 13 (day 0 = Onset of preceding oestrus). The hormone is usually injected intramuscularly, but may be given subcutaneously or intravenously. The onset of oestrus can be expected to occur about three days later, but considerable variation is encountered, and a proportion of ewes fail to exhibit oestrus. Ewes which come on heat earlier than two days or later than four to five days after PMSG generally have a poor superovulatory response. In those ewes that do come on heat, the duration of oestrus may be longer but is usually slightly shorter than the normal length of about 30h.

Better control of the timing of onset of oestrus is obtained if PMSG is followed 48 to 72h later by an injection of prostaglandin (Trounson et al 1976). Since prostaglandin causes luteal regression during the major part of the oestrous cycle in the ewe, the use of prostaglandin also widens the interval during the cycle in which superovulation may be obtained. However, the number of follicles responsive to PMSG seems to reach a peak during the latter part of the cycle.

The major drawback to the use of prostaglandin in PMSG treated ewes is the high incidence, following this procedure, of premature regression of the corpora lutea formed as a result of superovulation. Premature

Table 2 Superovulation of mature Welsh Mountain ewes

Season	Treatment[1]	No. ewes	No. ewes responding	Total No. ovulations	No. ovulations/ewe	No. ewes showing different ovulation response				
						≤5	6-10	11-15	16-20	>20
Weight of breeding season	1000 i.u.PMSG D.12-13	45	40	317	7.9	16	13	8	2	1
"	1250 i.u.PMSG D.12	25	23	242	10.5	4	9	6	3	1
"	1250 i.u.PMSG D.13	38	34	308	9.1	7	19	3	2	2
"	1500 i.u.PMSG D.11 125µgPG D.13[2]	24	20	232	11.6	3	6	8	2	1
"	1250 i.u.PMSG D.12 125µgPG D.14 500 i.u.HCG at oestrus	13	13	232	17.8	1	3	3	1	5
"	1250 i.u.PMSG D.27-28 125µgPG D.29-30[2]	19	17	238	14.0	1	5	6	2	3
Late anoestrus	1250 i.u.PMSG D.4-5 125µgPG D.6-7	18	15	63	4.2	12	-	3	-	-
"	1250 i.u.PMSG D.12-13	25	24	179	6.9	13	5	3	2	1
Early breeding season	1250 i.u.PMSG D.7-8 125µgPG D.9-10	18	16	157	9.8	1	10	3	1	1
"	1250 i.u.PMSG D.12-13	24	23	228	9.9	6	7	6	4	-

[1]Day of treatment refers to preceding (induced) oestrus.

[2]500 i.u. HCG given to ewes which had already ovulated at the time of insemination.

luteal regression, which is also observed occasionally in PMSG treated ewes not injected with prostaglandin, occurs between day 5 and day 7 in some 50% or more of ewes treated with both PMSG and prostaglandin. Although it is apparent that the uterus is implicated (Willadsen & Moor, unpublished observation), the underlying causes are not understood in detail at present. The effect however is that the ewe returns to oestrus and any embryos present in the uterus are expelled. Embryo collection from prostaglandin treated donors should therefore not be postponed beyond day 4. Alternatively, the ewes should be treated with progestagen from day 2 onwards to the day of collection.

PMSG may also be used in conjunction with progestagens (Moore 1970). Progestagen priming is necessary if anoestrous ewes are to be used as donors (cf Lamond 1964) and is also essential if prepuberal ewes are to be superovulated (Trounson et al 1977). The animals should be kept on progestagen for at least ten days and injected with PMSG about 3 days prior to planned onset of oestrus.

Examples of methods of superovulation and the results obtained are listed in Table 2. It should be emphasized that this table refers to experiments in which mature, in most instances rather old, Welsh Mountain and Welsh Mountain Crossbred ewes (25-35kg) were used and that procedures (especially the dose of PMSG) must be adjusted to suit the particular type of animal which is being used.

The method which we currently use for superovulating groups of cycling Welsh ewes is as follows:

(i) Oestrus synchronization with subcutaneous progesterone implants (350mg Progesterone, Abbott Labs) for ten days or two intramuscular injections of 125 μg Cloprostenol (Estrumate, ICI Ltd) ten to twelve days apart. Practically all ewes come into oestrus 24 to 48h following progesterone withdrawal or second injection of prostaglandin.

(ii) On day 11 to 12 of the following cycle an intramuscular injection of 1250 i.u. PMSG (Folligon, Intervet Labs).

(iii) Forty-eight hours later an intramuscular injection of 125 μg Cloprostenol after which the ewes are penned without access to food and water, and checked regularly for onset of oestrus. The majority of ewes have come into oestrus 28h later, practically all by 36h.

(iv) When a ewe has been detected in oestrus, an intravenous injection of 500 i.u. Human Chorionic Gonadotrophin (HCG) (Chorulon, Intervet Labs) and as soon as possible surgical insemination with freshly collected ram's semen. Ovulation may be expected to occur 24 to 30 h following HCG injection.

With this treatment a proportion of ewes experience one or two premature ovulations even before the onset of oestrus. The subsequent progesterone production tends to block the release of LH one or two days later, so that multiple cystic follicles, rather than superovulation, result unless HCG is given. The superovulatory response was not increased when a dose of 1500 i.u. of PMSG was used, and after this dose the signs of oestrus were muted so that oestrus detection became imprecise.

MATING AND INSEMINATION

When natural mating is used, the ewe should be mated on several occasions during oestrus; but failure of fertilization occurs frequently in superovulated ewes. In heavily superovulated animals (>10 ovulations), fertilization after natural mating is the exception rather than the rule. The failure appears to be due mainly to decreased transport of sperm through the cervix uteri. If semen is introduced directly into the uterus, high fertilization rates result even in heavily supervoulated ewes (Killeen & Moore 1971, Trounson & Moore 1974). The stress of surgery and particularly the manipulations necessary to exteriorize the uterus tend to interfere with the functional relationship between the oviduct and the infundibular fimbria so that a varying proportion of the oocytes are not picked up by the infundibulum after ovulation. Therefore the embryo collection rate may be drastically reduced, particularly after careless surgical insemination late in oestrus. Insemination should be performed as soon as possible after the ewe has come into oestrus and the necessary manipulations should be done as gently as possible. The ovary and oviduct should not be touched or exteriorized.

SURGICAL PROCEDURES

Surgery is necessary for collection of embryos from the live donor and for transplantation of embryos. Deposition of semen into the uterus also necessitates surgery. The techniques we use are as follows:

Prior to surgery the ewe is penned without access to food and water for 12 to 36h. Pentobarbitone sodium (Sagattal, May & Baker Ltd) is given intravenously at a dose of 20 to 25 mg/kg body weight for basal anaesthesia. An endotracheal tube is inserted, and anaesthesia is maintained with an oxygen-fluothane mixture in a closed circuit system. Five mg of atropine sulphate may be given subcutaneously to prevent excessive salivation and regurgitation of ruminal contents.

The animal is positioned on its back, and the ventral abdominal area is sheared, washed and disinfected with 70% ethanol. Under aseptical conditions a ca 7cm skin incision is made parallel with but ca 1cm lateral to the midline (between the large median and the smaller lateral abdominal vein) ending 1 to 2 cm in front of the udder. Fascia and peritoneum are incised in the midline. The uterus is located in the pelvic cavity and exteriorized along with oviducts and ovaries, but kept moist with gauze swabs soaked with 0.9% NaCl. After insemination, egg recovery or egg transfer, the peritoneum and fascia are sutured with a continuous nylon suture. The skin wound is dusted with antibiotic powder, sutured with continuous or interrupted sutures of braided silk, and covered with an adhesive plaster.

After the operation the ewe is placed on its chest, slightly inclined towards the left. The endotracheal tube is left inserted until reappearance of the swallowing reflex. The skin sutures should be removed about ten days later.

Insemination

Ca 0.04ml freshly collected, undiluted ram's semen is drawn into a Pasteur pipette. The tip of the pipette is wiped with a sterile swab and inserted into the lumen of the uterine horn through the wall ca 5cm from the utero-tubal junction, the semen is expelled, and the pipette slowly withdrawn. The same procedure is performed on the opposite horn. Ovaries and oviducts should not be exteriorized during inseminations.

Egg recovery

A curved 0.2 x 10 cm glass or plastic tube is inserted 1.5 to 2 cm into the oviduct via the infundibulum and held in position by a first operator. A blunt 21g needle is inserted into the lumen of the uterine horn 2 to 3 cm from the bifurcation and the lumen posterior to the point of insertion occluded manually by a second operator. Ten ml of PBS at 38°C is injected into the uterine lumen and gently manipulated along the horn towards the tip, up through the oviduct and collected in a glass cup via the tube. If the collection is made on day 2, 3 or early on day 4 only the tip of the uterine horns and the oviducts are flushed. There is often considerable resistance to the passage of fluid through the utero-tubal junction. It may help to straighten the junction manually. In any event patience rather than force should be relied upon to recover the flushing fluid. Excessive pressure of fluid in the uterine lumen may damage the embryos and cause rupture of the uterine wall along the curvatura minor. Such ruptures generally heal spontaneously without affecting the subsequent fertility of the ewe.

Collection of eggs from superovulated ewes within the first few hours of ovulation generally results in low recovery rates. The best results are obtained if no tube is inserted into the oviduct. The fimbria are held over the edge of a glass cup and flushed both from the tip of the uterine horn and on the surface. If bleeding can be avoided the surface of the ovary should also be flushed.

For collection of day 12, 13 and 14 embryos, a curved large bore glass tube (0.5 x 15 cm) is inserted into the uterine lumen through a small incision at the base of the horn. Fifteen to twenty ml of medium is flushed from the tip of the uterine horn or from the oviduct to be collected in a glass cup via the tube. The incision is closed with an absorbable suture.

Bleeding from the corpora lutea in superovulated donors is virtually unavoidable and almost invariably leads to adhesions between ovary and oviduct unless it is stopped and all clotted blood carefully removed before the genital tract is returned to the abdomen.

Embryo transfer

The ovaries are inspected and the corpus luteum located. The embryo to be transferred is picked up with a Pasteur pipette in ca 0.04 ml of medium. For uterine transfers the tip of the pipette is inserted 2 to 3 cm into the uterine lumen through the wall 2 to 3 cm from the utero-tubal junction and the fluid containing the embryo gently expelled. It may be helpful to make the puncture of the uterus with a blunted 21g needle, before the pipette is inserted. For transfers to the oviduct the pipette containing the embryo in a minimum volume of fluid ($\leq$ 0.02ml) is inserted 3 to 5 cm into the oviduct via the infundibulum. While the pipette is slowly withdrawn, the fluid is gently expelled.

Collection of embryos

Sheep embryos for transplantation may be collected from a few hours after ovulation up to day 14 after onset of oestrus. Viable embryos may be collected from slaughtered donors in which case the genital tract should be isolated and flushed without delay as soon as the animal has been killed. Surgical collection from the live donor is to be preferred. In ewes which have been mated naturally collected rates of 75% or more of the eggs ovulated are readily obtained, whereas the average collection rate from surgically inseminated ewes is 50%, but may be improved significantly if the ewes are inseminated as soon as they come into oestrus and with minimal manipulation of the genital tract. Maximum egg recovery rates are most easily obtained if collections are made while the embryos are still lodged in the oviducts or soon after they have entered the uterus (day 3 to 4). As mentioned earlier collections from prostaglandin treated donors should be made before day 5.

Donors from which all embryos are not collected may remain pregnant. If an animal is to be used again as a donor, it is therefore advisable to give it an injection of prostaglandin on day 7 - 12 after superovulation.

A ewe may be used as an embryo donor several times. A period sufficient for normal oestrous cycles to be reestablished should be allowed between PMSG treatments. If surgical insemination is used, one can generally expect to be able to collect embryos at least twice, but in most instances less good results are obtained the second time. Even with careful surgery repeated operations tend to lead to fibrous adhesions in the ovarian region which may interfere with ovulation and oocyte transport, eventually leading to sterility. Also adhesions between the omentum majus and the body wall at the laparotomy site are practically unavoidable.

Evaluation of embryos

In the mated or inseminated ewe sperm penetration takes place within the first two to three hours of ovulation in the ampular region of the oviduct. The one-celled fertilized egg surrounded by its zona pellucida has a diameter of 120 to 130μ and is thus barely visible to the naked eye. Early embryonic development in the sheep has been described by Rowson & Moor (1966a). During the first five to six days after fertilization the egg undergoes a series of cleavage divisions without increasing its total volume. The first cleavage takes place some 15 to 18h after sperm penetration, i.e. late on day 1 or early on day 2 after the onset of oestrus, the second some 12h later, on day 2. After this the spherical blastomeres cleave more or less synchronously every 16 to 24h. Four- to eight-celled embryos are typical of day 3. On day 3 to 4 the embryo passes from the oviduct into the uterus. Twelve to sixteen celled embryos are typical of day 4, 24 to 32 celled morulae typical of day 5. On day 6 most embryos develop into compacted late morulae of some 40 cells which are difficult to distinguish individually, or early blastocysts of about 60 cells. On day 7 the majority are early or expanding blastocysts of 100 to 120 cells which have differentiated into trophectoderm and inner cell mass (embryonic disc), and on day 8 they are fully expanded blastocysts of 150 to 200 cells, some of which have hatched or are in the process of hatching from the greatly distended zona pellucida. On day 9 the majority of embryos are hatched blastocysts of some 200-300 cells. During the next three to four days the cell number and volume of the embryo both increase, but the spherical shape is maintained. On day 12 to 13 the embryo starts to grow at a phenomenal rate, and by day 14 it has generally developed into a threadlike sack, several centimeters long. All the developmental stages mentioned are candidates for transplantation, day 3 to day 7 embryos being those most frequently used. After day 7 the embryos may be difficult to distinguish from aggregations of oviduct cells in the flushing fluid, particularly on day 8 and day 9, due to the shedding of the zona pellucida and the tendency in recently hatched blastocysts to contract into a compact ball of cells. Superovulated donors are not suitable for collection of day 13 and day 14 embryos, partly because a wave of embryonic dgeneration seems to occur after day 12, but particularly because the threadlike embryos tend to get completely entangled in one another.

Eggs collected from superovulated ewes often constitute a considerably more heterogenous material than the preceding description suggets. Unfertilized or completely degenerate eggs are easily recognised. Unfertilized eggs generally remain one-celled but may show few signs of degeneration till after day 5 to 6. If the collection was made late on day 1 or early on day 2, the presence of two polar bodies or sperm attached to the zona pellucida may be taken as evidence that the egg is fertilized. In completely degenerate embryos the plasma membranes are disrupted and the individual cells only represented by a coarsely granular material. Partial cleavage and retarded development also occur relatively frequently. Embryos which are two or more cleavage divisions behind their more advanced litter mates should be considered retarded, and their viability is questionable. The same applies to embryos in which the size of individual blastomeres varies widely, suggesting that in some blastomeres cleavage has been arrested. Whereas it may be extremely difficult to assess the normality, let alone predict the viability of early cleavage stages, compaction, blastulation and the presence of a distinct inner cell mass may be accepted as indications that the embryo is viable. In compacted morulae and early blastocysts it is usual to observe some cells lodged between the zona pellucida and the embryo proper without participating in compaction or blastulation. In most instances such cells do not contain normal nuclei, but the phenomenon is generally considered to be insignificant in relation to the viability of the embryo. Since the majority of late morulae and blastocysts, when fixed and stained, are found to contain a few degenerate cells (Killeen & Moore 1971), it appears that an embryo will tolerate the loss of a proportion of its cells without reduction of its viability.

The basic equipment for locating, handling, examination and storage of embryos includes:

(i) A dissecting microscope with a magnifying power of up to 200x.

(ii) An incubator with a working temperature of 37 to 38°C.

(iii) Glass cups (ca 20ml capacity) with covers.

(iv) Pasteur pipettes (ca 15cm long, with an outer tip diameter of 1.0 to 1.5mm. The tip should be fire polished). A rubber adaptor may be used to fit a 1 ml syringe to the pipette.

(v) Glass test-tubes (e.g. 0.5 x 5 cm) with silicone rubber bungs.

All glassware should be siliconized, cleaned and sterilized according to standard tissue culture procedures.

Various media with osmolarities of 270 to 300 m.osm. and pH7 to 8 have been used for collection, storage, culture and transfer of embryos. Our standard medium is Dulbecco's phosphate buffered saline medium enriched with pyruvate, glucose and Bovine Serum Albumin (PBS, Whittingham 1971). The composition of this medium is listed in Table 3. The medium is prepared in 4 to 5 litre batches, filtered through a Millipore filter and stored in 100ml bottles at -20°C.

Sterile sheep serum, heated to 56°C for 30 min and stored at -20°C is an equally satisfactory medium for routine use.

STORAGE AND CULTURE OF EMBRYOS

As a general rule sheep embryos for transplantation should be transferred as soon as possible after collection. Between collection and transfer the embryos should be stored either at room temperature or at 37-38°C, and precautions should be taken to avoid contamination of the medium during storage. When unsealed containers are used, they should contain a reasonably large volume of medium.

Storage in PBS or sheep serum in covered glass cups or tissue culture dishes for a few hours does not adversely affect the viability of the embryos. If the embryos are to be stored for longer periods of time, they should be placed in stoppered test-tubes or sealed ampoules.

Morulae and blastocysts appear to tolerate storage in PBS at room temperature quite well for up to 24h, whereas younger embryos are generally thought to be more sensitive. When embryos are stored at room temperature, cleavage is arrested, whereas embryos, particularly morulae and

Table 3 Dulbecco's salt solution

NaCl	8.0 gm		
KCl	0.2 "		
$Na_2\ HPO_4$	1.15 "		
$KH_2\ PO_4$	0.20 "		
$CaCl_2$	0.10 "	or $CaCl_2.6H_2O$	0.1974 gm
$MgCl_2.6H_2O$	0.10 "		

Supplemented with

Na pyruvate	0.036 gm	
Glucose	1.0 "	
BSA	4.0 "	Bovine Serum Albumin (Fraction V) (Sigma)
Pen.	0.060 "]	or Kanamycin sulphate (Sigma) 25mg
Strep.	0.050 "]	
Phenol red	1.0 ml	

All in 1 litre of double distilled deionized water.

blastocysts, stored at 37-38°C in PBS often continue their development during storage. When 20% sheep serum is added to the PBS medium, the overwhelming majority of morulae and early blastocysts will grow at an essentially normal rate for two or three days at 38°C. The same result is observed when other equally simple culture methods are used (Moor & Cragle 1971). The success of attempts to culture early embryos (2-8 cells) in vitro has been limited (Moore 1970, Moor & Cragle 1971, Tervit 1973). Continued development in vitro is a good indication that the embryo is alive, and culture in vitro is therefore very useful for assessing the immediate effect of various experimental treatments on embryos. Even morphologically normal in vitro cultured embryos are often found, however, to have a reduced viability (Moor & Cragle 1971), and culture in PBS with 20% sheep serum should not be extended beyond one or two days if the embryos are to be transferred.

The oviduct of an oestrous phase rabbit is a very satisfactory substitute for culture in vitro (Lawson et al 1972) and is the only culture system devised so far, apart from the sheep oviduct, which will consistently support the continued development of one to eight celled embryos. Only embryos with an intact zona pellucida are suitable for culture in the ligated rabbit oviduct. Sheep embryos may be kept for three days - or, in the case of more advanced embryos, until hatching from the zona pellucida commences - in the ligated rabbit oviduct with only minor reduction of viability upon re-transfer to ewes.

The procedure we use is as follows: The rabbit is starved for 12 to 24h prior to surgery. Anaesthesia is induced with pentobarbitone sodium (100 to 120mg) and maintained with an oxygen-fluothane mixture administered with an open mask. The rabbit is positioned on its back with its lumbar region elevated. A ca 5cm incision is made in the midline stretching forwards from the caudal pair of mammary glands. A ligature is placed on the oviduct at the uterotubal junction, and the embryos are transferred in a minimal volume of medium via the infundibulum with a curved Pasteur pipette. Body wall and skin are sutured with braided silk. When the embryos are to be recovered, the rabbit is killed, and the genital tract removed aseptically. The oviduct and the tip of the uterine horn are dissected free of fat tissue, the ligature is cut, and the oviduct flushed via a blunt 20 or 22g needle inserted into the uterotubal junction from the tip of the uterine horn.

Sheep embryos are relatively resistant to cooling and have been successfully stored at temperatures between 5 and 12°C for up to five days (Kardymovicz & Kremer 1971, Moore & Bilton 1973). Deep-freezing promises to be the ultimate method of embryo storage, since it makes possible both long term storage and long distance transport. Lambs have now been born after transplantation of embryos stored for several years (Willadsen & Tischner 1978). The deep-freezing methods which have been developed so far (Willadsen 1977, Moore & Bilton 1977) are best suited for late morulae and early blastocysts collected on days 6 and 7. Although the results obtained after transplantation of frozen-thawed embryos have been less good than those obtained with freshly collected embryos, the use of deep-freezing may be advantageous under certain circumstances. The simplest method is as follows.

(i) Equilibration of embryos with PBS containing 1.5M Dimethylsulphoxide (DMSO) by three steps at room temperature; 0.5M DMSO for 10 min, 1.0M DMSO for 10 min, 1.5M DMSO for 20 to 40 min. While in 1.5M DMSO the embryos are pipetted into ampoules (prepared from 0.5 x 5 cm test-tubes). Four to six embryos per ampoule is suitable. The ampoules are sealed over a gas burner.

(ii) Cooling at 1°C/min to -7°C the wall of the ampoule is touched with a pair of forceps cooled in liquid nitrogen, so that ice formation is induced.

(iii) Cooling at 0.3°C/min from -7 to -30°C, and at 0.1°C/min from -30 to -33°C. From -33°C the ampoules are plunged directly into liquid nitrogen (-196°C).

(iv) After storage, the ampoules are rewarmed rapidly in water at 25 to 30°C.

(v) The embryos are transferred to a fresh lot of 1.5M DMSO in PBS, and DMSO is diluted in six steps at room temperature 1.5M DMSO for 5 min, 1.25M, 1.0M, 0.75M, 0.5M, 0.25M for 10 min each, PBS.

Only embryos which have maintained normal morphology, generally 50 to 75% of those originally frozen, should be transferred. The zona pellucida is often damaged even in embryos which have survived. This in itself does not affect their viability, provided they have reached the compacted morula stage. The viability of morphologically normal embryos appears to be unimpaired so that, with the method described, between 30 and 50% of the embryos originally frozen may be expected to develop into lambs under optimal conditions.

PRINCIPLES OF EMBRYO TRANSPLANTATION

Developmental stages ranging between one cell and elongated gastrula (day 14) have been successfully transferred in the sheep. If the main interest is to maximize the number of lambs produced, embryos ranging between eight cells (day 3-4) and early blastocysts (day 7) are most convenient to work with. The embryonic survival rates have varied, but under optimal conditions some 60% or more of the embryos transferred can be expected to give rise to lambs (Moore & Shelton 1964b, Rowson & Moor 1966b, Moor & Trounson 1977).

Differences with respect to breed between donor and recipient are of no consequence as long as the recipient is sufficiently large to carry and given birth to the type of lamb produced (Hunter et al 1955, Lawson & Rowson 1972). Healthy, mature ewes which have previously lambed should be chosen as recipients.

In the planning of embryo transplantations one should aim at precise synchrony of oestrus in donor and recipient. Although acceptable pregnancy rates are still obtainable if there is an asynchrony of ± 1 or even ± 2 days (Rowson & Moor 1966b), the best results are probably obtained when donor and recipient are at precisely the same stage of the oestrous cycle, (Moore & Shelton 1964b). Recipients that came into oestrus slightly later than the donor are preferable to recipients that came into oestrus earlier. Day 13 and day 14 embryos should be transferred to day 12 recipients, since luteal regression is initiated about day 13 in the non-pregnant ewe (Moore & Rowson 1966). Ideally, the developmental stage of the individual embryo, rather than the stage of oestrous cycle of the donor should govern the choice of recipient, e.g. a morula of some 30 cells should preferably be transferred to a day 5 or 6 recipient even if it was collected from a day 7 donor. Similarly, the stage of development rather than the age of the embryo should decide the choice of recipient for an embryo which has been cultured or stored in vitro.

If two embryos are to be transferred per recipient, one should be transferred to each side of the genital tract; a single embryo should be transferred to the side ipsilateral to the ovary carrying the corpus luteum. The overall embryonic survival rate is about the same whether one or two embryos are transferred per recipient.

Embryos less advanced than eight cells should be transferred to the oviduct, more advanced embryos to the uterus.

One cannot escape the conclusion that although relatively efficient methods of superovulation and embryo transplantation have been developed in the sheep, the practical application of these techniques would only be economically feasible in exceptional circumstances. It is equally evident that they are potentially extremely valuable to investigators of basic mechanisms and relationships in reproductive physiology, genetics and developmental biology.

REFERENCES

Falkenberg, J.A.;Hulet, C.V.;Kaltenbach, C.C. (1964) The effects of hormone combinations on oestrus, ovulation and fertility in ewes. Journal of Animal Science 32,1206-1211.

Haresign, W. (1973) Ovulation control in the sheep. In Control of Ovulation (Eds) D.B. Crighton, N.B. Haynes, G.R. Foxcroft & G.E. Lamming. pp435-451 Butterworths, London.

Hay, M.F.;Moor, R.M. (1978) Changes in the Graafian follicle population during the follicular phase of the oestrous cycle. In Control of Ovulation (Eds) D.B. Crighton, N.B. Haynes, G.R. Foxcroft & G.E. Lamming. pp177-196 Butterworths, London.

Hunter, G.L.;Adams, C.E.;Rowson, L.E.A. (1955) Inter-bred ovum transfer in sheep. Journal of Agricultural Science, Cambridge 46,143-149.

Kardymowicz, O.;Kremer, M. (1971) The transport of cleaved sheep ova by air and rail. Acta biologica Cracoviensia 2001,14.

Killeen, J.D.;Moore, N.W. (1971) The morphological appearance and development of sheep ova fertilized by surgical insemination. Journal of Reproduction and Fertility 24,63-70.

Lamond, R. (1964) Synchronization of ovarian cycles in sheep and cattle. Animal Breeding Abstracts 269-285.

Lawson, R.A.S.;Rowson, L.E.A. (1972) The influence of breed of ewe and offspring on litter size after egg transfer in sheep. Journal of Reproduction and Fertility 28,433-439.

Lawson, R.A.S.;Adams, C.E.;Rowson, L.E.A. (1972) Development of sheep eggs in the rabbit oviduct and their viability after re-transfer to ewes. Journal of Reproduction and Fertility 29,105-116.

Moor, R.M.;Cragle, R.G. (1971) The sheep egg: Enzymatic removal of the zona pellucida and culture of eggs in vitro. Journal of Reproduction and Fertility 27,401-409.

Moor, R.M.;Rowson, L.E.A. (1966) The corpus luteum of the sheep: Functional relationship between the embryo and the corpus luteum. Journal of Endocrinology 34,233-239.

Moor, R.M.;Trounson, A.O. (1977) Hormonal and follicular factors affecting maturation of sheep oocytes in vitro and their subsequent developmental capacity. Journal of Reproduction and Fertility 49,101-109.

Moore, N.W. (1970) Preliminary studies on in vitro culture of fertilized sheep ova. Australian Journal of Biological Science 23,721-724.

Moore, N.W.;Bilton, R.J. (1973) The storage of fertilized sheep ova at 5°C. Australian Journal of Biological Science 26,1421-1427.

Moore, N.W.;Bilton, R.J. (1977) Frozen storage of embryos of farm animals: progress and implications. In The freezing of mammalian embryos. Ciba Found. Symp. 52 (new series) (Elsevier-Excpt. Medic. - Nth. Holl.) 203-219.

Moore, N.W.;Shelton, J.N. (1964a) Response of the ewe to horse anterior pituitary extract. Journal of Reproduction and Fertility 7,79-87.

Moore, N.W.;Shelton, J.N. (1964b) Egg transfer in the sheep. Effect of degree of synchronization between donor and recipient, age of egg, and site of transfer on the survival of transferred eggs. Journal of Reproduction and Fertility 7,145-152.

Rowson, L.E.A.;Adams, C.E. (1957) An egg transfer experiment on sheep. Veterinary Record 69,849-850.

Rowson, L.E.A.;Moor, R.M. (1966a) Development of the sheep conceptus during the first fourteen days. Journal of Anatomy 100,777-785.

Rowson, L.E.A.;Moor, R.M. (1966b) Embryo transfer in the sheep: the significance of synchronizing oestrus in donor and recipient animal. Journal of Reproduction and Fertility 11,207-212.

Shelton, J.N.;Moore, N.W. (1967) Evaluation of alternative methods of administration of progestagens in the entire cyclic. In The control of the ovarian cycle in the sheep (Sydney University Press),59-75.

Tervit, H.R. (1973) Culture and transfer of sheep and cattle ova. Ph.D. Thesis, Univ. Cambridge.

Thibault, C. (1977) Are follicular maturation and oocyte maturation idependent processes? Journal of Reproduction and Fertility 51,1-15.

Trounson, A.O.;Moore, N.W. (1974) Fertilization in the ewe following multiple ovulation and uterine insemination. Australian Journal of Biological Science 27,301-304.

Trounson, A.O.;Willadsen, S.M.;Moor, R.M. (1976) Effect of prostaglandin analogue Cloprostenol on oestrus, ovulation and embryonic viability in sheep. Journal of Agricultural Science 86,609-611.

Trounson, A.O.;Willadsen, S.M.;Moor, R.M. (1977) Reproductive function in prepubertal lambs: ovulation, embryo development and ovarian steroidogenesis. Journal of Reproduction and Fertility 49,69-75.

Warwick, B.L.;Berry, R.O. (1949) Inter-generic and intra-specific embryo transfers. Journal of Heredity 40,297-303.

Willadsen, S.M. (1977) Factors affecting the survival of sheep embryos during deep-freezing and thawing. In The freezing of mammalian embryos. Ciba Found. Symp. 52 (new series) (Elsevier-Excpt. Medic.-Nth. Holl.) 175-189.

Willadsen, S.M.;Tischner, M. (1978) Successful long-distance transport by car of sheep embryos. Bull Acad. Pol. Sci. 26(10),725.

DISEASE AND REPRODUCTION IN SHEEP

DR. W.A. WATSON

MAFF, Hook Rise South, Tolworth, Surbiton, Surrey

INTRODUCTION

This review deals with infectious causes of foetal mortality, abortion and ram infertility. Since the middle of the 19th century abortion has been recognised as a major cause of loss to the sheep industry. Early records suggest that foot-rot, cold winds and the feeding of unripe watery roots were responsible and it is of interest that the first appreciation of the possible infectious nature of the condition followed the demonstration by Bang (1897) that Brucella abortus caused abortion in experimentally infected ewes.

The systemic reaction alone caused by a wide range of infectious agents, particularly viruses, may cause a pregnant ewe to abort, but other viruses, bacteria, parasites and fungi affect both the ewe and the foetus. If more than 1-2% of a lambing flock abort it is probable that one of these specific infections is the cause. Not only may infection result in abortion but the lamb may die early in gestation and the ewe is apparently barren. There may be death of the ewe, loss of wool, infertility in breeding males and spread of infection to other farm livestock and man. Infection with the same agents which cause abortion may result in the birth of stillborn lambs at full term or weak lambs which fail to survive.

The earliest survey of the incidence of abortion in sheep in this country was carried out by Wortley Axe (1885) who observed abortion in 12% of 51,000 ewes in 200 flocks in Lincolnshire. At the present time 4-5% of the national flock produce premature or stillborn lambs each year. The true incidence is difficult to assess as farmers tend to ignore small numbers of abortions and fail to seek a diagnosis, insufficient and unsuitable material being sent for laboratory examination. Not all laboratories use the full range of available techniques to identify the cause and no diagnosis is made on two thirds of the specimens examined. Each year the 3 most prevalent diseases are Campylobacteriosis, Chlamydiosis and Toxoplasmosis. These infections are also common in many other countries although in Germany and Eastern Europe Salmonella abortus-ovis is very important and in the Middle East Brucella melitensis.

CAMPYLOBACTER FOETUS (VIBRIOSIS)

Infection of the reproductive tract of both cattle and sheep with vibrios was first described in 1913. Earlier work on the disease was confused by the failure to appreciate the presence of saprophytic vibrios in the genital tract of cattle and sheep. These must be recognised and differentiated from the pathogenic strains. The relationship of strains is by no means clear and, with the isolation of campylobacters from the intestinal tract and faeces of normal sheep, pigs, man and domestic and wild birds, it is essential that comprehensive tests are used for their identification.

Epidemiology

The disease in sheep has been reported from many parts of the world. The most important factor influencing the spread of infection and producing quite a different pattern of disease from that in cattle is the oral rather than the venereal transmission of the organism. The available evidence suggests that the ram plays no part in the transmission of infection.

Abortion often occurs in closed flocks of sheep, to which hoggs or replacement ewes are not introduced. The presence of infection in such flocks may suggest introduction by some wild life vector; isolations of campylobacter strains from sparrows, carrion crows and ravens supports this hypothesis. Such isolates have been shown to produce abortion in ewes. Once infection is established in a flock transmission can take place by the contamination of feed and indirectly by the movements of personnel. The organism will remain viable in vaginal discharge on pasture or crops for several days. Carrier sheep excreting the organism in their faeces may be responsible for the transmission of infection from flock to flock. Serious losses due to the disease only occur in a flock during the lambing season when infection first becomes established, resulting in an abortion storm.

Symptoms

A feature of this disease is the occurrence of abortions, followed by stillborn or weak lambs, many of which fail to survive, from the ewes lambing at or near term. As many as 70% of the lamb crop may be lost but on average 15-20% of the ewes at risk lose their lambs. Normally losses are spread through all ages of ewes and not, as with certain other infectious causes of abortion, confined to the younger ewes.

Diagnosis

This depends upon the isolation of the organism from foetal or placental material. The existence of several serotypes limits the value of serological tests for the diagnosis of infection and there is a rapid fall in serum antibody titre following abortion.

Treatment

Although chlortetracycline fed continuously for several days before experimental challenge has proved effective in preventing abortion, the cost, the possible risk involved and the variable pattern of the disease does not justify treatment in this way. Other antibiotics used following a diagnosis of campylobacter infection have not proved effective.

Prevention

Ewes which abort should be retained in the breeding flock and not disposed of as is the common practice. Mature ewes which lamb normally in a flock during an outbreak also develop a satisfactory immunity.

Vaccination prior to mating has been shown to give good protection. The vaccine has also been used successfully in the face of an outbreak. The formalin-killed adjuvant vaccine is available commercially in the United States and the serotypes to which the sheep population is exposed must be incorporated in this. Aborting ewes should be isolated until all discharges have ceased, preferably for a period of 3-4 weeks, and foetuses and foetal membranes submitted for laboratory diagnosis or destroyed. The area in which the abortion occurred should be disinfected and steps taken to avoid the contamination of food and water supplies.

SALMONELLA SPECIES

Some 1,800 salmonella serotypes have now been identified most of which are capable in certain circumstances of infecting domestic animals and many of which are a hazard to public health. Certain serotypes are host specific and one of these, Salmonella abortus-ovis, is a cause of sheep abortion. The incidence of this disease in Great Britain is rapidly declining and other serotypes are assuming greater importance, particularly S.dublin and S.typhimurium.

SALMONELLA ABORTUS-OVIS

Epidemiology

The disease has been recorded from many parts of the world and in Germany, Hungary and Bulgaria it is the most common cause of ovine abortion. In Britain it is now confined to Devon and Cornwall.

There is some difference of opinion on how infection may spread. It is acquired by mouth from the products of conception but as the organism may occasionally be isolated from the genital tract of the ram venereal transmission has been postulated. It may persist in carrier ewes from one year to the next although it is considered that after abortion a ewe is usually immune.

Symptoms

Losses occur in the last 6 weeks of pregnancy with an average abortion rate of 10%, though this may range from a single animal to 60% of the flock. Mortality in lambs is a feature of some outbreaks and they may be born alive to die shortly after birth, or apparently healthy lambs less than a fortnight old may suddenly become ill and die. Others develop diarrhoea and rapidly lose weight. Some ewes may die from metritis.

Diagnosis

The organism must be isolated and typed serologically. The serum agglutination test has proved to be a useful aid in diagnosis but a high proportion of sera from flocks in areas where the disease does not occur have demonstrable specific agglutinins.

Treatment

This is of doubtful value.

Prevention and control

Vaccines are widely used in West Germany with apparent effect. Until more is known of the transmission of the disease and of the carrier state, sound advice cannot be given on eradication from a flock. The home rearing of replacement ewes and the maintenance of a closed flock will remove one possible source of infection. Aborting ewes should be isolated and the contaminated area disinfected.

OTHER SALMONELLAS

Salmonella dublin is frequently responsible for abortion in sheep and losses may be heavy; there is also a high mortality in young lambs and ewes. Ewes which abort commonly show malaise, diarrhoea and a purulent vaginal discharge. Many excrete the organism in faeces and the vaginal discharge but most cease to excrete in the faeces 4-5 months after abortion. Infection may be acquired from cattle.

Diagnosis depends upon isolation of the organism but serological tests may be helpful.

Treatment of an affected flock is of doubtful value and the use of the commercially available cattle vaccine cannot be recommended.

Many other salmonella serotypes have been associated with ovine abortion and losses in some instances can be serious. The risk of human infection in those handling sheep with disease due to these should be borne in mind.

LISTERIA MONOCYTOGENES

Epidemiology

Disease of both cattle and sheep due to infection with this organism occurs in all parts of the world.

Listeriosis may be characterised by:

i. abortion in late pregnancy,

ii. encephalitis or

iii. septicaemia in lambs with loss of condition and death in the first few weeks after birth.

It usually causes sporadic abortion in sheep but occasionally more severe outbreaks may occur.

Listeria monocytogenes is widely distributed in the environment, being found in soil, sewage of animal and human origin, bedding and feeding stuffs. In such locations it is able to withstand direct sunlight, freezing and drying for long periods. A relationship has been observed between listeriosis in cattle and sheep and silage feeding.

Symptoms

The majority of losses occur in the last 4 weeks of gestation and metritis or placental retention may follow abortion.

Diagnosis

Diagnosis is based on the isolation of the organism from the tissues of the foetus or placental membranes. Small yellowish or grey necrotic foci are often visible in the substance of the liver, lung and spleen. The placental villi are necrotic and there is an abundant brownish-red exudate on the membranes. Serological tests are of limited value.

Treatment

Antibiotic treatment in affected flocks has given equivocal results.

Prevention and Control

The possibility of silage as a source of the infection should be considered. All carcasses and contaminated feed should be destroyed.

There are varying reports on the value of vaccination against listeriosis, but no commercial vaccine is available. It is not advisable to introduce susceptible sheep into flocks in which the disease has been diagnosed.

BRUCELLOSIS

As a cause of ovine abortion *Brucella abortus* is unimportant. It has been isolated in sporadic cases in Britain but does not spread within the flock. The importance of the disease in sheep is in the context of brucella eradication in cattle and on any farm, where there is a brucella outbreak in cattle for no apparent reason, the possibility of infection in the sheep flock should be borne in mind.

The ewe can become a carrier and excrete the organism in the milk for several months.

Diagnosis is based on the recovery of *Brucella* on culture. Sheep give a variable serological response to infection with *Brucella abortus* and a full range of serological tests should be used to identify infected animals.

Brucella melitensis is of considerable importance in sheep and goats in many countries and is often a serious public health problem.

Following abortion, infected ewes excrete the organism in milk, vaginal discharge and urine for several weeks. Diagnosis is based on the isolation of Br. melitensis and serological tests.

Evaluation of the various vaccines prepared from killed, smooth and rough strains, Strain 19 and Rev 1, has shown the latter to give the best protection.

CHLAMYDIA

This group of organisms includes the causal agent of enzootic abortion of ewes (EAE), epizootic bovine abortion (EBA), sporadic bovine encephalitis (SBE), ovine polyarthritis and psittacosis. The group also includes agents capable of causing pneumonitis and others prevalent in the faeces of apparently normal animals. The interrelationship between these strains is as yet by no means clear.

Epidemiology

Greig in 1936 focussed attention upon outbreaks of abortion in Scotland in which no bacterial infection was demonstrated. Stamp and his colleagues succeeded in reproducing the disease experimentally with aborted material and considered organisms identified in the foetal membranes to be responsible for the disease. Since that time chlamydial abortion has been recognised in all parts of Britain and many other countries.

The agent is picked up from aborting ewes by uninfected animals at the time of lambing. Pregnant, non-pregnant ewes and lambs may be infected in this way. Although the most likely route of infection is oral it has been suggested that aerosol and tick transmission may occur. The infection remains inapparent until the new host is well advanced in pregnancy. Lambs may be born alive but infected, and these later produce infected foetal membranes at parturition. The agent may be excreted in vaginal discharge for 3 weeks after abortion.

It was thought that the ram played no part in the transmission of the disease but more recent evidence indicates that venereal transmission may occur and rams in infected flocks are frequently found to be serologically positive.

As infection persists in a flock from year to year and ewes develop immunity after abortion, it is the younger animals which tend to be clinically affected in the years after the initial abortion storm. This type of abortion does not occur in flocks which lamb on hill pastures but appears with increasing frequency the more closely the ewes are confined at lambing time. Introduction into clean flocks often follows the purchase of carrier animals.

Symptoms

The disease is characterised by abortion from about 100 days of gestation. Some ewes carry their lambs to full term and produce stillborn lambs often accompanied by a small mummified foetus. Weak lambs may be born which do not survive. When the disease first occurs there may be a high incidence of abortion but subsequently it becomes enzootic with between 1 and 5% of the ewes producing dead lambs each year.

Diagnosis

In severe cases the appearance of the foetal membranes is almost diagnostic, with thickening and oedema of the chorion and necrosis of the cotyledons. The membranes are covered with a thick yellowish deposit. Minute necrotic foci may be seen in the liver of aborted lambs and microscopically there are focal accumulations of reticuloendothelial cells in the tissues of most foetuses. Diagnosis is confirmed by the examination of suitably stained smears from the placenta and impression smears from the foetal liver. Confusion may arise microscopically between this agent, Q fever and *Brucella abortus*.

The complement fixation test is a useful additional diagnostic tool and antibody may be detected between 2 weeks and 4 months after abortion; several ewes in the flock must be tested.

Treatment

Antibiotics have been used but chemotherapy cannot be recommended as a practical means of control in the affected flock.

Control

Early vaccines were prepared from embryonated hens' eggs and from the infected membranes of slaughtered ewes. Both appeared equally effective in conferring protection. Ewes should be vaccinated before service with an oil adjuvant vaccine, available commercially and which produces a high level of immunity lasting for 3 years. If the animals vaccinated are already infected they may still abort at the first pregnancy after vaccination. The use of vaccine in the face of an outbreak has been suggested to be of value.

To reduce the possibility of infection it is advisable to keep a flock self contained. As non-pregnant sheep may carry the infection, aborting ewes must be isolated and not mixed with lambs or yearlings.

TICK-BORNE FEVER

The problem of ovine abortion due to infection with tick-borne fever is limited to those areas in which the sheep tick *Ixodes ricinus* occurs. The sheep tick transmits the agent by inoculation and on an infected farm almost all the ticks harbour the rickettsia.

Abortion occurs when pregnant ewes are exposed to infective ticks for the first time and the febrile reaction resulting from infection is probably responsible. The immunity which follows an attack is not absolute.

It is not feasible to treat affected sheep to prevent abortion and care should be taken to avoid the introduction of susceptible pregnant animals to tick infested pasture.

BORDER DISEASE

Border disease caused by an agent related to bovine virus, diarrhoea virus and swine fever has now been recognised in several breeds of sheep in Britain and abroad. Lambs may be aborted, stillborn or born to term with an abnormally hairy birth coat, abnormalities of the skeleton and damage to the central nervous system.

The route of transmission in field outbreaks is not fully understood but it is presumed to be horizontal from one adult to another, as well as vertical from dam to offspring.

Infection prior to conception confers only partial immunity against experimental challenge during pregnancy. In field outbreaks the ewes infected during their first pregnancy produce severely affected lambs but progeny from subsequent pregnancies are affected less severely.

Some control of the disease can be achieved by retaining those ewes which have produced affected lambs in the breeding flock and mixing replacement breeding stock with these in the hope that they will become infeted and immune before pregnancy.

MYCOTIC ABORTION

A wide range of fungi are capable of producing abortion in domestic animals, more particularly in cattle. Occasional cases of mycotic abortion have been recorded in sheep, especially associated with Aspergillus fumigatus.

'Q' FEVER

Although Coxiella burnetii has been associated with abortion in sheep, it is infrequently the cause of clinical disease in this species. It is important however that the presence of the agent is recognised when foetal material is examined so that the possibility of human infection may be avoided. The 'Q' Fever organism may also easily be mistaken microscopically for Brucella abortus or Chlamydia.

The placental and vaginal discharges are heavily contaminated and infection spreads at the time of parturition or later by aerosols from the fleece or dust in the lambing area. The organism will persist for weeks in buildings after removal of stock and is very resistant to drying. Incidents of abortion have been recorded especially in housed flocks. Personnel exposed to the agent may become infected.

Treatment of infected sheep is not practicable and vaccination has not been used.

TOXOPLASMOSIS

Epidemiology

Toxoplasma gondii is a parasite related to the coccidia first described in 1908 in a North African rodent and a rabbit in South America. Medical interest in the disease was stimulated by the association of infection with eye damage (choroidoretinitis) and the birth of children with congenital hydrocephalus and brain calcification.

Studies in many animal populations and man followed the development of an effective serological test and infection was shown to be widespread.

Infection in man is associated with the ingestion of undercooked meat or it may be acquired from domestic cats. Work in recent years has shown that the parasite undergoes both a sexual and an asexual life cycle in the intestine of the cat resulting in the production of very resistant oocysts which may remain infective in the environment or soil for many months. Only members of the cat family produce this form of the parasite.

The disease is of particular economic importance as a cause of abortion in sheep and in some years is the most important diagnosed cause of ovine abortion in Britain.

Sheep can readily be infected experimentally by various routes but natural transmission may occur orally at lambing time from the products of conception of infected ewes. How infection enters a flock in the first place is not fully understood but in some instances it may be derived from cats which may contaminate feeding stuffs and pasture. Outbreaks of disease frequently occur in flocks which are self contained. Recent experimental work has shown that, following infection of the ram, toxoplasma may be excreted in the semen for many days. The importance of this possible method of transmission in the spread of natural infection is not yet confirmed.

Symptoms

The effect of toxoplasmosis on the pregnant ewe is related to the stage of gestation at which infection takes place and the immune status. In early pregnancy it will produce foetal resorption. Those ewes infected between 70 and 120 days will abort or have still born lambs at full term but those acquiring toxoplasmosis in later pregnancy will give birth to apparently normal viable lambs. After abortion the ewe will be immune and should be retained in the breeding flock.

Diagnosis

Characteristic lesions are seen in the foetal membranes as multiple white flecks or in the cotyledons as small yellow foci, 1 to 2 mm in diameter. These often have a gritty consistency because of calcification. When the focal lesions are absent the cotyledons are paler than normal, firm and fleshy.

The parasite can be isolated from the foetal brain or the placenta in mice, which are kept for 5-6 weeks and a brain emulsion examined for cysts. Antibody in mice can be detected by the latex slide agglutination test 19 days after inoculation and the same test may also be used for screening sheep. The examination of placental or foetal tissues by immunofluorescence provides a rapid and reliable technique to complement mouse inoculation and histopathology. Serological tests are mainly of value for establishing the immune status of the flock. A difficulty in the interpretation of the dye test antibody results is the fact that levels remain high for many months after abortion and it is difficult to associate high antibody titres with clinical abortion in individual animals.

Treatment

Treatment is of no value in sheep.

Prevention

When abortions occur, lambs born alive and ewes which do not abort should be mixed with those which have aborted so that they may acquire the infection and immunity. This procedure must not be adopted if there is any suspicion that other agents are present in the flock. In practice it has been shown that newly purchased seronegative ewes exposed to a resident flock on a farm where toxoplasmosis exists may acquire infection before pregnancy. If infection is acquired in this way the animal is resistant to a challenge during pregnancy; this method of exposure has been found to be effective in preventing abortion in recently introduced sheep.

SUMMARY

1. Veterinary guidance should be sought when any number of lambs are aborted or still born.

2. Diagnosis of the cause of abortion is essential to enable preventive measures to be taken.

3. Diagnosis can only be made by a veterinary laboratory following the submission of suitable fresh specimens.

DISEASE AND REPRODUCTION IN THE RAM

Work on infectious infertiity in the ram has been done mostly in those countries with large ram studs. Only recently has it become the practice to check the fertility of rams in this country but there is increasing developmental work in this field, both by the East of Scotland College and the Meat and Livestock Commission. The two agents primarily associated with infertility are *Brucella ovis* and *Actinobacillus seminus.* The former does not occur in Britain.

BRUCELLA OVIS

Disease due to this agent is also known as infectious epididymitis. It was first reported in 1942 in Australia and the causal organism was isolated in 1955, simultaneously in Queensland and New Zealand. Since then the infection has been identified in numerous countries all over the world. The agent may infect both rams and ewes.

Epidemiology

Infection is spread by venereal transmission from ram to ewe, from ewe to ram or from ram to ram and the latter is thought to be the most usual means of spread in the field. Infection of ewes tends to be self limiting and only rarely carries over from one season to the next. The organism can be recovered from the vagina for up to 10 days after expulsion of an infected placenta and thus rams can become infected at post abortion oestrus.

Infection by mouth, conjunctiva or nasal mucosa can occur but these routes are not thought important. It is possible that congenital infection might occur in a small number of ram lambs.

Symptoms

Infection in rams is chronic. The organism localises in the tail of one or both epididymi with resulting testicular degeneration, fibrosis and calcification. Rams can excrete Br. ovis without developing detectable epididymal lesions and individual rams although infected may have satisfactory semen quality and be highly fertile.

Ewes exposed at mating may have a lower lambing percentage because infection interferes with ovulation, conception and implantation and may cause embryonic mortality. Vaginitis and cervicitis often develop in ewes mated to infected rams and such ewes may be less fertile at subsequent matings but in time the infection tends to clear itself.

Lambs become immunologically competent to Br. ovis early in foetal life and thus many are born alive and survive even though they are associated with a grossly diseased placenta.

Diagnosis

This disease may be diagnosed by:

i. palpation of the testicles, although not all affected rams have palpable lesions and conditions other than Br. ovis infection can cause epididymitis

ii. microscopical examination of stained smears and semen culture

iii. serology.

The complement fixation test is the best as it misses few positive cases but some flocks may have false positives in rams which have been vaccinated or ram lambs which have passively acquired antibodies with colostrum (the titres may persist for several months). A recently developed, indirect immunofluorescence test compares favourably with complement fixation and immuno-diffusion.

Control

Control has been attempted by diagnosis of the disease in rams and vaccination of those which showed no signs of infection. The recommended procedure is manual palpation of all rams, CFTs on sera, removal of rams positive to CFT or palpation and vaccination of all rams retained for breeding and sale. Immunisation of rams with Rev 1 vaccine, if available, or Strain 19 gives varying protection. The high cost of treatment precludes this except for notably valuable rams in the early stages of infection.

ACTINOBACILLUS SEMINIS

Following the elimination of *Brucella ovis* from certain Australian flocks, some rams continued to develop epididymitis which was clinically and pathologically distinguishable from that caused by *Br. ovis.* *Actinobacillus seminis* was isolated from the semen of these.

Both naturally and experimentally infected rams develop significant CF titres to infection with this organism and it has since been identified in the United States, South Africa and more recently in Scotland.

The effect on fertility may extend to complete sterility and is dependent upon the severity of infection, contamination by inflammatory and bacterial products, testicular involvement and semen production, patency of the genital ducts and whether one or both sides are affected.

It is generally assumed that transmission is venereal but observations indicate that transmission occurs via the mucous membranes of all the natural orifices of rams during direct contact and crowding, especially in the pre-mating period or through the ewe as an intermediate carrier. The possibility of ewe to lamb transmission cannot be excluded. There is evidence for the existence of antigenically different strains of the organism. The complement fixation test is as reliable for diagnosis as clinical examination, semen smears and bacteriological examination.

REFERENCES

Campylobacter

Frank, F.W.;Scrivener, L.H.;Bailey, J.W.;Meinershagen, W.A. (1958) Chlortetracycline as a preventative of vibrionic abortion in sheep. Journal of the American Medical Association 132,24-26.

Gilmour, N.J.L.;Thompson, D.A.;Fraser, J. (1975) Vaccination against Vibrio (Campylobacter) fetus infection in sheep in late pregnancy. Veterinary Record 96,129-131.

Jensen, R.;Miller, V.A.;Hammarlund, M.A.;Graham, W.R. (1957) Vibrionic abortion in sheep. I. Transmission and immunity. American Journal of Veterinary Research 18,326-329.

McFadyean, J.;Stockman, S. (1913) Report of the Departmental Committee of the Board of Agriculture and Fisheries on Epizootic Abortion. III. Abortion in sheep. H.M.S.O. London (Col.7156-7157).

Watson, W.A.;Hunter, D.;Bellhouse, R. (1967) Studies on vibrionic infection of sheep and carrion crows. Veterinary Record 81,220-225.

Salmonella

Baker, J.R.;Faull, W.B.;Rankin, J.E.F. (1971) An outbreak of salmonellosis in sheep. Veterinary Record 88,270-276.

Jack, E.J. (1971) Salmonella abortion in sheep. Veterinary Annual 12,57-63.

Watson, W.A. (1960) Salmonella dublin infection in a lambing flock. Veterinary Record 72,62-65.

Brucella

Allsup, T.N. (1969) Abortion in sheep associated with Brucella abortus infection. Veterinary Record 84,104-108.

Cameron, R.D.A.;Lauerman, L.H.,Jr. (1976) Characteristics of semen changes during Brucella ovis infection in rams. Veterinary Record 99,231-233.

Laws, L.;Simmons, G.C¨;Ludford, C.G. (1972) Experimental Brucella ovis infection in rams. Australian Veterinary Journal 48,313-317.

Shaw, W.B. (1976) Brucella abortus infection in sheep. II. Experimental infection of ewes. British Veterinary Journal 132,143-151.

Chlamydia

Foggie, A. (1973) Preparation of vaccines against enzootic abortion of ewes. A review of the research work at the Moredun Institute. Veterinary Bulletin 43,587-590.

Foggie, A. (1977) Chlamydial infections in mammals. Veterinary Record 100,315-317.

Stamp, J.T.;McEwen, A.D.;Watt, J.W.;Nisbet, D.I. (1950) Enzootic abortion in ewes. I. Transmission of the disease. Veterinary Record 62,251-254.

Toxoplasma

Beverley, J.K.A. (1976) Toxoplasmosis in animals. Veterinary Record 99,123-127.

Dubey, J.P. (1976) Reshedding of Toxoplasma oocysts by chronically infected cats. Nature UK 262,213-214.

Watson, W.A. (1972) Toxoplasmosis in human and veterinary medicine. Veterinary Record 91,254-257.

General

Axe, J.W. (1885) Journal of the Royal Agricultural Society of England 21,199.

Bang, B. (1897) The etiology of epizootic abortion. Journal of Comparative Pathology 10,125-149.

Tonder, E.M. van (1973) Infection of rams with Actinobacillus seminis. Journal of the Southern African Veterinary Association 44,235-240.

Watson, W.A. (1973) The prevention and control of infectious ovine abortion. British Veterinary Journal 129,309-314.

RESPONSE OF SHEEP TO COLD EXPOSURE IN RELATION TO SELECTION FOR SURVIVAL ABILITY

DR. J. SLEE

ARC Animal Breeding Research Organisation, West Mains Road, Edinburgh

Sheep kept outdoors in the British Isles and in much of Europe suffer from exposure to wind, rain, snow and low environmental temperatures. This paper deals with the sheep's physiological responses to cold, with the factors controlling heat loss and with the effects of cold on energy demand and viability. Measurements of cold resistance (the ability to maintain body temprature during cold stress) are used to assess the viability of sheep and lambs during cold exposure.

PHYSIOLOGICAL RESPONSES TO COLD

As a homeothermic mammal the sheep maintains a stable body temperature, usually within 2°C of the normal 39.5°C. The physiological and biochemical responses involved in balancing heat production and heat loss are complex (Figure 1) and include changes in insulation and heat production.

Obviously the responses of the sheep will vary according to the intensity of cold exposure. Responses commence when the ambient temperature falls below the "critical temperature" which is defined as the ambient temperature below which the animal is obliged to raise its metabolic rate in order to maintain its deep body temperature. In practice, there is a graded sequence of responses including vasoconstriction of the ears and then the feet, accompanied by a progressive increase in metabolic rate as the ambient temperature declines further (Figure 2).

Factors affecting the critical temperature

The critical temperature is as high as 28°C for a closely shorn adult sheep in average body condition. This indicates a cold-sensitivity similar to that of a naked man. The critical temperature however is considerably affected by plane of nutrition and fleece length and by environmental factors, particularly wind and rain. Figure 3 shows the relationship between some of these variables and the critical temperature, using data obtained by Blaxter and his colleagues in Scotland. It can be seen that the capacity of wind to change the critical temperature is greater as the fleece length increases. This may be because long fleeces which produce considerable insulation in still air can readily be parted by wind. Skin insulation, although relatively less important than the fleece, may also vary between breeds and influence heat loss.

The critical temperature for newborn lambs with normal birthcoats is higher than for adult sheep (Table 1) and almost all lambs born outdoors in a temperate climate will be subject to cold stress.

Figure 1

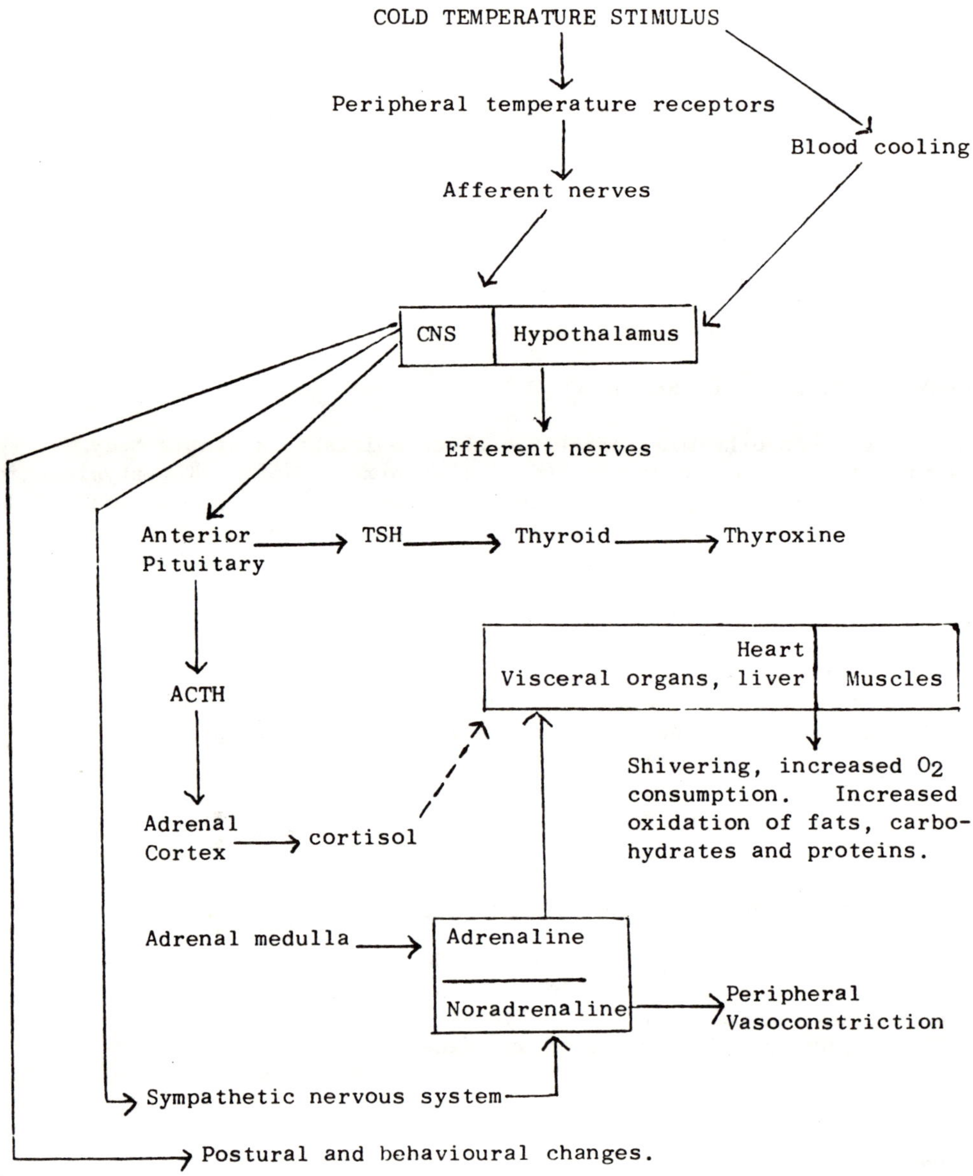

Figure 1. The pattern of response to cold exposure.

Table 1 Critical temperature of newborn lambs

Age	Birthweight	
	4kg	6kg
1-6 hours	37°C	32°C
14-20 hours	33°C	26°C
30-34 hours	31°C	22°C

HEAT LOSS AND PRODUCTIVITY

Despite the effects of physical and physiological insulation, cold exposure below the critical temperature will cause heat production to rise exerting an energy demand and producing effects on appetite and/or productivity. Cold exposure may initially depress appetite (Figure 4) perhaps due to the discomfort experienced but, over periods of a few weeks or more, voluntary food intakes will rise (Table 2) in the cold. The one exceptional breed (Southdown) showed the greatest degree of discomfort during the constant cold treatment.

When the food supply is limited, cold exposure would be expected to affect live weight. Over-winter losses of up to 5kg per head have been observed in some outdoor experiments involving food restriction.

Table 2 Changes in mean daily food consumption (grams) after temperature treatments in young wethers of four breeds.

	Intake and change in intake during the final 7 days of treatment			
	Treatment groups			
Breed	Thermoneutral	Constant Cold	Fluctuating Cold	Mean
Blackface	1615 (-84)	1643 (+175)	1916 (+267)	1724 +119
Welsh	873 (-275)	1359 (+201)	1206 (+77)	1146 +1
Southdown	1305 (+37)	1166 (-216)	1555 (+418)	1342 +80
Finn	1495 (-156)	1743 (+418)	1898 (+514)	1712 (+259)
Mean	1322 -120	1478 +144	1644 +319	1481 +115

There were four sheep of each breed in each treatment subset. The mean change in food consumption produced by treatment is shown in parenthesis. Treatments lasted for four weeks and were carried out in climate chambers.

Indicating the ambient temperatures causing physiological responses to cold and heat in the shorn sheep

Figure 2

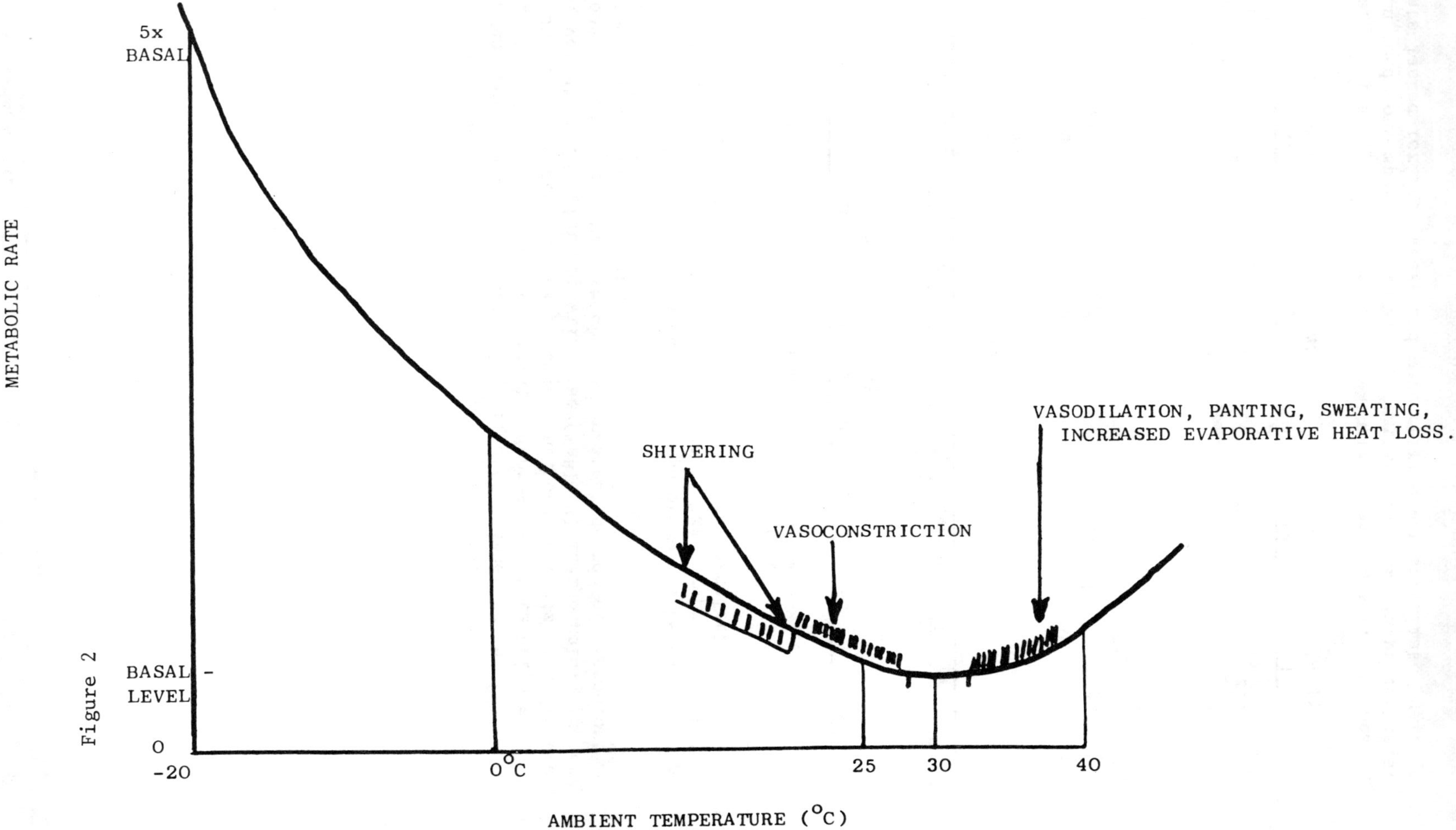

CRITICAL TEMPERATURE OF 100 lb BLACKFACE SHEEP
GAINING WEIGHT AT DIFFERENT RATES

FIGURE 3

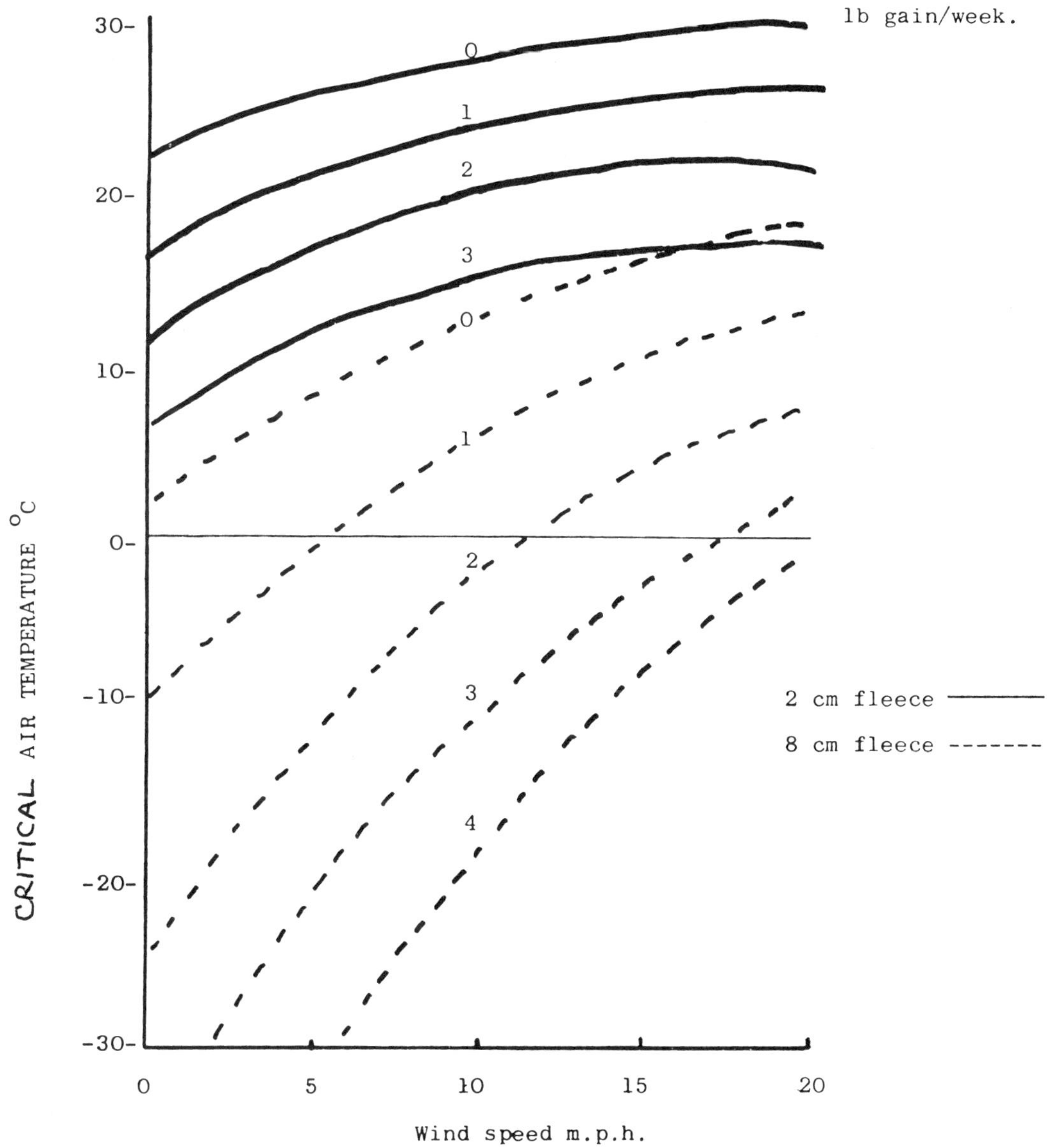

Fig. 3 The critical temperature of a Blackface sheep with either 2 or 8 cm. of wool kept in environments with different air velocities and differing food intakes as evidenced by different rates of gain per week. Critical temperature can vary from 30°C in a sheep given maintenance food with 2 cm. wool in a wind of 20 m.p.h., to -30°C in the same sheep with 8 cm. wool given food to support a gain of 4 lb. per week at a wind speed of 7 m.p.h.

- from Blaxter (1963)

NEWBORN LAMBS: COLD RESISTANCE AND SURVIVAL

On Scottish hill farms between ¼ million and ½ million lambs are lost each year. The effects of cold exposure are most crucial soon after birth. The main factors affecting survival are breed, birthweight, birthcoat type, weather, pre- and post-natal nutrition, maternal and lamb behaviour. Recent work at ABRO centres on breed differences in body temperature among newborn lambs in the field (Table 3). Such differences imply potential differences in mortality.

Table 3 Average Rectal Temperature One Hour After Birth of Lambs in the Field

Breed	Mean lamb temperatures (°C)	N	Mean adjusted ambient temperature (°C)[1]
Blackface	39.34	50	+1.4
Cheviot	39.81	19	-0.5
Welsh	39.96	25	0
Soay	39.52	24	+0.2
B. Leicester	36.73	10	+0.4
Southdown	37.15	10	-1.6
Finn	34.30	12	+0.4
Merino	35.14	13	-0.2
Oxford	38.60	14	-1.3

[1]Ambient temperatures were adjusted to allow for the cooling effects of wind if present.

Figure 5 shows the effect of some causal factors on the lamb's energy demand. If prolonged summit metabolism (about 18 k.cals/kg/hour) is required to maintain homeothermy, survival is unlikely for more than a few hours. When the energy demand of the environment is still high but does not call for maximum metabolic effort, survival will be extended but may then be limited by the available energy reserves. These reserves will depend largely on the prenatal nutrition of the ewe and may vary from 100-300 k.cals/kg, giving for a starved lamb about 5-12 hours of life in bad weather or 3-4 days in good weather. Superimposed on these variables are the effects of the lamb's speed in gaining its feet and suckling as soon as possible after birth and the mother's ability to assist.

Controlled experiments at ABRO, using climate chambers (Table 4) and water bath immersions (Table 5), have shown that tests can be devised which will distinguish between newborn lambs of different breeds with respect to their ability to resist body cooling. Significant breed differences have been demonstrated and the next step will be to look for genetic

FIGURE 4 FOOD REFUSAL DURING COLD EXPOSURE

Indicating the extent of inappetence in shorn sheep during cold exposure at 8°C. Before exposure began there was complete clearance of the same ration.

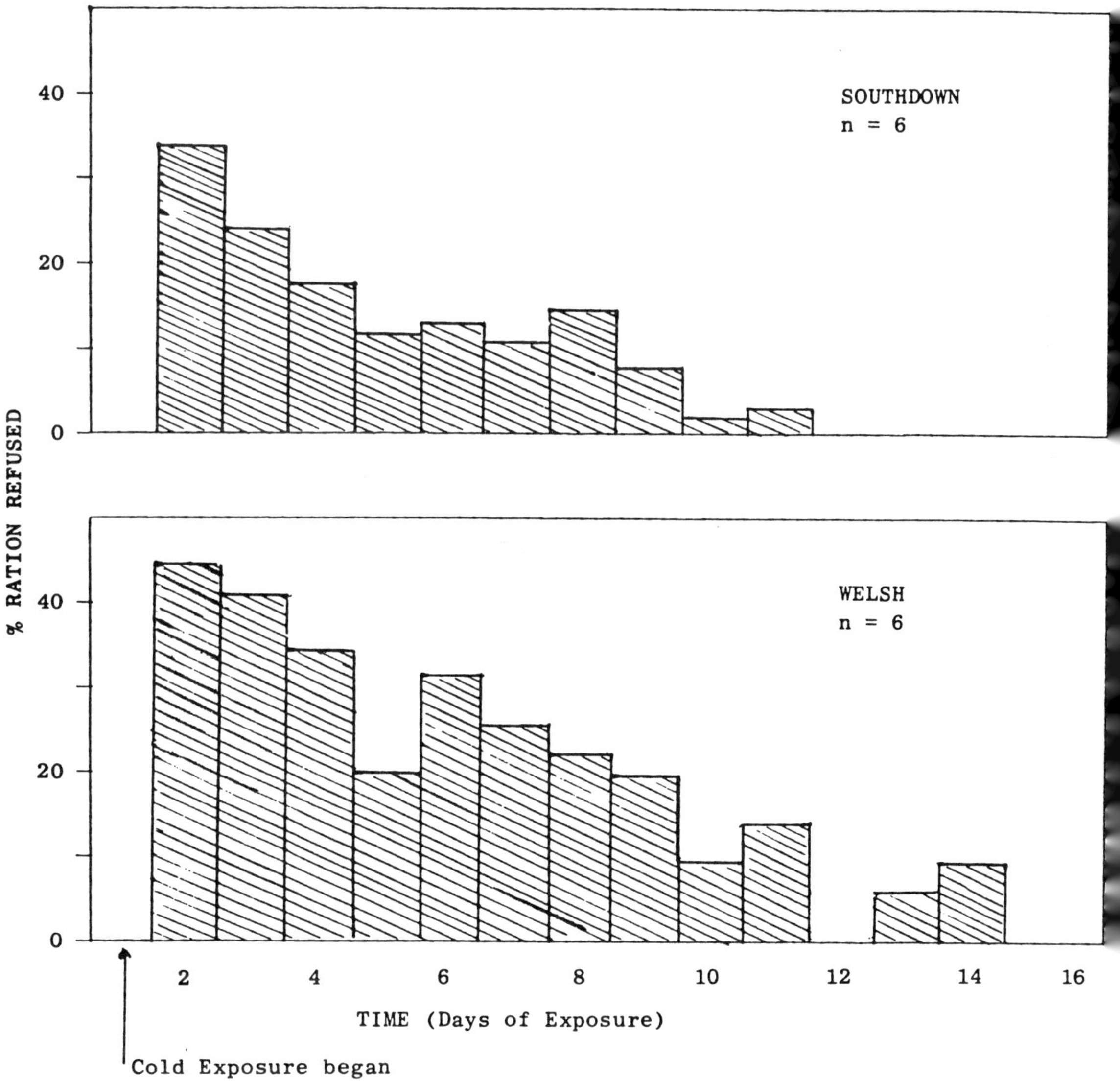

FIGURE 5

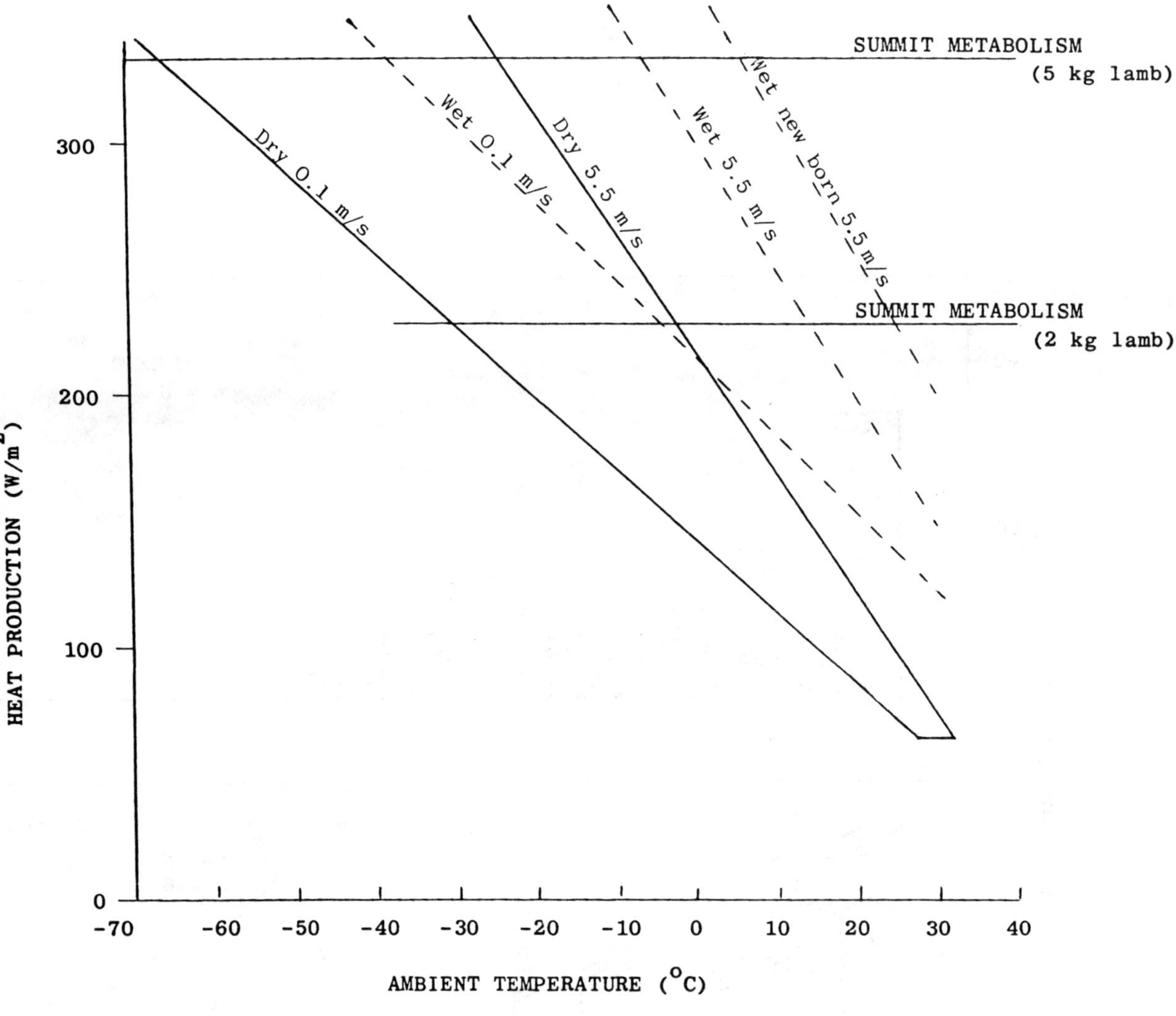

Fig.5 Relationship between metabolic rate and ambient temperature in young Merino lambs with short coats, in still air or in a wind of 5.5 m/s and with the coat thoroughly wet. Summit metabolism of 2 kg and 5 kg lambs is included. Metabolic rates below minus 10°C were estimated by extrapolation. Lambs with long hairy coats show 25-30% less heat loss.

- from Alexander (1962)

variance within breeds which might be used for selection towards increased cold resistance.

Table 4 Resistance to body cooling expressed as the number of minutes of cold exposure required to produce 1°C decline in rectal temperature of unclipped lambs in a climate chamber

	BREED			
	Blackface	Welsh (long birthcoats)	Welsh (short birthcoats)	Merino
Number of lambs	10	9	10	10
Mean cold resistance (minutes)	216	139	64	44
Mean birth-weight (kg)	4.0	2.6	2.6	3.8
Mean age (hours after birth)	23	22	20	39

Table 5 Average Water Bath Cold Resistance of Lambs of Nine Breeds

	n	weight (kg)	Mean CR (min)
Cheviot	16	3.6	96
Welsh	14	3.3	83
Blackface	22	3.7	79
B. Leicester	8	5.3	78
Southdown	8	3.8	47
Merino	9	3.9	46
Boreray[1]	10	2.8	41
Finn	18	3.0	37
Soay[1]	24	2.3	35

[1]Primitive breeds

Cold resistance (CR) was equivalent to the time in minutes required to induce rectal temperature to decline by 4°C in a bath water temperature falling from 37°C to 12°C.

CHANGES IN RECTAL TEMPERATURE DURING ACUTE COLD EXPOSURE IN 2X2 GROUPS OF SCOTTISH BLACKFACE SHEEP

FIGURE 6

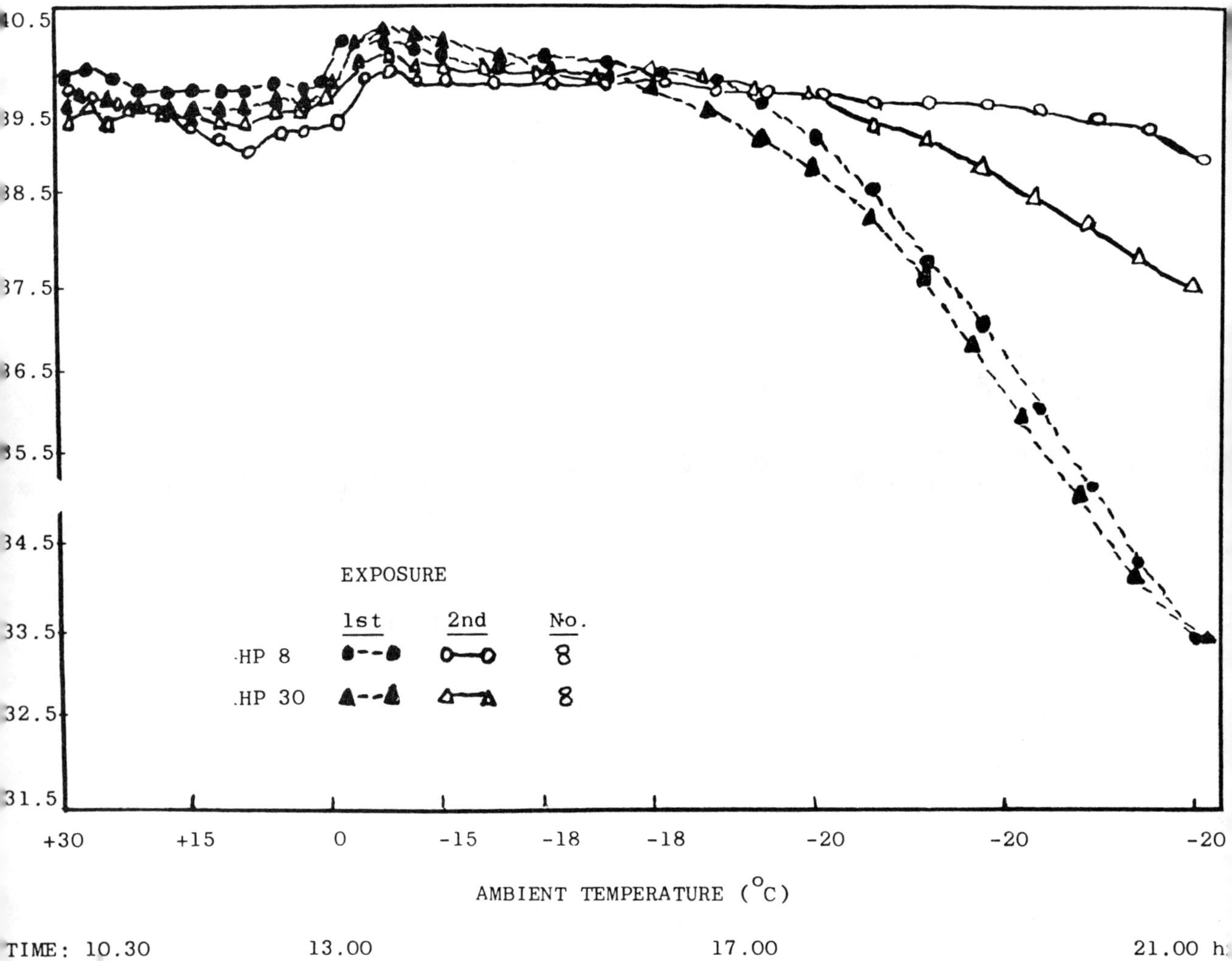

Fig 6

Mean rectal temperature changes in two groups each of 8 shorn sheep during two acute cold exposures before and after acclimatization. The HP 30 sheep were kept at +30°C (thermoneutral) between the first and second exposures; the HP 8 sheep were kept at +8°C (about 20°C below the critical temperature). Note that the rate of fall of rectal temperature was similar for both groups at the first exposure - ~~before~~ acclimatization. After acclimatization, resistance to cooling was greatly increased, especially in the HP 8 group which received the greatest dosage of cold.

The acute exposures involved a fall in ambient temperature first to zero and then subsequently to -20°C with a wind of 4 mph. Individual sheep were removed when rectal temperature fell to 36°C, but the group mean lines are extrapolated.

COLD RESISTANCE IN ADULT SHEEP

Although cold weather is less hazardous for adult sheep than for lambs, serious losses can occur. In the exceptional winter of 1962-63, up to a million sheep were lost in Scotland as a result of severe weather and undernutrition. In Australia up to a million adult sheep per year die as a result of cold wet weather immediately after shearing.

The cold resistance of shorn sheep can be measured in climate chambers. The sheep are exposed to a temperature of -20°C with a 4mph wind and the rate of body cooling is assessed by measuring the time taken for rectal temperature to decline from 39.5°C (normal) to 36°C. This character of cold resistance was found to be variable within and between breeds of sheep but had a high repeatability within individuals. Preliminary measurements in hill sheep suggest that cold resistance may have a fairly high heritability of about 30%.

Unfortunately sire-offspring tests carried out by ABRO have not demonstrated a clear relationship between the cold resistance of rams (measured in climate chambers) and the survival rate of their lambs in the field (Table 6). Progeny from selected high cold resistance rams did show slightly better survival rates than those from low cold resistance rams but the difference was not significant. This work is continuing.

It has been found that cold resistance is strongly influenced by plane of nutrition and inbreeding (Table 7), by breed (Table 8) and by previous acclimatization to cold (Table 8).

Table 6 Sire-offspring Comparisons of Cold Resistance and Lamb Mortality

Sires	Offspring	
	Lamb mortality rate (%)	
	Period after birth	
Ram selection (160 tested)	0-2 weeks	0-16 weeks
10 high-cold resistance rams	3.1	16.7
10 low cold-resistance rams	4.6	17.0

Table 7 Average cold resistance of outbred and inbred rams

Time in minutes for rectal temperature to fall to 36°C during acute exposure. There were 14 sheep in each treatment group.

Plane of nutrition	Outbred	Inbred
High	389	356
Low	267	161

Table 8 Resistance to body cooling in four breeds of sheep

Average time in minutes for body temperature to fall to 36°C during acute cold exposure

	Scottish Blackface	Southdown	Welsh Mountain	Tasmanian Merino
No previous cold experience	449	294	230	125
Cold-acclimatized	557	374	303	208

ACCLIMATIZATION TO COLD

One of the most powerful factors influencing cold resistance is acclimatization. Previous exposure to either chronic moderate cold or acute cold stress consistently produces a marked improvement in cold resistance (Figure 6 and Table 8) and apparently results from an increased ability to achieve and sustain a high metabolic rate. Other physiological changes which result from cold acclimatization include increased heart rate and resting metabolism at thermoneutral temperatures and increased skin temperature.

These physiological responses demand an increased energy expenditure but at least one other type of cold adaptation can be produced in some sheep. This involves progressive habituation to short, acute cold exposures resulting in passive body cooling (Figure 7) and a diminished metabolic response to cold. Since habituation and acclimatization differ in their energy demands, they may also differ in their effectiveness for specific situations. For a hill sheep exposed to prolonged cold stress with undernutrition, energy-sparing adaptation would be desirable. But for newly-shorn sheep, or newborn lambs with an assured milk supply, increased energy expenditure during a short cold stress would be acceptable if it permitted greater comfort and mobility and preservation of the drive to feed or suckle.

Figure 7

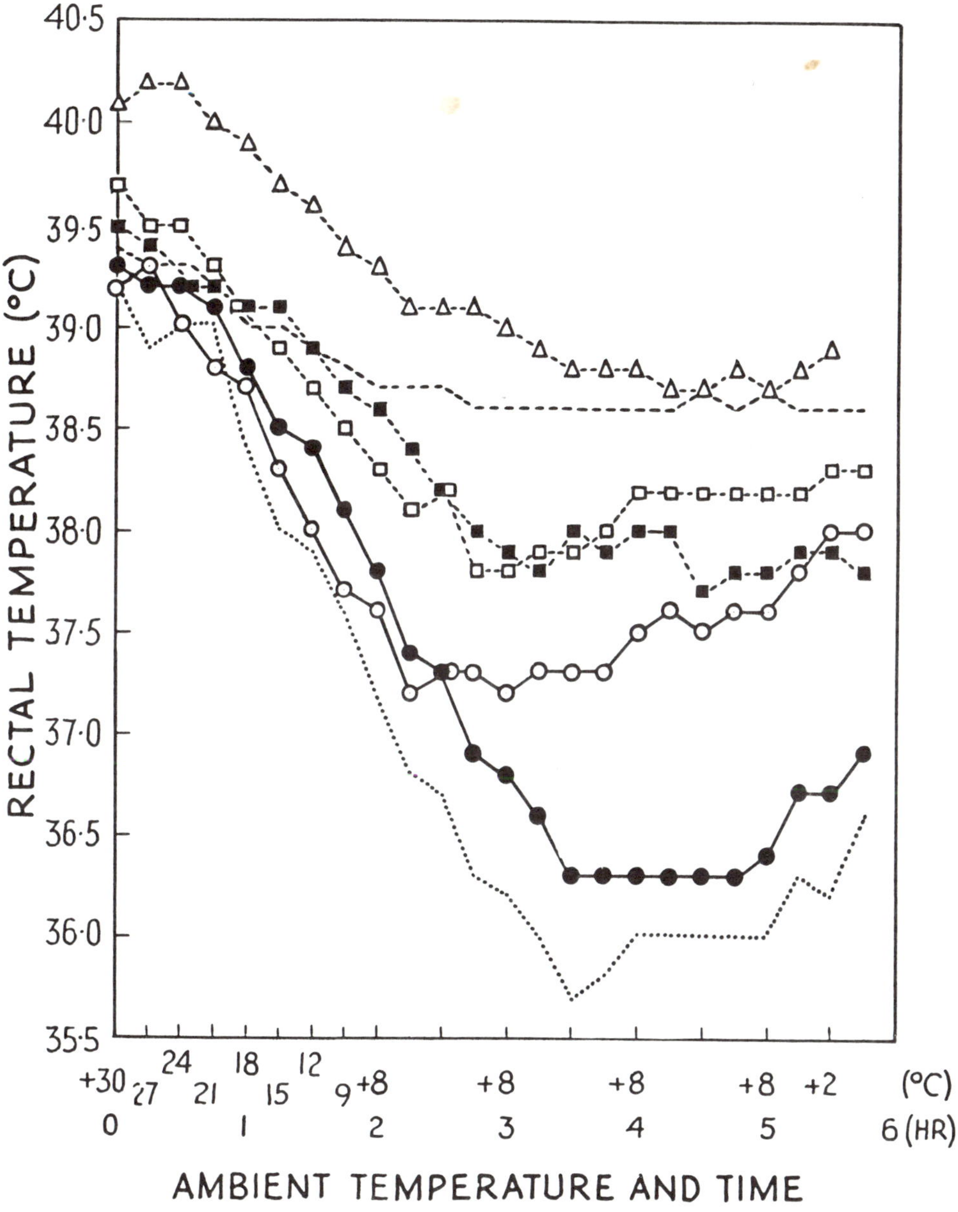

Figure 7 Voluntary hypothermia after cold-habituation assessed by decline in mean rectal temperature in shorn sheep exposed to a slow fall in ambient temperature from 30°C to 2°C. Six different groups of sheep had previously been either acclimatized to cold (△), habituated to cold (■ , □ ● , ○) or untreated (– – –). The different effects of acclimatization (raising rectal temperatures) and habituation (permitting lower rectal temperatures) are clear. One cold habituated sheep (.....) showed considerable hypothermia during mild cooling.

REFERENCES

Alexander, G. (1962) Temperature regulation in the new-born lamb. V. Summit metabolism. Australian Journal of Agricultural Research 13, 100-121.

Alexander, G. (1973) Heat loss from sheep. In: Monteith, J.L.;Mount, L.E. (Eds) 'Heat loss from animals and man', 173-204. Proceedings 20th Easter School in Agricultural Science, University of Nottingham.

Blaxter, K.L. (1963) Problems of cold and wet environments. Proceedings 6th International Congress of Nutrition. p197.

Mercer, J.B. (1974) Ph.D. Thesis. Trinity College, Dublin.

Slee, J. (1974) The retention of cold acclimatization in sheep. Animal Production 19,201-210.

Sykes, A.R.;Slee, J. (1969) Cold exposure of Southdown and Welsh mountain sheep. Animal Production 11,65-75.

Sykes, A.R.;Griffiths, R.G.;Slee, J. (1976) Influence of breed, birth weight and weather on the body temperature of newborn lambs. Animal Production 22,395-402.

Slee, J. (1977) Cold exposure and survival in newborn lambs. Animal Breeding Research Organization Report pp11-16.

GENETIC AND PHYSIOLOGICAL VARIATION IN REPRODUCTIVE PERFORMANCE

DR R.B. LAND

A.R.C. Animal Breeding Research Organisation, West Mains Road, Edinburgh

INTRODUCTION

We have two methods to improve the performance of domestic animals. We may intervene to change the system of husbandry or alter the normal physiology of the animal, or we may breed for genetic improvement. Although these may be considered as alternatives, they are not exclusive; in practice it is desirable to combine them optimally, the relative balance between them being determined by their relative costs and advantages.

The use of husbandry or physiological intervention has the advantage of immediate response, but the response may be only temporary. The use of pregnant mare serum gonadotrophin (PMSG) may raise the number of eggs shed at a particular ovulation, but not at any other. Improved nutrition in early life may lead to greater body size and reproductive performance, but it is necessary to continue the better diet throughout an individual's life; it is necessary to repeat the extra input in the next generation. Intervention may be compared to the donkey driving the water wheel; it goes round and lifts some water, but at the end of the circuit the opportunities for raising water are no greater than they were when it started. Research may lead to the development of improved pharmacological products which will simplify intervention techniques and lead to greater cost effectiveness, but this should not be relied upon; although knowledge is increasing rapidly, so too is the difficulty of marketing new products. The restrictions on the introduction of new animal products are increasing at such a rate that drug companies may not consider it worthwhile to develop and market them. The case for genetic improvement is likely to increase relative to that for improved intervention, but what are the possibilities?

CLASSICAL GENETIC IMPROVEMENT

Genetic improvement of reproduction is dependent upon increases in the frequency of favourable genes. For a chosen population, two alternative approaches to improvement have been considered. Firstly, and much more generally, superior individuals have been chosen and bred on the basis of their and/or their relatives' performance. Secondly, attempts have been made to identify superior genes directly by biochemical or immunological means and thus increase their frequency by breeding from individuals carrying these genes. Neither technique has been a resounding success.

Studies of the genetic improvement of one component of reproductive performance, litter size, have been reviewed by Turner (1969) and Bradford (1972), and illustrate the difficulties involved. Both authors report that litter size can and has been changed by selection within flocks, but that the response to selection has been very slow. In the long term selection experiments of Wallace and Turner, referred to by Bradford, Romney and Merino sheep have shown a response of 0.017 lambs per ewe per year between the selected Romneys and an unselected control, and a divergence of 0.023 lambs per ewe per year between high and low twinning Merino lines. Four main factors are responsible for this slow response; the heritability of number born is low at around 5 to 10% (Bradford 1972), the trait is sex limited and can only be measured in females, even in the females the trait can often not be measured until they are 2 to 3 years of age, and the low reproductive rate limits the selection pressure which can be applied. Although the heritability is low, progress is not quite as slow as this alone might suggest because the phenotypic standard deviation for reproductive traits is high. The product of $h^2 \sigma_p$ in the genetic response equation $R = ih^2 \sigma_p$ is of the same order of magnitude for reproductive traits as it is for productive traits. The low heritability of reproductive traits may arise from high environmental variation rather than an absence of genetic variation. It is necessary to consider changes both in selection intensity and in $h^2 p_p$ in any programme to improve rates of response to selection. As a first step it is necessary to increase our understanding of the sources of variation in reproductive traits.

THE PHYSIOLOGICAL COMPONENTS OF REPRODUCTION

The display of oestrus and the number of eggs shed and fertilised are dependent upon the response of the ovary to the stimulus of gonadotrophic hormones, principally follicle stimulating hormone (FSH) and luteinizing hormone (LH), secreted by the pituitary gland. Ovarian activity therefore determines the fertility of an individual and the upper limit to prolificacy. If the ovaries are inactive an individual will not ovulate or show heat and, except for monozygous twins, no more lambs are born than eggs fertilised. The number born is clearly influenced by the losses between fertilization and parturition, but superovulation (e.g. Newton et al 1970), shows that ewes can carry more lambs than the number of eggs normally shed. Furthermore, the introduction of fertilised eggs in excess of the normal ovulation rate by embryo transfer leads to an increase in prolificacy. Selection for litter size has led to positively correlated changes in ovulation rate (Packham & Triffit 1966, Bindon et al 1971, Trounson & Moore 1972). At the present level of performance increases in the number of fertilised eggs available lead to increases in litter size. The key to the improvement of flock performance is the improvement of the ovarian activity of the individual females.

VARIATION IN OVARIAN ACTIVITY

The improvement of overall reproductive performance by increasing ovarian activity is dependent upon the possibility of measuring it, the existence of genetic variation and the relationship between ovulation rate and other components of reproductive performance.

Seasonal variation in the incidence of ovulation and the ovulation rate as estimated by the number of corpora lutea has been studied extensively by laparoscopy (e.g. Thimonier & Mauleon 1969, Land et al 1973,

Wheeler & Land 1977). These have shown that both the duration of the breeding season and the maximum ovulation rate vary markedly among breeds. Adult Ile de France ewes showed an increase in the incidence of ovulation in June and then a decline in February, eight months later, whereas in the Romanov ovarian activity did not rise until August but continued at a high level until April. The maximum ovulation rate recorded for the Romanov was around 2.9, whereas that for the Solognote breed was only 1.1. Bradford (1972) also refers to breed differences in ovulation rate recorded at single observations. There is considerable breed variation both in the duration of ovarian activity over the year, and in the level of activity.

The correlated change in ovulation rate following selection for prolificacy, referred to previously, and the variation in the date of first oestrus between Welsh Mountain ewes of different sire groups (Purser 1972) also indicate the presence of individual variation in the activity of the ovaries of unmated, non-lactating sheep.

There is therefore little doubt of the existence of genetic variation in ovarian activity, but its utilisation is dependent upon a convenient method of measurement and furthermore the advantages of using estimates of ovarian activity are dependent upon the correlations between different aspects of ovarian activity. If for example variations in the age at which ovarian activity commences, the duration of activity during the year, the maximum ovulation rate and the incidence of ovulation and oestrus during lactation were all physiologically and genetically independent of each other, selection for one would not be expected to change the others, and selection for one aspect of ovarian activity would offer little advantage over direct selection of 'end' traits. If the different aspects of ovarian activity however are genetically correlated, selection for an individual parameter would be expected to change to some extent the correlated components of reproductive performance.

COVARIATION AMONG COMPONENTS

The relationships between different aspects of reproductive performance have been studied both within and between breeds, and in general it has been found that the different aspects of ovarian activity are not independent of each other.

Comparative studies have been helped greatly by the availability of breeds such as the Finnish Landrace and the Romanov, both of which give birth to large litters. Their litter sizes are in the region of 3 lambs (Maijala 1966, Donald & Read 1967, Desvignes 1972) and both breeds have correspondingly high ovulation rates. Comparative studies of Finnish Landrace and British breeds of sheep have shown that several aspects of ovarian activity are higher in the Finnish Landrace and its crosses. The proportion of developing oocytes at birth in Finn cross Welsh and Finn cross Blackface lambs is greater than that of the pure bred Welsh and Blackface lambs. Similarly, the duration of oestrus is greater in Finn crosses than in pure Blackfaces, which in turn is greater than that of Merino crosses; the incidence of oestrus during lactation is greater in Finnish ewes than in Dorset Horns (Land 1971).

The regularity of ovarian activity during the breeding season and the mean ovulation rate have been found to be much higher in Finnish ewes than in Tasmanian Merinos (Wheeler & Land 1977), and the incidence of oestrus in Merino ewe lambs is much lower than that in Finns (Land et al 1974).

Within breeds, Chang & Rae (1971) found that the duration of the first breeding season "number of hogget oestruses" was correlated with ewe fertility in Romney sheep (genetic correlation = 0.5-0.9, depending on the method of estimation). The litter size of Welsh Mountain ewes however was found to be genetically independent of the onset of the breeding season (Purser 1972). Within the breeding season the incidence of ovulation in Romanov ewes was correlated (phenotypic correlation = 0.8) with the mean ovulation rate of those ovulating at that time (Land et al 1973).

There is now considerable evidence to show that the repeatabiity of ovulation rate differs among breeds and that it may be related to the mean value of a breed. Hanrahan (1976 personal communication) reports the repeatability of ovulation rate to be 0.7, 0.8, 0.2 and 0.2 for Finnish Landrace, "High Fertility", Galway and Fingalway ewes, with mean ovulation rates of 3.8, 2.6, 1.6 and 2.6 respectively. Wheeler & Land (1977) report repeatabilities of 0.4, 0.1 and 0.1 for Finnish Landrace, Scottish Blackface and Tasmanian Merino ewes with ovulation rates of 3.0, 1.3 and 1.1 respectively. There is therefore evidence for the hypothesis that ovarian activity may be considered as a quantitative character which is expressed in several ways, such as ovulation rate and the duration of cyclical activity, the characteristics of which vary from one breed to another.

Although ovarian activity is more difficult to measure than litter size or the incidence of oestrus, it has the advantage that it can be measured in relatively young individuals. It does not however overcome the limitations of low selection intensity. These difficulties may however be overcome by the development of indirect measures.

QUANTITATIVE ENDOCRINOLOGICAL ESTIMATES OF OVARIAN ACTIVITY

The activity of the ovaries is controlled by the gonadotrophic hormones from the anterior pituitary gland. It is logical therefore to consider the measurement of the gonadotrophic stimulation of the ovaries as an indication of their potential activity. The results of the study of such a possibility are reviewed by Land & Carr (1978).

It has been shown that young lambs of highly prolific breeds or crosses have higher concentrations of LH in the plasma of peripheral blood, but that there are two problems. At any one time, the concentration of LH is very variable, the repeatability over a range of ages and breeds being only 0.12 (Land & Carr 1978), so that any single estimate is a very poor estimation of the concentration at any other time, let alone the potential reproductive merit of that individual. Secondly, the concentration first increases with age then declines, the peak occurring at an earlier age in lambs of highly prolific types than in those of low prolificacy. This led to the suggestion of a model to explain apparently anomalous results; at young ages highly prolific lambs exceed those of low prolificacy and the ranking then becomes the same and finally reverses.

It may therefore be concluded that the level of activity of the ovaries appears to be quantitatively related to differences in an endocrinological parameter known to be qualitatively involved in their control, and that these differences are present in young animals of both sexes. It may therefore be possible to base genetic selection on them (Land 1971). Such an opportunity would have a marked effect on the selection intensity, for in a species such as the sheep with a low reproductive rate much more intense selection can be applied to males than to females. Similar endocrinological criteria may be used to try to predict the reproductive performance of a male e.g. sperm production or libido. This however would be dependent upon the phenotypic correlation within males rather than the genetic correlation between sexes.

A relationship between the fertility of males and of females may at first be thought surprising, but consideration of the physiological basis of reproductive activity shows that this might be expected. It has been known for many years that the same gonadotrophic hormones, FSH and LH, control the development of follicles and ovulation in the female and also spermatogenesis and testosterone production in the male (e.g. Cole 1969). This relationship is illustrated in Figure 1. Furthermore, with the exception of one sex chromosome the genotype of an individual is independent of its sex, and all autosomal genes controlling gonadotrophin production are common to both sexes.

The critical step is to translate this knowledge into a suitable selection criterion, a male characteristic which has both a high repeatability and a high genetic correlation with the female trait to be improved.

POSSIBLE SELECTION CRITERIA

Two possibilities have been considered; the response to LRF and testis growth. Of these, Bindon & Piper (1976) concluded that testis growth was the more promising, but it is relevant to consider both.

Response to LRF

The philosophy behind the "LRF test" is that the use of a small dose of the releasing factor responsible for the release of LH (LRF) may overcome the short term variation in LH concentration while still reflecting the sensitivity of the pituitary gland to endogenous LRF. Early studies (Carr et al 1976) show that the response of male Finnish Landrace lambs is greater than that of Blackfaces. Further studies show that Finn x Blackface lambs exceed Merino x Blackface lambs. The differences between breeds and crosses are in the same direction as the differences in unstimulated hormone concentration and the differences in female performance. The critical relationship however is that within populations; this is now being tested at ABRO.

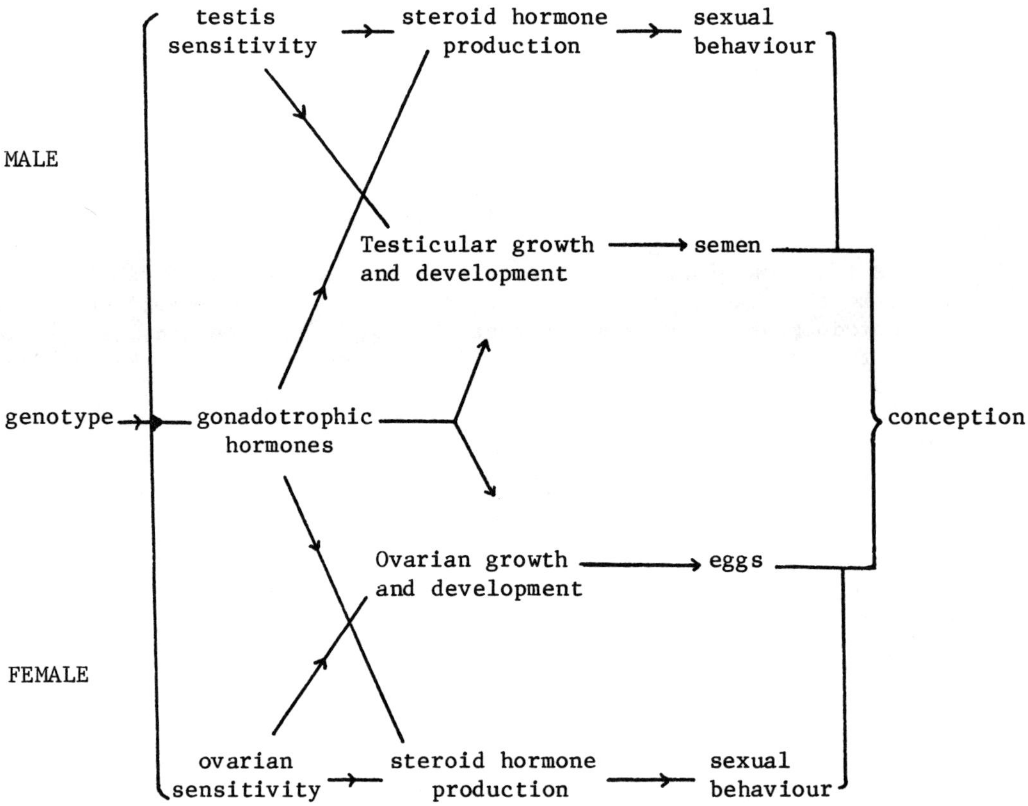

Figure 1 The central role of the gonadotrophic hormones in the pathway from genotype to conception in males and females.

Testis Growth

The comparison of testis growth in breeds and crosses of sheep with varying female reproductive characteristics provided the first information (Land 1973). The most extreme of the breeds compared were again the Finnish Landrace and the Tamanian Merino; their characteristics and those of the cross between them are summarised in Table 1.

Table 1 The reproductive characteristics of females of the Finnish Landrace and Tasmanian Merino breeds and of the crosses between them

	Finnish Landrace	Finn-Merino Crossbred	Tasmanian Merino
Showing heat at 7 months old (%)	97	94	5
Lambing at 1 year of age (%)	83	80	0
Lambs born per ewe lambing at 2-4 years of age	2.7	1.7	1.0
Ovulation rate	3.0	2.0	1.1

Testis growth was rapid in the Finn and Finn-Merino males, both groups reaching approximately 80% of their "adult" diameter by 7 months of age, whereas the testes of the Merino rams remained at only 40% of their adult diameter. Sexual maturation appeared to be more rapid in Finn and Crossbred males and females than in Merinos and, at this level of measurement, to be genetically dominant in the Finn. Such a conclusion may however be superficial for it is known that the two parent breeds show different reproductive responses to the same change in the breeding season. The dominance may have arisen from the time of birth of the lambs in the experiment reported, the Merino lambs missing the period of photoperiod stimulation in the first year of their lives (Land 1978). As with the response to LRF, the comparison of different breeds and crosses indicates that testis growth in the male may be associated with the reproductive characteristics of the female.

Critical to animal breeding is the nature and extent of the relationships within populations. The relationship between testis growth and ovulation rate has been studied in an experimental mouse population, and the genetic correlation between the two traits found to be 0.5 (Islam et al 1976); the heritability of testis weight was also 0.5. The characteristic is being examined within a population of sheep by divergent-selection for testis size (Land & Lee 1976). Early results indicate that the ovulation rates of females of these lines may diverge by 20% after 4 years of selection among males, and that there may be a similar divergence in number of lambs born. This result must be treated with caution until the results of subsequent generations of selection are available.

In addition to the indicated correlated response in ovulation rate and litter size, the onset of the breeding season of 18-month-old females of the two lines has been found to differ in favour of the earlier onset of females of the high line. It is argued (Land 1978) that this is the result of selection during an adverse photoperiod; males which grew their testes more rapidly during an adverse photoperiod, i.e. when daylength was increasing, sire daughters which are in turn less dependent on a stimulatory photoperiod, i.e. the period when daylength is declining. It may therefore be possible to control the extent to which testis growth in the male is genetically correlated with ovulation rate, as distinct from the time of occurrence of ovulation, by varying the environment in which the testes grow by altering the time of birth of the lambs.

IMPLEMENTATION

Techniques are available for the genetic improvement of reproductive performance, and these may be supplemented by the development of new quantitative physiological criteria. In the meantime, it is relevant to consider improvement programmes and the development of strains of extreme merit will facilitate the improvement of locally adapted breeds throughout the world. The impact of the Finnish Landrace and the Romanov would have been considerably greater had their difference from other sheep also been greater. With additive inheritance, sheep with an ovulation rate of for example six would be twice as "potent" as the Finnish Landrace, so that the same improvement in the reproductive performance of any other breed could be attained with only half the proportion of introduced genes.

The development of biologically extreme types lends itself to international cooperation and should be considered in such a framework.

CONCLUSIONS

The investigation of physiological sources of genetic differences in the reproductive performance of the sheep has led to the hypothesis that many aspects of reproductive activity in males and female are controlled at least in part by the same underlying variables and has indicated that it may be possible to predict the performance of females, and of a male's potential daughters, from measurements of hormone levels or the products of hormone action in young lambs. It may therefore be possible to incorporate such a criterion in a genetic improvement scheme. Similar ideas are being pursued in the dairy cow. The development of quantitative physiological criteria may have a considerable impact on the design of future breeding programmes.

REFERENCES

Bindon, B.M.;Ch'ang, T.S.;Turner, H.W. (1971) Ovarian response to gonadotropin by Merino ewes selected for fecundity. Australian Journal of Agricultural Research 22,809-20.

Bindon, B.M.;Piper, L.R. (1976) Assessment of new and traditional techniques of selection for reproductive rate. In 'Sheep Breeding'. (Eds) G.J. Tomes;D.E. Robertson;R.J. Lightfoot. West Australian Institute of Techology 357-371.

Bradford, G.E. (1972) Genetic control of litter size in sheep. Journal of Reproduction and Fertility Suppl.15,23-41.

Carr, W.R.;Land, R.B.;Sales, D.I. (1976) The effect of gonadotropin releasing hormone (GnRH) on plasma levels of luteinizing hormone (LH) in lambs of breeds with high and low ovulation rates. Annales de Biologie animale, Biochemie, Biophysique 16,167 (Abst.)

Ch'ang, T.S.;Rae, A.L. (1972) The genetic basis of growth, reproduction and maternal environment in Romney ewes. II. Genetic covariation between hogget characters, fertility and maternal environment of the ewe. Australian Journal of Agricultural Research 23,149-165.

Cole, H.H. (1969) Physiological characterization of gonadotropins. In 'Reproduction in domestic animals'. (Eds) H.H. Cole and P.T. Cupps. Academic Press 1969. pp17-45.

Desvignes, A. (1972) La race ovine Romanov. Annales de Zootechnie 20,353-270.

Donald, H.P.;Read, J.L. (1967) The performance of Finnish Landrace sheep in Britain. Animal Production 9,471-476.

Islam, A.B.M.M.;Hill, W.G.;Land, R.B. (1976) Ovulation rate of lines of ewes selected for testis weight. Genetical Research, Cambridge 27, 23-32.

Land, R.B. (1971) The incidence of oestrus during lactation in Finnish Landrace, Dorset Horn and Finn-Dorset sheep. Journal of Reproduction 24,345-352.

Land, R.B. (1973) The expression of female sex-limited characters in the male. Nature, Lond 241,208-209.

Land, R.B. (1978) Reproduction in young sheep: some genetic and environmental sources of variation. S.S.F. symposium 'Puberty'. Journal of Reproduction and Fertility 52,427-436.

Land, R.B.;Carr, W.R. (1978) Reproduction in Domestic Mammals. In 'Genetic variation in hormone systems'. Ed. J.G.M. Shine. C.R.C. Press. In press.

Land, R.B.;Lee, G.J. (1976) Testis growth, a possible genetic predictor of female reproductivity. Animal Production 22,137 (Abstr.)

Land, R.B.;Pelletier, J.;Thimonier, J.;Mauleon, P. (1973) A quantitative study of genetic differences in the incidence of oestrus, ovulation and plasma luteinizing hormone concentrations in the sheep. Journal of Endocrinology 58,305-317.

Land, R.B.;Russell, W.S.;Donald, H.P. (1974) The litter size and fertility of Finnish Landrace and Tasmanian Merino sheep and their reciprocal crosses. Animal Production 18,265-271.

Land, R.B.;Thimonier, J.;Pelletier, J. (1970) Possibilite d'induction d'une decharge de LH par une injection d'oestrogen chez l'agneau femelle en fonction d'age. Comptes Rendus de l'Academie des Sciences, Paris D.371,1549-1551.

Newton, J.E.;Betts, J.E.;Large, R.V. (1970) Increasing litter size in three breeds of sheep by superovulation. Journal of Agricultural Science, Cambridge 75,335-360.

Maijala, K. (1966) Causes of variation in litter size of Finn-sheep ewes. 9th International Congress on Animal Production, Edinburgh. Abstr.29.

Packham, A.;Trifitt, L.K. (1966) Association of ovulation rate and twinning in Merino ewes. Australian Journal of Agricultural Research 17,515-520.

Purser, A.F. (1972) Variation in date of first oestrus among Welsh Mountain ewes. Proceedings, British Society of Animal Production. Abstr.133.

Thimonier, J.;Mauleon, P. (1969) Variations saisonnieres du compartement d'oestrus et des activites ovarienne et hypyphysaire chez les ovines. Annales de Biologie animale, Biochemie, Biophysique 9,233-250.

Trouson, A.O.;Moore, N.W. (1972) Ovulation rate and survival of fertilised ova in Merino ewes selected for and against multiple births. Australian Journal of Agricultural Research 23,851-858.

Turner, H.N. (1969) Genetic improvement of reproductive rate in sheep. Animal Breeding Abstracts 37,545-563.

Wheeler, A.G.;Land, R.B. (1977) Oestrus and ovarian activity of Finnish Landrace, Tasmanian Merino and Scottish Blackface ewes. Animal Production 24,363-376.

FACTORS AFFECTING PUBERTY AND MATING BEHAVIOUR IN SHEEP

J.L. LEES

Department of Agriculture, University College of Wales, Aberystwyth

FEMALE LAMBS

The advantages of early attainment of puberty

Early breeding of female lambs is an important and relatively simple method of intensifying sheep production in the self-replacing flock. It will:

(i) increase annual flock output;

(ii) halve the period of non-productivity with its attendant overhead costs;

(iii) facilitate selection programmes by reducing the generation interval and supplying earlier information in the case of progeny tested rams (Donald 1962, Donald et al 1968, Gjedrem 1969, Joakimsen 1969).

Early attainment of puberty may also increase total lifetime productivity.

(i) It allows an increase in the number of lamb crops (Bowstead 1930, Briggs 1936, Longriff 1961).

(ii) Female lambs reaching puberty during their first year have a higher potential prolificacy, even if they are not bred, than those failing to do so (Wiggins 1955, Hulet et al 1969).

(iii) Among lambs attaining puberty in their first year those that do so earliest, and thereby exhibit a greater number of heat periods in their first breeding season, are likely to be more prolific in later life than those slower to attain puberty and showing fewer heats (Ch'ang 1967, 1968). Subsequent prolificacy may not be directly dependent on early sexual maturity for both may be affected by the prepuberal nutritional or growth pattern.

(iv) Female lambs which give birth in their first year of life acquire their mothering experience earlier and usually with the advantage of having to care for only single lambs. Thus they become better mothers sooner and before they have the responsibility of larger litters. Good mothering at parturition reduces neo-natal mortality and shepherding requirements (Eltan 1974, Lees & Eltan 1978).

Determination of the attainment of puberty

Most observers determine puberty in female lambs in terms of age at first behavioural oestrus (Joubert 1963, Dyrmundsson 1973a). In few cases the onset of puberty has been estimated from a study of the formation of ovarian follicles and the development of the reproductive tract (Voss 1950, Allen & Lamming 1961, Downing & Lees 1974). Ovulation may be determined in the living animai by laparotomy (Hulet & Foote 1968, Southam et al 1971). Monitoring of blood hormone levels may also be carried out (Downing et al 1974).

Factors affecting the attainment of puberty

Genetic effects Breed and strain differences in the age and body weight at first oestrus are well established (Roux 1936, Voss 1950, Hafez 1952, 1953, Tierney 1969, Foote et al 1970, Southam et al 1971, Laster et al 1972, Dickerson & Laster 1975). There may also be considerable variability among individuals of the same breed (Hafez 1952, 1953, Ch'ang & Raeside 1957, Ch'ang 1968, Dyrmundsson 1972, 1973a).

Cross-bred female lambs tend to have a better reproductive performance than pure-breds (Fox et al 1964, Lang & Hight 1967). Sefidbakht (1969) and Riera et al (1970a and b) reported that a higher proportin of cross-bred than pure-bred female lambs attained puberty in their first year when subjected to various hormone treatments. Hulet et al (1969) showed that range female lambs which attained puberty in their first winter had lower inbreeding coefficients than those not exhibiting oestrus. Thus heterosis may contribute to earlier sexual development.

Few attempts however appear to have been made on any flock basis to select for the early attainment of puberty, although Southam et al (1971) concluded that this was a practical proposition. Indeed breeding from female lambs is not widely practised and the great majority of sheep breeders throughout the world breed ewes for the first time at the yearling stage. Dyrmundsson (1973a) has fully reviewed reports of early breeding practices but even this information does not give a complete indication of breed potential.

In Britain comparatively little information is available on the potential for early breeding from female lambs. Work at Cambridge (Hammond Jr 1944, Hafez 1952) included mostly Suffolk but also Romney Marsh and Scottish Blackface lambs, both pure and cross-bred, but some of the results were based on small numbers. British breeds certainly differ in their degree of sexual precocity but female lambs of the Clun Forest breed will mate fairly readily in their first autumn and in certain favourable localities advantage is taken of this fact. (Williams 1954).

With April-born lambs in the University College of Wales Clun Forest flock at Aberystwyth, 98% of the females, studied over several seasons, reached puberty by the end of the year; 80% did so by November 30th and 95% by December 15th (Dyrmundsson 1972, Downing & Lees 1977). These figure do not imply that every female lamb mated will satisfactorily conceive and carry a lamb to full term but they do mean that female Clun Forest lambs can make a significant contribution to flock productivity within the time limits of the normal lambing season acceptable in Britain.

Over a 14 year period in the Cockle Park flock of the University of Newcastle, 89% of nearly 2000 female Clun Forest lambs achieved puberty by mid-December (Bichard et al 1973).

There are little similar data concerning the early breeding potential of other breeds in Britain. If there were, the situation might be surprising. For example female lambs of the Welsh Mountain breed, if managed as lowland sheep, may exhibit oestrus in their first autumn to an extent almost comparable to that of the Clun Forest breed (Lucas 1973). According to an MLC breed survey, about 15% of Welsh Mountain female lambs are exposed to the ram usually when they are being away wintered.

Whatever the nature and extent of genetic effects on the attainment of puberty in female sheep, environmental factors tend very largely to obscure them (Cupps et al 1969).

<u>Age and body weight at puberty</u> Female lambs which exhibit oestrus tend to be heavier than those not doing so in the first year (Spencer et al 1942, Williams 1954, Ch'ang & Raeside 1957, Bowman 1966, Lang & Hight 1967, McGuirk et al 1968, Winn & Cumberland 1974, Keene 1976). This situation for the UCW flock is shown in Appendix Figure 1 (from Dyrmundsson & Lees 1972a).

Thus faster growth during rearing will normally favour an earlier onset of oestrus, i.e. at a lower age and heavier body weight; lambs growing at slower rates, and this usually includes twin lambs, will normally exhibit oestrus later in the season at a higher mean age but lower mean body weight (Hafez 1958, Mounib et al 1956, Southam et al 1970, Dyrmundsson & Lees 1972a and e, Downing & Lees 1977).

<u>Nutrition</u> The literature shows that age and weight at puberty are strongly influenced by level of nutrition, which therefore also affects sexual development in female lambs (Casida 1959, Red 1960, Allen & Lamming 1961, Tassell 1967, Lamming 1969, Moustgaard 1969, Allden 1970). Southam et al (1971) found that the incidence of puberty could be changed from 12% to 88% simply by providing additional feed following early weaning.

While it is possible in fully grown adult sheep to measure the direct effects of nutrient variation on the reproductive processes, in sexually immature females it is very difficult to discriminate between the effects of varying the plane of nutrition on growth and on sexual maturation independent of growth.

Recent experiments in the University College of Wales Clun Forest flock at Aberystwyth include a study of varying planes of nutrition to the point of sexual maturity in early born (February) and late born (April) female lambs. Some results are given in Tables 1 and 2 (Downing & Lees 1977). A close relationship between growth rate and the proportion of lambs reaching puberty was found. At higher growth rates, plane of nutrition made little difference but at slower growth rates differences in nutrition were more important, particularly for lambs born later in the year. In 1973 all planes of nutrition were higher than in 1974 because of the contrasting local patterns of grass growth in these two years.

Both early and late lambs in 1973 and early lambs in 1974 may have satisfied sufficient thresholds so that only non-nutritional factors remained decisive but, with the 1974 April born lambs, a higher early plane of rearing increased the proportion exhibiting oestrus and a higher later plane of rearing decreased it. Thus a HL pattern of rearing gave the highest proportion and the LH pattern the lowest proportion exhibiting oestrus in the first autumn.

It is difficult to explain this contrasting effect except in terms of an 'initial' metabolic signal yet to be defined which triggers sexual maturation; for seasonal breeders such as the female sheep this would need to operate within the scope permitted by seasonal influences. A recent study (Wilen & Naftolin 1977), based on directly measured body composition within a homologous group of puberal rats, showed that, although the average body weight or the proportion of total body fat or water could change, the percentage of total protein stayed the same no matter what the age at first oestrus. Since lean body mass or total body protein is the greatest contributor to the basal metabolic rate, the attainment of similar body protein proportions in both early and late maturing females at puberty may indicate the existence of such a metabolic signal. If precise proportions of body components are important, this points to a weakness in using total body weight as a parameter of development in this type of study and may explain why, in Tables 1 and 2, among those lambs which reached puberty, level of nutrition itself, though it resulted in large differences in total body weight at puberty, had little effect on the mean age at which first oestrus was observed.

Lamming (1963) illustrated the need however to discriminate between age at puberty, i.e. first behavioural oestrus, and age at first conception. In the work of Downing & Lees (1977), lambs on a low plane of nutrition during the breeding season were older at first conception, even though they tended to be younger at puberty, than lambs on a high plane at this time. Overall a high proportion of mated female lambs conceived. Conception rate was affected by nutrition but this effect was related to body weight, more of the heavier lambs conceiving and lambing at full time than lighter ones. Higher planes of nutrition at both the earlier and the later stages of rearing to puberty increased the conception rates but the effects of high earlier rearing were more consistent.

Failure to conceive, particularly in lambs younger and lighter at first oestrus, may also be related to endocrine imbalance and/or ovulation failure. Nutrition was shown to affect the progesterone status of female lambs so that those on a low plane of nutrition during the breeding season tended to have unstable blood levels. Some lambs also failed to ovulate at first oestrus. Thus overt oestrus does not necessarily imply ability to conceive and breed successfully.

Earlier evidence suggested that nutritional flushing prior to mating does not have a clear effect upon female lambs (Williams 1954, Allen & Lamming 1961), but Downing & Lees (1977) showed a clear flushing effect among better reared lambs. It appeared that the effects of nutrition on ovulation rate interacted with those of body weight, heavier lambs being more responsive. Keane (1974, 1975) reported a similar response.

Table 1 The effects of different planes of rearing on the attainment of puberty in Clun Forest female lambs

Group	No. in group	No. exhib-iting 1st oestrus	Mean growth rate (kg/wk)	Mean date at 1st oestrus	Mean age at 1st oestrus (days)	Mean live-weight at 1st oestrus (kg)
1973						
Feb. born						
HH	18	18	1.18	11 Oct.	229	43.9
HL	18	18	1.05	3 Oct.	222	41.6
LH	17	17	0.96	13 Oct.	232	35.4
LH	17	16	0.82	7 Oct.	225	32.0
April born						
HH	13	13	1.25	24 Oct.	193	39.3
HL	13	13	1.12	28 Oct.	195	36.1
LH	12	12	1.08	21 Oct.	190	32.9
LL	12	11	0.98	26 Oct.	197	30.5
1974						
Feb. born						
HH	33	33	0.99	7 Nov.	251	38.8
HL	33	33	0.89	30 Oct.	243	35.8
LH	31	26	0.81	11 Nov.	257	31.7
LL	31	24	0.69	19 Nov.	264	28.1
April born						
HH	27	14	1.01	29 Nov.	226	33.6
HL	27	23	0.88	14 Nov.	214	30.2
LH	25	7	0.82	4 Dec.	235	29.6
LL	25	16	0.74	24 Nov.	223	25.9

Table 2 The effects of different planes of rearing in conception rates in Clun Forest female lambs

Group	No. in group	No. exhibiting oestrus	No. known to conceive	Lambing % of those lambing
1973				
Feb. born				
HH	18	18	18	139
HL	18	18	18	100
LH	17	17	16	119
LL	17	16	16	113
April born				
HH	13	13	13	115
HL	13	13	9	100
LH	12	12	10	118
LL	12	11	7	100
1974				
Feb. born				
HH	33	33	24	109
HL	33	33	28	107
LH	31	26	22	105
LL	31	24	12	100
April born				
HH	27	14	10	100
HL	27	23	9	100
LH	25	7	3	100
LL	25	16	5	100

More work needs to be done if nutritional effects at different stages of rearing are to be fully determined. It may be however that total growth over the whole period is more important in ensuring high conception and lambing rates from female lambs. Moreover, Gunn (1977) has shown that high plane rearing to 12 months of age results in greater subsequent ovulation rates and that the advantage over low plane rearing to this stage shows up particularly where adult nutrition is low.

Daylight and date of birth Much evidence indicates that seasonality is an important factor in the attainment of puberty in female lambs. A relationship between date of birth and date of first oestrus has been demonstrated by Hammond Jr (1944), Hafez (1952) and Dyrmundsson & Lees (1972a and e).

Although lambs born early tend to attain puberty earlier in the season they normally do so at a greater age and heavier bodyweight; similarly later born lambs attain puberty at a younger age and lighter body weight. Weight effects are frequently confounded with differences in date of birth but Dyrmundsson & Lees (1972e) concluded that, where growth rates of early and late born lambs were similar, the daylight environment experienced by the lamb could be a critical factor in regulating the onset of breeding activity.

The more recent work of Downing & Lees (1977) has shown that in practice date of birth and growth rate are likely to interact. Although early born lambs tended to exhibit first oestrus earlier in the season, for lambs with faster growth rates date of birth had little effect on when the first oestrus was exhibited. Where lambs had slower growth rates, time of birth exerted a greater influence, with a greater number of late born lambs exhibiting first oestrus later in the breeding season or failing to show oestrus at all. Keane (1975) reported similar results.

Thus there is neither a fixed age, body weight nor time of year at which female lambs exhibit their first oestrus, owing to the complex interaction between these factors and time of birth. This situation for Clun Forest lambs at Aberystwyth is shown in Table 3 and Figure 2. Whether it operates through variations in accumulated light pattern, seasonal nutritional differences, body weight effects or any combination of these factors, date of birth is of fundamental importance to the onset of breeding activity in female lambs. The highly significant relationship between day of birth and day of first oestrus is illustrated in Figure 3.

Table 3 The effect of month of birth on the attainment of puberty (mean date for each month)

Birth		Attainment of puberty			
Date	Day No.	Date	Day No.	Age (days)	Wt. (kg)
Jan. 18	18	Oct. 10	283	265	41.4
Feb. 16	47	Oct. 24	297	250	36.1
Mar. 16	75	Nov. 2	306	231	34.3
Apr. 11	101	Nov. 15	319	218	33.8
May 10	130	Dec. 4	338	208	32.2

Figure 1

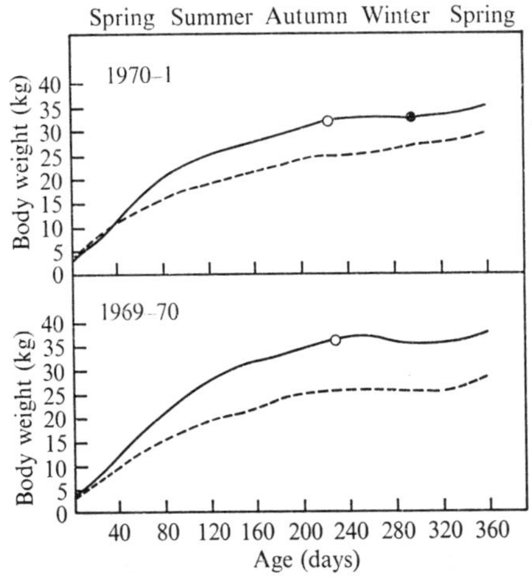

Fig. 1. Mean growth rates of ewe lambs from birth to 12 months of age. ————, Ewe lambs which attained puberty; – – – – –, ewe lambs which failed to attain puberty; ○, mean onset of breeding season; ●, mean cessation of breeding season.

Fig. 1

Figure 2

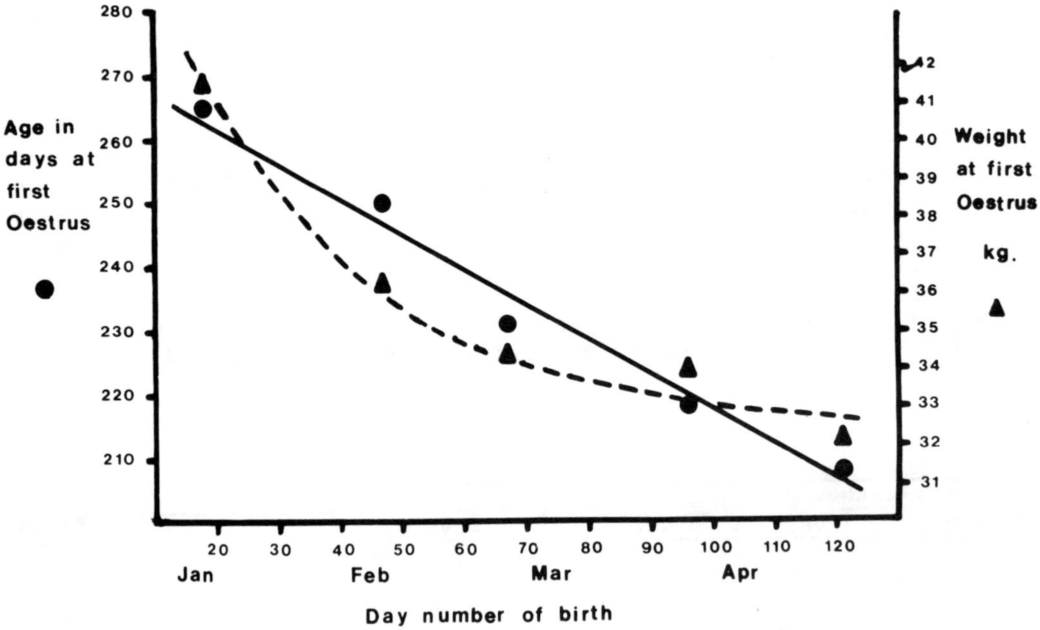

Age and body weight at first oestrus in relation to day of birth in Clun Forest female lambs

Little attention has been given to the photoperiodic regulation of breeding activity in female lambs under controlled conditions. Neither Radford (1961) nor Smith (1967) could demonstrate any convincing effect of extended light patterns. Ducker (1971) found that various artificial light treatments modified the occurrence of oestrus but that the breeding pattern was basically seasonal in nature irrespective of light treatment.

Use of hormones There is the possibility of inducing sexual behaviour by hormonal treatments but the results of Southam et al (1971) show only small advantages over control lambs receiving adequate feed. It is likely that when compared with older ewes or even with those lambs which breed naturally, a larger proportion will need assistance at lambing and that the lambs will be smaller at birth.

Cyclic activity

First oestrus is experienced later and the breeding season is shorter in female lambs than in yearling or adult ewes (McKenzie & Terrill 1937, Hammond Jr 1944, Hafez 1952, Mounib et al 1956, Ch'ang & Raeside 1957, Watson & Camble 1961, Wiggins et al 1970, Dyrmundsson & Lees 1972a).

Female lambs attaining puberty early usually have a more extended breeding season than those attaining it later (Hafez 1952, Ch'ang & Raeside 1957, Dyrmundsson & Lees 1972a). Thus the number of oestrus cycles may vary from 1 to 10 depending on breed, environment and sexual development of the lambs. This situation is illustrated in Figure 4 from Dyrmundsson & Lees (1972a).

Female lambs exhibit less regular cycles and tend to have slightly shorter ones than adult ewes (Hafez 1952, Mounib et al 1956), though Dyrmundsson (1972) observed that in Clun Forest lambs mean cycle lengths were fairly consistent throughout the breeding season and did not appear to be influenced by seasonal factors. Among individual lambs, those exhibiting few cycles during the season show greater variation in cycle length and in the incidence of oestrus than those with longer breeding seasons (Sefidbakht et al 1966, Hulet et al 1969, Foote et al 1970, Dyrmundsson & Lees 1972a).

The duration of oestrus is usually shorter in lambs than in yearlings and adult ewes (McKenzie & Phillips 1930, Hafez 1952, Mounib et al 1956, Wiggins et al 1970) but there may be breed effects; Land (1970) and Osborne (1970) reported that oestrus was of similar duration in Finnish Landrace ewes and lambs.

Mating behaviour

Sexual behaviour is less pronounced in female than in male lambs during the pre-puberal period (Scott 1945, Fraser 1968).

At puberty the behavioural signs especially of the first oestrus are usually weak and the intensity of oestrus is less marked than in older sheep (Hafez 1952, Hafez & Scott 1962, Anderson 1969). Thus female lambs in oestrus appear to make little or no effort to seek out the ram and

Figure 3

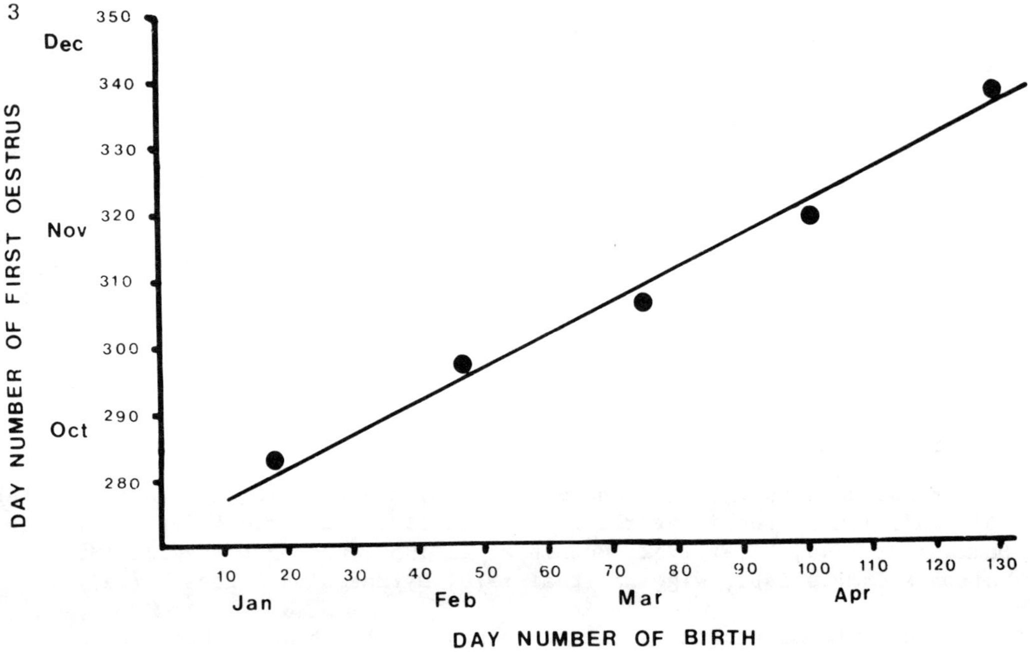

Relationship between day number of birth and day number of first oestrus in Clun Forest female lambs

Fig. 3

Figure 4

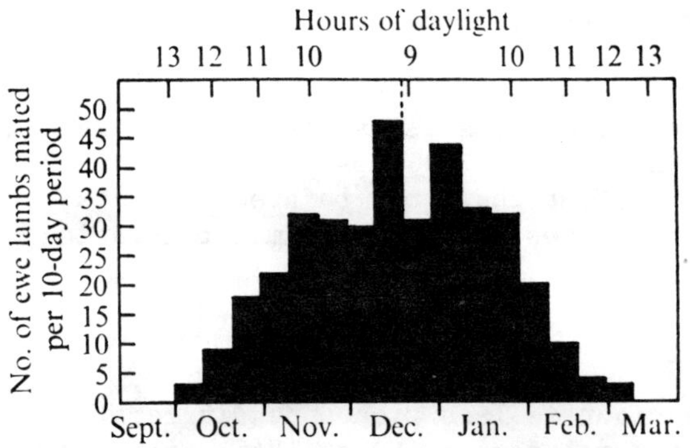

Frequency distribution of matings in ewe lambs during the breeding season 1970–1. The broken vertical line shows the mid-breeding season.

may be reluctant to accept service when the latter makes his sexual advances. However in experiments with tethered rams this reluctance was largely overcome perhaps because mating could then coincide with the brief period of maximum receptivity by the lamb and unwelcome premature attention by the ram was eliminated (Lees 1971). At the same time, with a choice of rams available, Clun and Kerry lambs showed a preference for mating with sires of their own breed, but under ordinary flock conditions such female mating preferences would normally be masked.

Practical considerations

The less intense and shorter oestrus of the lamb may lower mating efficiency, particularly if rams are used which are themselves less efficient at detecting a weak oestrus (Lees 1971, Dyrmundsson & Lees 1972a, Downing & Lees 1977). Lambs are likely to receive only scant attention from rams if adult ewes exhibiting stronger heats are present. At the same time adult oestral ewes may physically compete for the attention of a ram to the considerable exclusion of the immature lambs. Female lambs are thus better separated from ewes and preferably joined with mature and experienced rams of proven efficiency at detecting oestrus. This choice of rams is particularly important in experimental work requiring the testing of female lambs for first behavioural oestrus. Mating will also be more effective if it takes place in small enclosures rather than in very large ones or in open or broken country.

The spread of the onset of puberty may lead to a prolonged mating season and hence to a lengthy and, at the end, very late lambing period. The introduction of the rams could be delayed to compress lambing but this might still tend to give a late lambing. Hormonal synchronisation would similarly lead to delay in introducing the rams as well as the likelihood of some difficulty in the insertion of progestagen pessaries. A compromise might be to turn the rams to the lambs towards the end of the adult tupping period and to withdraw them about 5 weeks later (after 2 cycles), regardless of the numbers covered. In the case of Clun Forest sheep some 80-90% of female lambs would have mated, and failure to mate could be used as one of the criteria for culling surplus female lambs. A further possibility would be to use the effect of the unaccustomed presence of the ram. This would require some prior knowledge of the expected time of onset of cyclic activity but it has been shown by Dyrmundsson & Lees (1972c) to be effective (Figure 5).

This paper is primarily concerned with puberty and mating but it is not possible readily to dissociate these phenomena from subsequent reproductive performance, and the consequences of early breeding must be one of the factors that affect the exploitation or otherwise of the early attainment of puberty. The occurrence of early behavioural oestrus and mating in the female lamb is no guarantee that it will carry a foetus to full term. In the UCW Clun Forest flock the percentage of mated female lambs which eventually experience parturition is about 85%. Bichard et al (1973) reported corresponding percentages which, over 14 seasons, varied from 36% to 77%. These workers considered that some of this large variability may have been due to differences from year to year in the plane of nutrition. Downing & Lees (1977) have shown a large variability from

Figure 5

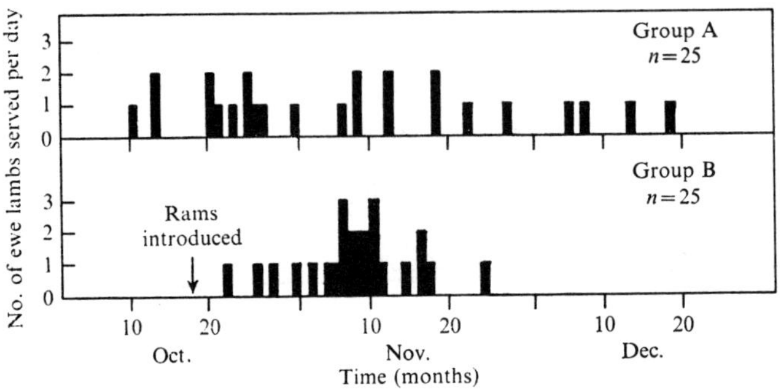

Frequency distribution of first matings in ewe lambs in relation to the introduction of rams.

Fig. 5

Figure 6

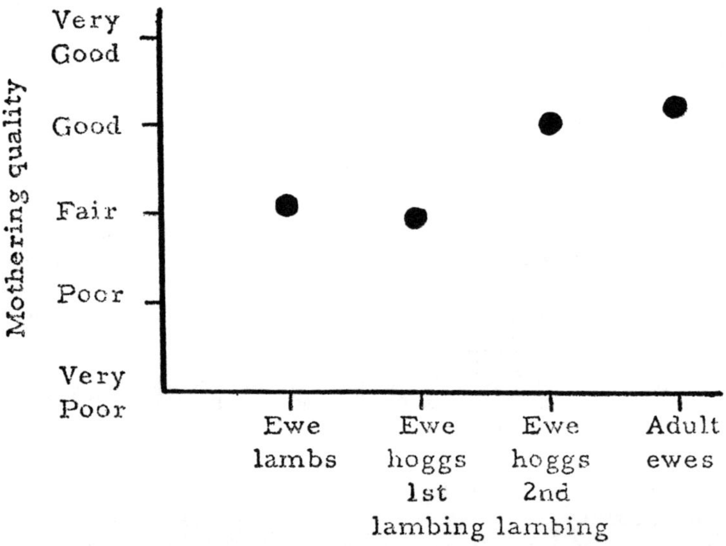

12% to 100% in the percentage of mated lambs actually experiencing parturition, due to differences in experimental planes of nutrition. MLC data (Read 1974) indicate that, where early breeding takes place in lowland flocks, 66 rams are reared for every 100 female lambs mated; in the uplands the corresponding figure is 51. Information within these recorded flocks emphasises the importance of body weight in this situation.

Clearly it is necessary to feed pregnant lambs at a level that will allow sufficient growth to continue, though some measure of retardation may be both inescapable and acceptable, as well as to ensure the proper development of the foetus and udder tissue. It may also be important not to overfeed immature female lambs during the very latest stages of pregnancy since their single lambs may grow too large and cause lambing difficulties (Lees & Eltan 1978). Thus the proper level of late pregnancy nutrition may be even more critical in the lamb than in the adult ewe.

It is generally assumed that lamb mortality is higher in sheep lambing at 1 year of age. Bichard & Cooper (1966) gave mortality figures for the Cockle Park Clun flock as 23.7% at 1 year, 16.6% at 2 years and 12.9% for ewes over 2 years. In the UCW Clun flock the corresponding figures, taking 1972 and 1973 together, are 16.5%, 14.4% and 6.8% respectively. However Laster et al (1972) reported that lambing difficulty was not a large problem among the female lambs of the 19 breeds and cross-breeds that they studied, and neither did it appear to have any significant effect on the numbers of lambs weaned.

Poor mothering behaviour may be exhibited by female lambs following parturition but this is mainly because it is their first experience. The same difficulty is likely to arise at first lambing at whatever age it takes place (Eltan 1974). Figure 6 compares the mothering quality scores of different classes of dam. It shows that ewe hoggs at their first lambing are on average only fair mothers, as are female lambs. Ewe hoggs, which have already lambed as lambs, and adult ewes are usually good mothers. Table 4 gives as an example the average intervals from the commencement of labour to delivery and from delivery to first successful suckling for different classes of sheep. This illustrates the effect of the age of the mother. Following a more protracted parturition single lambs from lambs take on average twice as long as single lambs from adult ewes to achieve their first successful suckling. This results from both the reduced vigour of a smaller lamb and the inexperience of the mother. On the other hand Table 5 shows that the percentage of cross-mothering and mis-mothering in lambs may be much less than that which occurs among older sheep. On balance it seems that yearlings previously bred as lambs tend to be more reliable breeders, better mothers and to have fewer lambing troubles (Apps 1953, Boaz 1957, 1960, Lees & Eltan 1978).

Table 4 Average time intervals

	Commencement of labour to delivery (mins)	Delivery to first successful suckling (mins)
Adult ewe with single	50.6	33.6
Adult ewe with twins	58.1	45.1
Ewe hogg with single	59.2	45.4
Ewe hogg with twins	62.5	64.8
Ewe lamb with single	87.6	70.5

Table 5

Class of dam	% cross and mis-mothering	
	Indoor lambing (group pens)	Outdoor lambing (heavily stocked lambing paddock)
Adult ewes	23.9	20.3
Ewe hoggs	23.9	6.9
Ewe lambs	5.9	11.1

It is often suggested that lambing female lambs requires a disproportionate amount of shepherding. Detailed records kept at lambing time in the UCW flock (Downing & Lees 1977) show that more lambs needed assistance during and after parturition than did other sheep (see Table 6). Nevertheless, less time was required per assisted parturition than for older sheep and in terms of average time spent per animal lambing the amount of shepherding has been the same regardless of the age of the mother. Thus the involvement of female lambs at lambing does not necessarily mean more shepherding except in so far as there may be more parturitions to deal with.

Table 6

Class of dam	% assisted lambing	
All adult ewes	9.7	
Adult ewes with singles		16.4
Adult ewes with twins		6.3
All ewe hoggs	9.8	
Ewe hoggs with singles		7.4
Ewe hoggs with twins		21.4
All ewe lambs	19.2	

With Clun Forest sheep there would seem little justification for advancing the date of birth simply to increase female lamb age in the autumn. March/April lambing in Britain will already allow a high proportion of female lambs to breed in their first year and to advance their date of birth to any marked extent will not only increase flock feeding and management costs but is also likely to lower very substantially the prolificacy of the adult flock (Figure 7).

It is now generally accepted that, provided they are adequately managed and fed, the use of female lambs for breeding need not have any adverse effects on their subsequent growth and development. Palsson (1953) found that ewes bred as lambs and those bred as yearlings reached the same body weight at about 2 to 2½ years, and similar results have been reported by Spencer et al (1942), Williams (1954) and Boaz (1957). It seems equally well accepted that with adequate nourishment there need to be no detrimental effects on subsequent reproductive efficiency (Boaz 1960, Donald 1962, Yalcin & Bichard 1964, Cottier 1965, Holland & Ruttle 1966, Suiter & Croker 1970). On the contrary there is mounting evidence that early breeding may result in enhanced subsequent fertility and a worthwhile increase in the overall lifetime productivity of the ewe (Bowstead 1930, Griswold 1933, Briggs 1936, Hume 1939, Spencer et al 1942, Sharafeldin 1960, Longrigg 1961, Mair 1963, Cannon & Bath 1969).

MALE LAMBS

The advantages of early attainment of puberty

The advantages of early puberty in male lambs are relatively less important than in female lambs. Rams account for only about 3% of total flock numbers and the exploitation of early sexual maturity in male lambs cannot give the same economic advantages as breeding from female lambs. Nevertheless the satisfactory attainment of puberty in male lambs is important in husbandry practice for, in Britain at least, a very high proportion of male lambs kept for breeding begin work in their first autumn. Moreover where selected male lambs are used in breeding

Figure 7

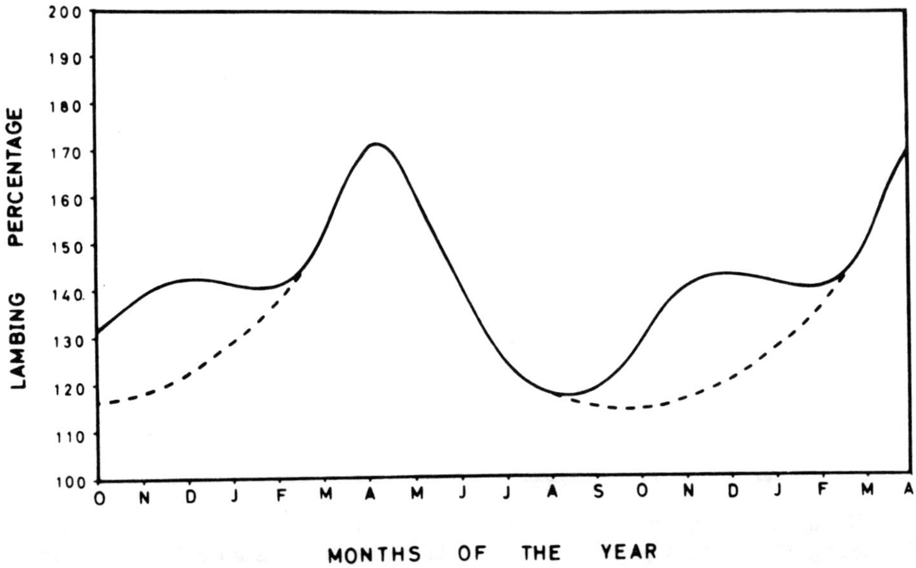

Seasonal variation in prolificacy in
Clun Forest sheep

Fig. 7

Figure 8

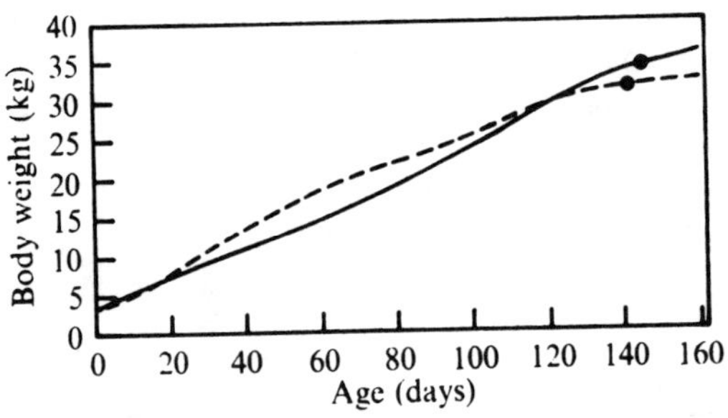

Mean growth rates of early and late-born ram lambs from birth to 160 days of age. ——, Early-born; ----, late-born; ●, mean age and body weight at puberty.

programmes on performance and progeny testing, the generation interval is reduced and information obtained on the ram at the earliest possible age, especially if his daughters are in turn bred as lambs. Such a practice may facilitate the earlier selection and use of outstanding sires and consequently the rate of genetic improvement is likely to be accelerated (Terrill 1938, Sjodin 1966, Gjedrem 1969, Joakimsen 1969, Naerland 1970).

Determination of the attainment of puberty

In the male lamb puberty is associated wth a marked increase in endocrine function, the onset of spermatogenesis and subsequently with the manifestation of full sexual behaviour. These developments are accompanied by certain anatomical changes in the reproductive organs.

There is little information concerning the attainment of puberty in male lambs in Britain. Skinner et al (1968) studied post-natal development of the reproductive tract of the Suffolk ram and changes in the gonadotrophin content of the pituitary, while Skinner & Rowson (1968) reported on puberty in pure and cross-bred Suffolk rams.

In studies on male lambs in the UCW Clun Forest flock (Dyrmundsson 1972, Dyrmundsson & Lees 1972b and d) the attainment of puberty was defined in several ways but firstly in terms of the stage of development when mature and viable spermatozoa were first produced and released in the gonads. This was done both by serialised castration and by semen collection by electro-ejaculation. Secondly, observations were made on the development of the male reproductive organs. Certain anatomical changes must take place before spermatozoa are released; the testes first descend into the scrotum, and the glans penis and the processus urethrae, which in the immature lamb are completely adherent to the prepuce, gradually become detached until at puberty the penis is freely movable in the sheath. Thirdly, sexual maturity was defined in terms of the lambs' willingness to mate with adult ewes known to be strongly oestrus.

These phenomena are not contemporaneous and physiological puberty preceded the first copulation often by several weeks. Thus the attainment of puberty in the male cannot be measured as simply and as accurately as in the female lamb. Dyrmundsson (1973b) has emphasised the importance, when comparing reports, of considering the criteria on which the definition of puberty may have been based.

Factors affecting the attainment of puberty

Genetic effects The literature clearly shows that there are marked differences in puberal age and body weight between male lambs of various breeds of sheep (Dyrmundsson 1973b) and genetic factors may account for some of this variation. Wiggins & Terrill (1953) showed that penis development tended to be retarded by inbreeding.

Genetic factors may also account for some of the variation within breeds. Dyrmundsson & Lees (1972b) demonstrated large differences among individual Clun Forest male lambs in age and in both testicular and epididymal weights at puberty, even where birth dates and growth rates were carefully standardised.

No direct attempts appear to have been made to select for earliness of breeding in male lambs.

Age and body weight at puberty Growth rate and body weight are a better guide than chronological age to physiological puberty in male lambs (Dun 1955, Watson et al 1956, Symington 1961, Couret 1962, Pretorius & Marincowitz 1968, Ortavant et al 1969, Dyrmundsson & Lees 1972b).

Testicular weight and volume as well as the diameter and volume of the seminiferous tubules also show a closer relationship with body weight than with age (Carmon & Green 1952, Watson et al 1956, Courot 1962, Skinner et al 1968, Courot et al 1970, Colyer 1971).

Similarly the freeing of the penis is more closely associated with the growth rate than with the chronological age of the lamb (Johnstone 1948, Wiggins & Terrill 1953, Dun 1955, Watson et al 1956, Belonje 1965, Skinner & Rowson 1968, Pretorius & Marincowitz 1968, Dyrmundsson 1972).

Thus singles are likely to attain puberty earlier than twins in all the above respects.

Data for early and late-born male lambs in the UCW Clun Forest flock are given in Table 7 and Figure 8 (from Dyrmundsson & Lees 1972b). Marked variation is shown in the age and body weight at which spermatozoa were found in the epididymis at castration. While the majority of lambs reached this stage of development at 4½ - 5 months the age range for 52 lambs was 99 - 176 days. The mean body weight at puberty was 32.5 kg with a range of 24.5 to 38.0 kg or 35 - 45% of adult body weight.

Table 7 A comparison between 28 early and 26 late-born male lambs (mean date)

Parameter	Early born	Late born
Birth date	Jan. 14	April 3
Date of puberty	June 9	Aug. 23
Age	145 days	142 days
Body weight	33.7 kg	31.1 kg
Testes weight	149.1 g	175.9 g
Epididymis weight	20.6 g	20.9 g

Correlation coefficients between various parameters of anatomical sexual development and both age and body weight similarly showed such development to be more closely related to body growth than to chronological age and this was emphasised by multiple regression analyses.

Though chronological age seems less important than growth rate it cannot be ignored. In some cases male lambs, which had grown at a fast rate and which had well developed testes, had no spermatozoa present in the epididymis when castration took place at heavy body weights but early ages. This suggests a certain limit of chronological age below which puberty is not attained, irrespective of body and testicular weights etc.

Nutrition An adequate plane of nutrition is clearly important for the early sexual development of the male lamb. Nutritive deficiency, especially low energy intake, will retard sexual development and delay the attainment of puberty (Marshall & Hammond 1952, Mann 1964, Donovan & Ten Bosch 1965, Foote 1969). There is a lack of gonadotrophin from the hypophysis though the testes will usually continue to produce testosterone (Mann 1964); thus the androgenic function of the testes is retarded more markedly than is spermatogenesis (Casida 1959, Mann et al 1967, Skinner & Rowson 1968). Conversely in very well fed and rapidly growing lambs it would appear that androgenic function is preferentially advanced (Dyrmundsson & Lees 1972).

Studies under controlled feeding conditions by Ragab et al (1966) and Pretorius & Marincowitz (1968) demonstrated that rams reared on higher planes of nutrition normally attain puberty at lower ages and heavier body weights than those on lower feeding levels.

Daylight and date of birth There appears to be no direct evidence of a relationship between photoperiodism and puberal development in male lambs. Skinner & Rowson (1968) suggested that such a relationship might exist and the results obtained by Dyrmundsson (1972) and Dyrmundsson & Lees (1972b and d) do not preclude the possibility that, apart from seasonal fluctuations in nutritional conditions during rearing, other seasonal factors such as the natural light environment might have influenced the sexual development of their male lambs. Indeed as can be seen in Table 7, late born lambs had a much heavier mean testes weight at puberty than early-born lambs despite their lower mean age and body weight at this stage of development. With the data corrected for age and body weight at puberty so that the only known difference was month of birth, a highly significant difference was confirmed.

Semen production at puberty Dyrmundsson & Lees (1972b) have shown that, as would be expected, the concentration of sperm in smears taken at puberty from epididymides after castration is very low, and even lower in the ejaculates collected by electro-ejaculation, which had a clear or slightly cloudy appearance.

Symington (1961), Louw & Joubert (1964) and Skinner & Rowson (1968) have demonstrated marked increases in both the quantity and quality of ram semen after puberty is attained. Thus the fertilizing potential of the semen is likely to rise at a steady rate with advancing age and more mature body development (Terrill 1938, Dun 1955, Watson et al 1956, Symington 1961, Louw & Joubert 1964, Skinner & Rowson 1968, Skinner 1971).

Mating behaviour

Pre-puberal mounting of both a hetero- and a homosexual nature occurs commonly among male lambs (Symington 1961, Banks 1964, Louw & Joubert 1964, Ragab et al 1966, Dyrmundsson 1972). Very little information is available on the true mating ability of puberal male lambs of British breeds.

Dyrmundsson & Lees (1972d) studied the mating behaviour of 37 Clun Forest male lambs from 140 days of age introduced individually at 14-day intervals to adult Clun ewes exhibiting high quality oestrus. After successful matings had been achieved, the penis and scrotum were examined and their development assessed. The testes were removed by surgical castration on the day following the successful mating.

At each introduction to ewes in oestrus the male lambs exhibited varying degrees of libido, ranging from a complete lack of sexual interest to intense mounting activity which eventually resulted in successful matings taking place. A behaviour pattern of frequent occurrence was the nosing of the udder region and sometimes the nuzzling of the teats of the ewe. According to Banks (1964) such behaviour is not seen in adult rams and it has not been observed among adult rams in the UCW flock.

Only a few of the rams were able to copulate successfully at their first introduction to ewe in oestrus, but generally mating dexterity increased gradually as the number of encounters increased. This is shown in Table 8.

The mean values and ranges of date, age and body weight at first complete mating are summarised in Table 9. These data show great variation in the age and body weight at which copulation first took place but the majority of lambs copulated when they had reached approximately 40-50% of adult body weight.

Table 8 Copulation success in relation to the number of encounters of puberal male lambs with adult ewes in oestrus

No. of encounters	% of male lambs successful for the first time
1	16.2
2	29.7
3	29.7
4	13.5
5	8.2
6	2.7
	100.0

Table 9 Date, age and body weight at first mating of 37 Clun Forest male lambs

Parameter at first mating	Mean	Range
Date	Sept. 16	July 29 - Nov. 10
Age	174 days	142 - 223 days
Body weight	34.8 kg	24.0 - 49.0 kg

The male lambs were always kept as a group so that homosexual tendencies could be observed. Such activity seems common among male lambs and there is some evidence to suggest that it may interfere with the establishment of normal heterosexual mating behaviour, especially when the male lambs are reared in large groups (Hulet et al 1964, Banks 1964, Hulet 1966, Marincowitz et al 1966, Pretorius 1967, Mattner et al 1971). Dyrmundsson & Lees (1972d) observed that homosexual behaviour was more common among the very well grown and early-born lambs but it did not interfere with the establishment of normal heterosexual mating.

Observations on anatomical development indicated that physiological puberty, marked by the detachment of the penial adhesions and the release of viable spermatozoa, preceded complete physical mating ability to a widely varying extent but on average by 4-5 weeks.

Approximately 16% of the lambs exhibited abnormal testicular development at first mating. In two thirds of these only one of the testes was functional while in the other third testicular weights were exceptionally low, the testes were poorly descended into the scrotum and elongated in shape, and spermatozoa were completely absent in the epididymides. Such individuals, although partly or completely sterile, had normal body growth and fully developed penes and mated apparently successfully with ewes in oestrus.

Male lambs, reaching behavioural maturity earlier in the year in late summer, seemed to have a somewhat weaker sex drive at first mating than those reaching this stage in autumn. However definite conclusions cannot be drawn from this limited evidence of a possible seasonal influence on libido in the male lamb, although a rising trend in the mating capacity of adult rams during the period May to September in the same flock has been reported by Lees (1965, 1978). Skinner & Rowson (1968) noted a greater accessory gland development at autopsy in late-born (summer) than in early born (spring) male lambs at Cambridge which was suggestive of a seasonal influence on sexual development, but the number of lambs studied was very small.

Practical considerations

In the very rapidly growing male lamb, complete anatomical development does not always ensure that mature sperm cells are being released. Occasional shortcomings in testicular development at puberty emphasise the need to pay proper attention to this point when selecting male lambs for breeding. Routine scrotal palpation and careful examination of the penis may reveal some of these abnormalities.

Young rams have a lower breeding capacity than mature rams (Lightfoot 1968, Lightfoot & Smith 1968, Croker & Lindsay 1972), and the common practice of breeding a smaller number of ewes to male lambs than to older rams appears to be highly desirable. There appears to be no information however on differences in breeding capacity between individual male lambs. Marked differences in breeding capacity have been shown among adult Clun rams, given a series of controlled mating loads, and in certain cases sub-normal breeding capacity has almost certainly been inherent in the animal since puberty (Lees 1978). Frequently such differences cannot be detected by physical examination, including scrotal palpation, or by semen evaluation; in multi-sire pens they may go unnoticed.

Should puberal male lambs be sexually inhibited this can usually be overcome by confining such a lamb in a small pen with a mature ewe very strongly in oestrus.

In the work of Dyrmundsson & Lees (1972d) no opportunity was given to puberal male lambs to select among a group of ewes in oestrus. However Rouger (1969) noted that male lambs introduced to ewes in oestrus showed preferences for individual ewes and it appeared that the more sexually active male lambs were also the more selective. Especially in single-sire pens such mating preferences might lower mating efficiency.

Male lambs, being relatively smaller than mature ewes of their own breed, may sometimes experience difficulty in achieving successful copulation with them (Dyrmundsson 1972). At the same time they are better mated with adult ewes because of the more intense heat which the latter are likely to exhibit.

Since well-grown early-born lambs tend to reach puberty earlier in the autumn, selection of these for breeding will mean that less attention can be paid to the prolificacy of the dam which, in the early part of the year, is less able to reveal her reproductive potential (see Appendix Figure 7). Single male lambs usually reach puberty earlier than twin males. Thus selection for the best grown males may indirectly amount to selection for lower prolificacy.

The use of male lambs for breeding does not appear to have any adverse effects on their subsequent growth and development (Gaal 1964, Dukin 1966). However the extent to which early breeding may affect their reproductive performance throughout adult life does not seem to have been examined.

REFERENCES

Allden, W.G. (1970) Nutrition Abstracts Review 40,1167-1184.

Allen, D.M.;Lamming, G.E. (1961) Journal of Agricultural Science, Cambridge 57,87-95.

Anderson, L.L. (1969) In 'Reproduction in domestic animals'. 2nd edn. (Eds) H.H. Cole and P.T. Cupps. UK: Academic Press. pp541-568.

Apps, A.W.J. (1953) New Zealand Journal of Agriculture 87,112.

Banks, E.M. (1964) Behaviour 23,249-279.

Belonje, P.C. (1965) Journal of the South African Veterinary Medical Association 36,381-383.

Bichard, M.;Cooper, M.M. (1966) Animal Production 8,401-410.

Bichard, M.;Younis, A.A.;Forrest, P.H.;Cumberland, P.H. (1973) Proceedings, British Society of Animal Production 2,81 (abstr.)

Boaz, T.G. (1957) Spillers Livestock Magazine 64,18-21.

Boaz, T.G. (1960) Journal of the Yorkshire Agricultural Society 111,25-36.

Bowman, J.C. (1966) Animal Breeding Abstracts 34,293-319.

Bowstead, J.E. (1930) Scientific Agriculture 10,429-459.

Briggs, H.M. (1936) North Dakota Agricultural Experimental Station Bulletin 285,2pp.

Cannon, D.J.;Bath, J.G. (1969) Australian Journal of Experimental Agriculture and Animal Husbandry 9,467,477-481.

Carmon, J.L.;Green, W.W. (1952) Journal of Animal Science 11,674-687.

Casida, LE. (1959) In: Reproductive physiology and protein nutrition. Ed. by J.H. Leatham, New Brunswick, New Jersey, USA. Rutgers University Press 34-40.

Ch'ang, T.S. (1967) Massey University. Sheep Farming Annual 1967, 7-13.

Ch'ang, T.S. (1968) Wool Technology and Sheep Breeding 15,11-13.

Ch'ang, T.S.;Raeside, J.I. (1957) Proceedings, New Zealand Society for Animal Productin 17,80-87

Colyer, R.J. (1971) Journal of Agricultural Science 76,433-441.

Cottier, M. (1965) Mouton 20,87-88.

Courot, M. (1962) Annales de Biologie animale, Biochemie, Biophysique 2,25-41.

Croker, K.P.;Lindsay, D.R. (1972) Australian Journal of Experimental Agriculture and Animal Husbandry 12,3,13-18.

Cupps, P.T.;Anderson, L.L.;Cole, H.H. (1969) In: Reproduction in Domestic Animals. Ed. by H.H. Cole and P.T. Cupps. London, UK. Academic Press, 2nd ed.,217-250.

Dickerson, G.E.;Laster, D.B. (1975) Journal of Animal Science 41,1-9.

Donald, H.P. (1962) Animal Production 4,369-377.

Donald, H.P.;Read, J.L.;Russell, W.S. (1968) Animal Production 10,413-421.

Donovan, B.T.;Bosch, J.J. Van Der W. Ten (1965) Physiology of puberty London, UK. Edward Arnold, vii + 206pp.

Dukin, Yu, P. (1966) Ovtsevodstvo 12,23-24.

Dun, R.B. (1955) Australian Veterinary Journal 31,104-106.

Dyrmundsson, O.R. (1972) Ph.D. Thesis, University of Wales, UK.

Dyrmundsson, O.R. (1973a) Animal Breeding Abstracts 41,275-289.

Dyrmundsson, O.R. (1973b) Animal Breeding Abstracts 41,419-430.

Dyrmundsson, O.R.;Lees, J.L. (1972a) Journal of Agricultural Science, Cambridge 78,39-45.

Dyrmundsson, O.R.;Lees, J.L. (1972b) Journal of Agricultural Science, Cambridge 79,83-89.

Dyrmundsson, O.R.;Lees, J.L. (1972c) Journal of Agricultural Science, Cambridge 79,269-271.

Dyrmundsson, O.R.;Lees, J.L. (1972d) Animal Production 14,259-262.

Dyrmundsson, O.R.;Lees, J.L. (1972e) Animal Production 15,311-314.

Eltan, O. (1974) Ph.D. Thesis, University of Wales.

Foote, R.H. (1969) In: Reproduction in domestic animals. (Eds) H.H. Cole and P.T. Cupps, London, UK. Academic Press, Ed.2, 313-353.

Foote, W.C. (1968) Wool Technology and Sheep Breeding 15,87-97.

Foote, W.C.;Sefidbakht, N.;Madsen, M.A. (1970) Journal of Animal Science 30,86-90.

Fox, C.W.;Eller, R.;McArthur, J.A.B.;Shelton, M. (1964) Journal of Animal Science 23,591 (abstr.)

Fraser, A.F. (1968) Reproductive behaviour in ungulates. London, UK. Academic Press x + 202pp.

Gaal, M. (1964) Kiserletugyi Kozlemenyek Allatten - yesztes (1962) No.1, 91-98.

Gjedrem, T. (1969) Acta Agriculturae Scandinavica 19,116-126.

Griswold, D.J. (1933) Proceedings, American Society of Animal Production 25th Annual Meeting 1932, 181-183.

Gunn, R.G. (1977) Animal Production 25,155-164.

Hafez, E.S.E. (1952) Journal of Agricultural Science, Cambridge 42,189-265.

Hafez, E.S.E. (1953) Empire Journal of Experimental Agriculture 21,217-225.

Hafez, E.S.E.;Scott, J.P. (1962) Chapter 11 in The behaviour of domestic animals. Ed. by E.S.E. Hafez, London, UK. Bailliere, Tindall & Cox Ltd.

Hammond, J. Jr. (1944) Journal of Agricultural Science, Cambridge 34, 97-105.

Holland, L.A.;Ruttle, J.L. (1966) Journal of Animal Science 25,583 (abstr.)

Hulet, C.V. (1966) Journal of Animal Science 25, Suppl.,5-20.

Hulet, C.V.;Blackwell, R.L.;Ercanbrack, S.K. (1964) Journal of Animal Science 23,1095-1097.

Hulet, C.V.;Foote, W.C. (1968) Journal of Animal Science 27,142-145.

Hulet, C.V.;Wiggins, E.L.;Ercanbrack, S.K. (1969) Journal of Animal Science 28,246-252.

Hume, D.L. (1939) Bi-monthly Bulletin, North Dakota Agricultural Experiment Station 1,7-8.

Joakimsen, O. (1969) Acta Agriculturae Scandinavica 19,175-177.

Johnstone, L.L. (1948) Australian Veterinary Journal 24,86-88.

Joubert, D.M. (1963) Animal Breeding Abstracts 31,295-306.

Keane, M.G. (1974) Irish Journal of Agricultural Research 13,263-274.

Keane, M.G. (1975) Irish Journal of Agricultural Research 14,91-98.

Keane, M.G. (1976) Farm and Food Research 7,10-12.

Lamming, G.E. (1963) Animal Production 5,225 (abstr.)

Lamming, G.E. (1969) In: The science of nutrition of farm livestock. Ed. by D. Cuthbertson, London, UK. Pergamon Press, Pt.1,411-453.

Land, R.B. (1970) Journal of Reproduction and Fertility 23,49-53.

Lang, D.R.;Hight, G.K. (1967) Proceedings, New Zealand Society for Animal Production 27,48

Laster, D.B.;Glimp H.A.;Dickerson, G.E. (1972) Journal of Animal Science 35,79-83.

Lees, J.L. (1965) Nature (Lond.) 207,221-222.

Lees, J.L. (1971) Veterinary Record 88,86-95.

Lees, J.L. (1978) Veterinary Record (in Press)

Lees, J.L.;Eltan, O. (1978) In preparation.

Lightfoot, R.J. (1968) Australian Journal of Agricultural Research 19, 1043-1057.

Lightfoot, R.J.;Smith, J.A.C. (1968) Australian Journal of Agricultural Research 19,1029-1042.

Longrigg, W. (1961) Prog. Rep. Exp. Husb. Fms. Exp. Hort. Stas. pp8-10.

Louw, D.F.J.;Joubert, D.M. (1964) South African Journal of Agricultural Science 7,509-520.

Lucas, I.A.M. (1973) Paper read at a meeting of the University College of Wales Agricultural Debating Society, November 1973.

Mair, R.B. (1963) Agriculture 70,531-534.

Mann, T. (1964) The biochemistry of semen and of the male reproductive tract. London, UK. Methuen & Co. Ltd., xxiii + 493pp.

Mann, T.;Rowson, L.E.A.;Short, R.V.;Skinner, J.D. (1967) Journal of Endocrinology 38,455-468.

Marincowitz, G.;Pretorius, P.S.;Herbst, S.N. (1966) South African Journal of Agricultural Science 9,971-980.

Marshall, F.H.A.;Hammond, J. (1952) Bulletin, Ministry of Agriculture, Fisheries and Food UK. No.39, Ed. 6, iv + 50pp.

Mattner, P.E.;Braden, A.W.H.;George, J.M. (1971) Journal of Reproduction and Fertility 24,149-150 (abstr.)

McGuirk, B.J.;Bell, A.K.;Smith, M.D. (1968) Proceedings, Australian Society of Animal Production 7,220-222.

McKenzie, F.F.;Phillips, R.W. (1930) Proceedings, American Society of Animal Production 23rd Annual Meeting 138-143.

McKenzie, F.F.;Terrill, L.E. (1937) Research Bulletin Missouri Agricultural Experimental Station No.264, 88pp.

Mounib, M.S.;Ahmed, I.A.;Hamada, M.K.O. (1956) Alexandria Journal of Agricultural Research 4,85-108.

Moustgaard, J. (1969) In: Reproduction in domestic animals. Ed. by H.H. Cole and P.T. Cupps, London, UK. Academic Press. Ed.2, 489-516.

Naerland, G. (1970) Veterinary Record 86,129-133.

Ortavant, R.;Courot, M.;Hochereau, M.T. (1969) In: Reproduction in Domestic Animals. Ed. by H.H. Cole and P.T. Cupps, London, UK. Academic Press, Ed. 2, 251-276.

Osborne, H.G. (1970) Australian Veterinary Journal 46,605.

Palsson, H. (1953) Atvinnudeild Haskolans, Rit Landbunadardeildar, B-Flokkur No.5, 84pp.

Pretorius, P.S. (1967) Proceedings, South African Society of Animal Production 6,208-212.

Pretorius, P.S.;Marincowitz, G. (1968) South African Journal of Agricultural Science 11,319-334.

Ragab, M.T.;Sharafeldin, M.A.;Khalil, I.A. (1966, publ. 1968) Journal of Animal Production U.A.R.6, 89-94.

Read, J. (1974) Quoted by Johnson in First Service, Farmers Weekly, 22 Feb. 1974, pp(v),(vii),(viii).

Reid, J.T. (1960) Journal of Dairy Science 43,Suppl.,103-122.

Riera, S.G.;Mathews, D.H.;Svejda, A.J.;Foote, W.C. (1970a) Journal of Animal Science 30,1029 (abstr.)

Riera, S.G.;Mathews, D.H.;Svejda, A.J.;Foote, W.C. (1970b) Journal of Animal Science 30,1029 (abstr.)

Rouger, Y. (1969) VI Congr. int. Reprod. Anim. Insem. Art., Paris, 1968 Vol.II, 1663-1665.

Roux, L.L. (1936) Onderstepoort Journal of Veterinary Science and Animal Industry 6,465-717.

Scott, J.P. (1945) Comparative Psychology Monographs 18, 29pp.

Sefidbakht, N. (1969) Dissertation Abstracts B 29, 4461.

Sefidbakht, N.;Madsen, M.A.;Foote, W.C. (1966) Journal of Animal Science 586 (abstr.)

Sharafeldin, M.A. (1960) Mededelingen van de Landbouwhogeschool te Wageningen 60, 1-61.

Sjodin, E. (1966) 9th International Congress on Animal Production, Edinburgh, 1966. Scient. Progr. Abstr. Eng. Ed., 54.

Skinner, J.D.;Booth, W.D.;Rowson, L.E.A.;Karg, H. (1968) Journal of Reproduction and Fertility 16,463-477.

Skinner, J.D.;Rowson, L.E.A. (1968) Journal of Reproduction and Fertility 16,479-488.

Southam, E.R.;Hulet, C.V.;Botkin, M.P. (1971) Journal of Animal Science 1282-1287.

Spencer, D.A.;Schott, R.G.;Phillips, R.W.;Aune, B. (1942) Journal of Animal Science 1,27-33.

Suiter, R.J.;Croker, K.P. (1970) Journal of Agriculture, Western Australia 11,92.

Symington, R.B. (1961) Journal of Agricultural Science, Cambridge 56, 179-181.

Tassell, R. (1967) British Veterinary Journal 123,257-264.

Terrill, C.E. (1938) Proceedings, American Society of Animal Production 31st Annual Meeting 308-310.

Tierney, M.L. (1969) Wool Technology and Sheep Breeding 16,33-35.

Voss, G. (1950) In: Neue Ergebnisse und Probleme der Zoologie. Ed. by W. Herre. Leipzig, German Democratic Republic, Akademische Verlagsgesellschaft. 1028-1056.

Watson, R.H.;Gamble, L.C. (1961) Australian Journal of Agricultural Research 12,124-138.

Watson, R.H.;Sapsford, C.S.;McCance, I. (1956) Australian Journal of Agricultural Research 7,574-590.

Wiggins, E.L. (1955) Journal of Animal Science 14,1260 (abstr.)

Wiggins, E.L.;Miller, W.W.;Barber, H.B. (1970) Journal of Animal Science 974-977.

Wiggins, E.L.;Terrill, C.E. (1953) Journal of Animal Science 12,524-535.

Wilen, R.;Naftalin, F. (1977) Pediatric Research 11,701.

Williams, S.M. (1954) Journal of Agricultural Science, Cambridge 45,202-228.

Winn, G.W.;Cumberland G.L.B. (1974) Proceedings, New Zealand Society for Animal Production 34,100-101.

Yalcin, B.C.;Bichard, M. (1964) Animal Production 6,73-84.

NUTRITION AND THE REPRODUCTIVE FUNCTION IN FEMALE SHEEP

DR J.M. DONEY

Hill Farming Research Organisation, Bush Estate, Penicuik, Midlothian

Any discussion of the influence of nutrition on reproductive function must take into consideration interactions with basic physiological regulators of reproduction, genetic variation and environmental factors. Much relevant experimental work has been carried out on other species and, although the basic principles are probably the same, these results cannot be extrapolated to sheep without direct evidence. Most of the earlier published work on sheep was observational and the considerable differences in objectives, breeds of sheep, methods and results led to a potential conflict in interpretation and application (Moule 1962).

For convenience the dietary effects on reproduction can be considered in relation to levels of nutrition before the onset of sexual maturity and to the general nutritional state of the adult during the whole of the annual reproductive cycle.

NUTRITION BEFORE MATURITY

Nutrition of the female from the foetal stage until the onset of sexual maturity may influence its subsequent reproductive performance by affecting the time or age of onset of puberty and the fertility at that oestrus, or by residual effects on reproductive performance during the rest of the reproductive life.

Effects of nutrition on the onset of puberty

Effects of nutrition have been studied indirectly by growth rate and development pattern by many workers. All the results suggest that the age of onset of puberty and the fertility of mated lambs is very dependent on genotype. Within breeds it appears that lambs must reach a threshold liveweight or associated stage of development at the time of the normal onset of seasonal oestrus in adult ewes of that breed. Beyond this threshold there appears to be an interaction between date of birth, pattern of growth and actual weight achieved at 6-7 months of age (e.g. Dyrmundsson & Lees 1972, Dickerson & Laster 1975) such that earlier born lambs reach puberty earlier in the season but at a later age and heavier weight than late-born lambs. Cedillo et al (1977) suggest that early, well-grown lambs reach the threshold weight before the changing daylight length permits oestrus activity. Lambs which fail to reach the characteristic weight for the breed by the middle of the normal breeding season, whether due to lateness of birth or to slow growth rate, are unlikely to come to oestrus in their first year. Fertility is generally much lower than in older ewes and again appears to be weight-dependent.

Even in breeds which are not normally mated in their first year a proportion of lambs may show oestrus activity. For example results from the Scottish Blackface breed suggest that relatively small increases in mean liveweight at 6 months of age may have a significant effect on the proportion of lambs reaching puberty in their first year.

Effects of early nutrition and growth on adult performance

There is no firm evidence of a direct effect of in utero nutrition on subsequent adult reproductive performance but indirect effects have been suggested. Lambs born as twins are likely to be more undernourished and have smaller birth weights than lambs born as singles even when ewes are in similar nutritional states (Russel et al 1967). Growth rates of lambs reared as twins are likely to be slower than those of singles because of lower birth weight and the restricted intake of milk which is likely to occur unless the ewes are exceptionally well fed during lactation. Similarly, it is known that severe undernourishment of ewes during late pregnancy can depress birth weight and vigour of lambs irrespective of the number carried. In the normal sequence such lambs would continue to be undernourished during the early growth period, partly because of their own size and partly because of the consequentially reduced milking capacity of their dams. Several reports (e.g. Allden 1970) suggest that there are no long-term, direct effects of foetal restriction on reproductive performance and that compensatory growth may eliminate any reproductive effect of nutrition in utero. Similarly, compensatory growth following inadequate nutrition during the lactation period and over the first winter suggests that early growth restriction may not necessarily inhibit long-term reproductive potential. Some observations however indicate that lambs with slower growth up to 12 months of age may show a lower adult reproductive rate than well grown lambs, long after the growth differential has been compensated (Gunn 1977). These results for the Scottish Blackface breed show that restricted nutrition, characteristic of most hill farms, may have a permanent inhibitory effect on a lifetime reproductive performance, even when the adult nutritional environment is less restricting and allows some compensatory growth. The importance of good early growth and the consequently increased adult size has been clearly shown by the high lambing percentages obtained from hill breeds reared under lowland conditions (Wiener 1967).

The depression of reproductive potential by restricted nutritional provision in the growth period is particularly noticeable in relation to the first lamb crop. Many earlier experiments were concerned only with the nutrient provision during the second 6 months of life (i.e. the winter period). These have shown conflicting results in terms of the first lamb crop at 2 years of age but very little permanent effect. The differences would appear to depend on interaction between the degree of restriction imposed and the opportunity for compensatory growth between 12 months and first mating at 18 months of age.

Existing knowledge suggests therefore that growth, and hence nutrition, during the whole juvenile period may have a significant effect on lifetime performance. It is far from certain how this variation in growth causes the associated differences but for purposes of sheep management it can be assumed that nutritional provision during development must be related to the provision during the productive years and to the level of performance expected or desired under the given management system.

ADULT NUTRITION

Reproduction in sheep is a cyclic phenomenon and in general the pattern of nutrient intake also follows an annual cyclic pattern. Hence it is hardly surprising that the efficiency of reproductive performance depends not only on the average level of intake over the year but on the actual level of intake at different stages in the annual cycle. In year-round grazing systems such as those of the Scottish hill regions, the annual cycle depends on the interaction of factors such as the sward composition, changes in herbage digestibility, duration of grazing season, grazing pressure, etc. Hence the intake pattern may be modified by changes in pasture management or by supplementary feeding. Under these conditions there is a reasonable balance between the nutritional provision and the requirements for single lamb production but there are gross deficiencies in the case of requirements for twin production. These deficiencies affect the ability to rear twins satisfactorily (early summer nutrition) but also affect the ability to conceive twins in the following season (late summer and autumn nutrition).

Flock reproductive rate, expressed as mean number of lambs born per ewe alive at lambing time, depends on two factors. The number of eggs per ewe (ovulation rate) which are available for fertilisation at the proposed time of mating sets the maximum potential lambing percentage. This number will be reduced by egg wastage. It is convenient to consider ovulation rate and egg wastage separately. Ovulation rate is completely determined by factors which operate up to the time of mating. Nutritionally, the most influential period is that between the previous parturition, or more conveniently the end of lactation, and mating. This can be described as the "recovery" period when body tissue reserves, utilised during pregnancy and lactation, are replenished. Egg wastage is also affected by nutrition during the recovery period but it may be further influenced by factors which operate during the whole of pregnancy.

Nutrition during recovery. The static and dynamic effects on ovulation rate

Two interrelated aspects of nutrition in this period must be considered. The first is the effect of feeding during the whole period which determines the body condition and weight of the ewe at mating time. A direct link between body conditiion and ovulation rate was established by Clark (1934). Subsequent work tended to ignore this factor and concentrate on the second or short-term aspect, which is the actual level of nutrition or rate of change in liveweight during the period immediately before mating. The latter became known as the "flushing" period. The terms static and dynamic effects were used by Coop (1962) to describe these separate factors. The static effect of nutrition is represented by the combination of body condition, liveweight and size of the ewe. Body condition measures the proportion of usable energy reserves possessed by an animal, irrespective of its physical size, and is therefore determined by the total nutritional pattern during the recovery period. Liveweight however takes both condition and size into account and is an imprecise measure of condition. In addition liveweight variation at equivalent conditions reflects differences in size due to early growth rate or to genetic causes, both of which may have indirect effects on fertility. Hence body condition, size or liveweight may be found to influence reproductive performance independently in different experimental situations.

The concept of flushing, the dynamic effect, has been accepted since at least the 19th century. Generally it is now taken to mean that the ewe should be in a rapidly rising body condition at the time of ovulation. Until recently, experimental designs and the interpretation of results in terms of practical recommendations have ignored the possibility of genetic differences and have also ignored the confounding of rapid change in weight or condition with the actual level achieved by mating time (Moule 1962). Coop (1966) discussed this latter interaction and concluded that both the static and dynamic aspects of nutrition influenced the reproductive rate, as measured by lambing percentage, mainly in terms of the proportion of multiple births. One of the difficulties of these earlier experiments was in the interpretation of the significance of quite small increases in twinning rate, where differential effects of egg wastage might have influenced the results. There is no adequate physiological hypothesis to account for the proposed effects either of body condition or of nutritional state (Casida 1963) and this limits the design of really critical experiments.

Following the developments of suitable laparoscopy techniques and the direct examination of the ovaries to establish actual ovulation rates, the interrelations of the effect of size, body condition, body weight and change in body weight or condition have been investigated in many breeds and under a wide range of nutritional patterns (e.g. Lindsay 1976, Doney & Gunn 1974, Ducker & Boyd 1977, Cumming 1977). Despite differences in detail and interpretation the results confirm the highly significant effects of one or other nutritional pathways on ovulation rate. It is apparent that the developmental history of the individual, the level of body reserves at mating time and the changes in nutritional state during the recovery period, especially in the period just before mating, are all involved to a varying extent in the achievement of the maximum ovulation rate for any given genotype. Beyond this it is impossible to generalize and this is likely to remain the case until the underlying relationship between nutritional and reproductive physiology is established. The current situation is summed up by Cumming (1977) in his statement that "... much investigation is still required to elucidate the neural and hormonal mechanisms relating mature skeletal size, tissue fat or protein reserves, energy and protein balance and photoperiod with ovulation rate".

<u>The effects of specific dietary components on ovulation rate</u>

The influence of minor elements in total reproductive performance has been reviewed by Cloete (1972), Lindsay (1976) and Edey (1976). In general it is accepted that these effects are indirect and are more likely to be associated with the basic health status of the ewe. Recently there has been renewed interest in the specific short-term dietary effect of high protein supplements on ovulation rate, following the benefits obtained in Australia from lupin-feeding (Lindsay 1976). Earlier studies contrasting the effect of high-energy or high-protein supplements had shown no consistent effects. Corbett & Edey (1977), who review the subject, found no improvement in protein status by feeding protected protein. Further developments are needed before the importance of dietary protein content and the limits of its influence on ovulation can be confirmed.

The effect of pre-mating nutrition on embryo wastage

Delay in or failure to show behavioural oestrus has been reported by a number of authors for ewes with exceptionally low body weights, in very poor condition or severely undernourished in the pre-mating period (Guerra et al 1977, Gunn & Doney 1975). Such gross effects of poor nutrition are uncommon in normal husbandry and the causes may be quite unrelated to those responsible for variation in the survival and development of fertilised ova within a more realistic range of nutritional states. There is little evidence available to link either under-nutrition or low body weight and condition with fertilization rate but results presented by Killeen (1974) suggest that both factors may depress fertilization in some circumstances.

Interpretation of information on the influence of pre-mating nutrition, expressed either as achieved weight and condition or as the short-term nutritional state, on early embryonic wastage is complicated by the effect of nutrition on ovulation rate. Basal embryonic mortality is directly related to ovulation rate (Edey 1969) such that ewes in higher body condition are expected to produce more eggs but to show a higher proportionate basal loss. The difference in viability of ova shed as singles or multiples depends partly on the location of the corpora lutea and on the extent of uterine migration. Location and migration are not directly affected by nutrition although there may well be significant breed factors which create differences in the relative loss of single and multiple ova (Doney et al 1973b). In addition to the basal loss there is evidence to suggest the existence of an opposite effect of nutritional state on embryo survival such that lower body condition and/or pre-mating nutritional state increases the rate of embryonic loss within both single- and multiple-ovulators (Gunn & Doney 1975). There are also suggestions that ewes in good body condition suffer less wastage associated with non-nutritional stress (Doney et al 1973a).

The effect of post-mating nutrition on embryo wastage

The direct effects of nutrition in the immediate post-mating period have been comprehensively reviewed by Edey (1969, 1976). The majority of lossess occur within the first 30-40 days after mating and such losses as may be found in the mid-pregnancy period are more likely to be associated with disease states than with nutrition. Peri-natal mortality, which can be a very important factor in the reduction of surviving lambing percentage, is outwith the scope of the present paper.

The potential nutritional influence on embryonic loss has been studied in relation to both over- and under-feeding during the post-mating period.

Early work on high levels of feeding gave rise to the suggestion that these may be associated with relatively high levels of embryonic loss. Many of these results however were confounded by the pre-mating feed level and by associated differences in ovulation rate (Edey 1976). The possibility of a deleterious effect of overfeeding in some situations was supported by Gunn et al (1972), who found a higher level of wastage in ewes very well fed after mating than in those which remained on restricted rations. This differential effect was not evident in ewes already in good body condition. Cumming et al (1975) found a difference in the response to over-feeding between two breeds of sheep. Over-feeding may therefore result in higher losses but the magnitude and significance of the effect must depend on interaction with pre-mating nutrition, ovulation rate and genotype.

Under-nutrition after mating has long been considered as a likely source of increased embryonic loss. The recent critical studies, reviewed by Edey (1976), show that this is indeed a factor to be considered. The extent and significance of the losses are influenced by the timing and duration of the undernourishment period, the severity of the treatment, the interrelationship of pre- and post-mating levels of nutrition and other factors such as ovulation rate and genotype. As with over-feeding there is some evidence to suggest that ewes subjected to change from an unrestricted level before mating to a low level after mating are likely to suffer greater losses than those which are maintained at a low level throughout the whole period.

Lindsay (1976) suggested that the drastic nature of the nutritional treatments employed in most of the studies on embryo loss indicates that normal nutritional fluctuations are unlikely to have a significant practical effect. The questions which must be answered for commercial sheep husbandry are what are the 'normal' levels of pre-mating nutrition for the given breed and how often does management create deviations outside the 'normal' range? The practical answer must be as imprecise, but probably as effective, as it is in relation to the nutritional regulation of ovulation rate. Early embryonic loss may be minimized by the avoidance of extremes in body condition, in levels of current nutritional provision or in changes in condition and nutrition in both the pre-mating and post-mating periods. The lower extremes of both condition or nutrition are likely to be more deleterious to both ovulation rate and embryo loss than the upper levels.

Although the loss of all potential embryos resulting from a single mating should allow the ewe to return for a second service, this return may be delayed beyond the mating period by extended cycles. Even if re-mating does occur, the conception rates at such a second service are considerably lower than at first service, irrespective of any nutritional factors (Edey 1976). This factor may be a much more significant source of barren ewes than the more extreme failure of oestrous behaviour and mating.

Many aspects of nutrition, from its effects on pre-pubertal growth, through variation in the annual cycles of food intake to the level of feeding or the diet composition in the mating period itself, can have a significant effect on the reproductive performance of sheep. Both genetic and non-nutritional environmental factors interact with and may even override the direct nutritional effects. It is essential therefore that the nutritional management should not create an unfamiliar 'stress' situation which could negate the expected gains. The general conclusions can be applied successfully to designed management practice but an accurate regulation of reproductive performance by nutritional manipulation must await more critical studies on the under-lying physiological mechanisms and the development of a unifying hypothesis.

REFERENCES

Allden, W.G. (1970) The effects of nutritional deprivation on the subsequent productivity of sheep and cattle. Nutrition Abstracts Review 40,1167-1184.

Casida, L.E. (1963) The level of fertility in the female as influenced by feed level and energy intake. Proceedings, 6th International Congress on Nutrition 366.

Cedillo, R.M.;Hohenboken,W.;Drummond, J. (1977) Genetic and environmental effects on age at first estrus and on wool and lamb production of crossbred ewe lambs. Journal of Animal Science 44,948-957.

Clark, R.T. (1934) Studies on the physiology of reproduction in the sheep. I. The ovulation rate of the ewe as affected by the plane of nutrition.

Cloete, J.G. (1972) Nutrition and reproduction in sheep. Journal of the South African Veterinary Medical Association 43,147-153.

Coop, I.E. (1962) Liveweight-productivity relationships in sheep. 1. Liveweight and reproduction. New Zealand Journal of Agricultural Research 5,249-264.

Coop, I.E. (1966) Effect of flushing on reproductive performance of ewes. Journal of Agricultural Science, Cambridge 67,305-323.

Corbett, J.L.;Edey, T.N. (1977) Ovulation in ewes given formaldehyde-treated or untreated casein in maintenance-energy rations. Australian Journal of Agricultural Research 28,491-500.

Cumming, I.A. (1977) Relationships in the sheep of ovulation rate with live weight, breed, season and plane of nutrition. Australian Journal of Experimental Agriculture and Animal Husbandry 17,234-241.

Cumming, I.A.;Blockey, M.A. de B.;Winfield, C.G.;Parr, R.A.;Williams, A.H. (1975) A study of relationships of breed, time of mating, level of nutrition, live weight, body condition and face cover to embryo survival in ewes. Journal of Agricultural Science, Cambridge 84,559-565.

Dickerson, G.E.;Laster, D.B. (1975) Breed, heterosis and environmental influences on growth and puberty in ewe lambs. Journal of Animal Science 41,1-9.

Doney, J.M.;Gunn, R.G. (1974) 6th Report Hill Farming Research Organisation Edinburgh. 1971-73,69-73.

Doney, J.M.;Gunn, R.G. (1976) Factors affecting reproductive performance in sheep. Veterinary Annual 1976 16,70.

Doney, J.M.;Gunn, R.G.;Griffiths, J.G. (1973) The effect of premating stress on the onset of oestrus and on ovulation rate in Scottish Blackface ewes. Journal of Reproduction and Fertility 35,381-384.

Doney, J.M.;Gunn, R.G.;Smith, W.F. (1973) Transuterine migration and embryo survival in sheep. Journal of Reproduction and Fertility 34, 363-367.

Ducker, M.J.;Boyd, J.S. (1977) The effect of body size and body condition on the ovulation rate of ewes. Animal Production 24,377-385.

Dyrmundsson, O.R.;Lees, J.L. (1972) Attainment of puberty and reproductive performance in Clun Forest ewe lambs. Journal of Agricultural Science, Cambridge 78,39-45.

Edey, T.N. (1969) Prenatal mortality in sheep: a review. Animal Breeding Abstracts 37,173-190.

Edey, T.N. (1976) Embryo mortality. In 'Sheep Breeding', Proceedings 1976 International Congress Muresk, Australia. Perth, Western Australia; Western Australia Institute of Technology, 400-410.

Edey, T.N. (1976) Nutrition and embryo survival in the ewe. Proceedings of the New Zealand Society for Animal Protection 36,231-239.

Guerra, J.C.;Thwaites, C.J.;Edey, T.N. (1972) The effects of components of body weight on reproductive efficiency in the Merino ewe. Journal of Agricultural Science, Cambridge 73,289-299.

Gunn, R.G. (1977) The effects of two nutritional environments from 6 weeks prepartum to 12 months of age on lifetime performance and reproduction potential of Scottish Blackface ewes in two adult environments. Animal Production 25,155-164.

Gunn, R.G.;Doney, J.M. (1975) The interaction of nutrition and body condition at mating on ovulation rate and early embryo mortality in Scottish Blackface ewes. Journal of Agricultural Science, Cambridge 85,465-470.

Gunn, R.G.;Doney, J.M.;Russel, A.J.F. (1972) Embryo mortality in Scottish Blackface ewes as influenced by body condition at mating and post-mating nutrition. Journal of Agricultural Science, Cambridge 79,19-25.

Killeen, I.D. (1974) Effects of liveweight of Border Leicester x Merino ewes and breed of ram on fertilization in flock mated sheep. Journal of Reproduction and Fertility 36,464-465.

Lindsay, D.R. (1976) The usefulness to the animal producer of research findings in nutrition on reproduction. Proceedings of the Australian Society for Animal Production 11,217.

Moule, G.R. (1962) Field trials in retrospect: Flushing. Proceedings of the Australian Society for Animal Production 4,195-200.

Russel, A.J.F.;Doney, J.M.;Reid, R.L. (1967) The use of biochemical parameters in controlling nutritional state in pregnant ewes and the effect of undernutrition during pregnancy on lamb birth-weight. Journal of Agricultural Science, Cambridge 68,351-358.

Wiener, G. (1967) A comparison of body size, fleece weight and maternal performance of five breeds of sheep kept in one environment. Animal Production 9,177-195.

THE PHYSIOLOGY OF PREGNANCY

DR D.J. MELLOR

Moredun Research Institute, 408 Gilmerton Road, Edinburgh

INTRODUCTION

I intend to concentrate on an outline of those maternal-foetal interactions which may eventually lead to an understanding of causes of lamb death or reduced viability. Areas that will receive attention include methods of study, foetal fluid balance, endocrinology and the initiation of parturition and influences on the foetus of changes in maternal physiological state during dehydration and changes in ambient temperature and nutrition.

ANATOMY OF THE CONCEPTUS

The conceptus consists of the foetus, the placenta, the foetal membranes and the two foetal fluids. The foetus is connected by its umbilical cord to the cotyledonary placenta, a major area of contact with the mother. The foetal membranes, the chorion, amnion and allantois, form the walls of the two fluid sacs; (i) the amniotic sac which encloses the foetus and amniotic fluid and is connected to the foetal bladder by the urethra, and (ii) the separate allantoic sac which contains allantoic fluid and is connected to the foetal bladder by the urachus which passes along the umbilical cord. The conceptus is enclosed by the uterus.

METHOD OF STUDY

The 'chronic' approach has become popular in recent years because it allows continuous observation of the conceptus for periods of 40-90 days after insertion of catheters (or electronic devices) into different parts. Catheters are preferred in this institute. Changes in composition of amniotic fluid (AMF), foetal urine (FU), allantoic fluid (ALF) and foetal plasma (FP) have been followed for variable periods to term (about 147 days gestational age). Most of the last half of pregnancy can be covered using this approach (Table 1).

Table 1 Longest periods of observation of foetal fluids, urine and plasma using chronic catheterization at the Moredun Research Institute

AMF for up to 90 days, from 60 to 150 days gestational age

FU for up to 80 days, from 70 to 150 days gestational age

ALF for up to 70 days, from 80 to 150 days gestational age

FP for up to 55 days, between 80 and 150 days gestational age

A 1.25 ml sample of foetal fluid or urine, or 2.5 ml of foetal blood, is taken daily through each catheter. These volumes represent a small proportion of the volume of the fluid spaces being sampled (Table 2). Sampling itself does not appear to alter gestational trends in fluid, urine or plasma composition as the same trends are observed with regular daily or weekly sampling, irregular sampling and when no samples are taken for periods of up to 3 weeks.

Table 2 Volumes (ml) of fluid spaces in the conceptus at different foetal ages

Fluid	Gestational age (days)				
	60	80	115	130	145
AMF	180	550	630	-	360
ALF	180	200	450	-	750
FP	50	80	200	360	620
FU	-	flow rate of 10-45 ml/hr			

FOETAL FLUID BALANCE

The rate of urine flow into each fluid sac is as yet technically too difficult to measure directly. Since however the compositions of amniotic fluid, allantoic fluid, and foetal urine are distinct, gestational trends in foetal fluid composition allow changes in urine flow into each sac at different stages of pregnancy to be assessed. Before 80 days gestational age, urine flow along the urethra into the amniotic sac is not sufficient to alter amniotic fluid composition but thereafter urine entry into this sac increases. Until about 100 days urine entry into the allantoic sac decreases but urine continues to flow into both fluid sacs until term.

Although urine is a major source of both foetal fluids it is not the only one, the presence of each fluid being due to a complex interaction of a variety of factors, listed here.

Amniotic Fluid

(i) Relative impermeability of the amnion as a whole to the passage of solutes.

(ii) Active or passive exchange of solutes and water between amniotic fluid and foetal blood in the amnion or foetal skin vessels until wool growth is significant.

(iii) Entry of foetal urine via urethra (Total for both sacs of 200-600 ml/day).

(iv) Foetal swallowing (250-480 ml/day).

(v) Foetal long fluid production (100-200 ml/day).

(vi) Foetal nasopharyngeal and buccal cavity secretions.

The addition of solute and water in foetal urine and nasopharyngeal and buccal cavity secretions, their equilibration with or without active transport between amniotic fluid and foetal blood, and swallowing, are all responsible for the presence of amniotic fluid.

Allantoic Fluid

(i) The chorioallantois as a whole seems to be more permeable to solutes than the amnion.

(ii) Solute and water exchange between allantoic fluid and maternal blood in the endometrium.

(iii) Solute and water exchange between allantoic fluid and foetal blood in the choriollantois.

(iv) Active transport of sodium from allantoic fluid into foetal blood and potassium in the opposite direction, apparently under the control of foetal adrencortical hormones.

(v) Active transport of amnio acids into allantoic fluid from maternal or foetal blood, or both.

(vi) Entry of foetal urine via the urachus.

The passive entry of solute and water into the allantoic sac in foetal urine and by equilibration with maternal and foetal blood, and active modification of the fluid by membrane pumping mechanisms under foetal control, are major factors in allantoic fluid formation.

FOETAL ENDOCRINOLOGY AND THE INITIATION OF PARTURITION

The foetal pituitary-ACTH-adrenal axis plays a major role in the initiation of parturition. Prolonged pregnancy follows foetal pituitary stalk section, hypophysectomy and adrenalectomy, and premature delivery may be induced by foetal infusion of adrenocorticotrophin (ACTH) or glucocorticoids but not mineralocorticoids. In addition, the corticosteroid concentrations of foetal plasma markedly increase during the last 3-7 days before delivery at full-term or before delivery induced prematurely by foetal ACTH infusion.

Before normal delivery there is a marked reduction in the progesterone levels and a rise in the oestrogen concentrations of maternal plasma, and during the last few hours a rise in myometrial and maternal cotyledonary prostaglandin F_2 levels. Oxytocin levels do not change until dilation of the cervix begins.

The currently accepted mechanism which relates these observations is as follows. In response to one or several unknown stimuli, which may include foetal hyperthermia and compression, the hypothalamus becomes increasingly sensitive towards the end of pregnancy and increases the release of ACTH. This hormone causes a marked increase in foetal plasma levels of corticosteroids which decrease progesterone and later increase oestrogen secretion by the placenta resulting in parallel changes in their respective levels in maternal plasma. These hormonal changes promote the synthesis of prostaglandin F_2 in maternal cotyledons and myometrium and this in turn sensitises the uterus to the action of oxytocin which induces labour. Once established labour is reinforced by dilatation of the cervix which elicits reflex abdominal muscle contractions and release of oxytocin. These changes end after delivery.

Apart from the rise in foetal plasma levels of corticosteroids during the last 3-7 days before birth, plasma levels of antidiuretic hormone (ADH) increase during the same period, causing a marked rise in the osmolality of foetal urine during the same period. Apart from the labour-inducing role the rising levels of corticosteroids may be important for survival of the newborn lamb by preparing it for the traumas of labour and delivery and by aiding indirectly the initiation of respiration. At birth the change from a fluid to a dry environment may also be accompanied by haemorrhage when the umbilical cord is severed. The preparturient increase in foetal ADH secretion with its associated renal water retention may therefore be a further essential preliminary to these events.

FOETAL RESPONSES TO:

(i) Maternal dehydration after eating dry feed.

A rise in maternal plasma osmolality is accompanied by a similar rise in the osmolality and a fall in the volume of foetal plasma. These changes cause an increase in the secretion of foetal ADH which acts on the foetal kidneys to reduce urine output. A reversal of maternal osmolality changes after drinking is associated with decreasing foetal plasma osmolality, increasing blood volume and increasing foetal urine production.

(ii) Nutrition of the ewe.

In well nourished, unstressed ewes, there is little change in the sodium and potassium concentrations of allantoic fluid until the last 10 days before delivery, when sodium levels decrease and potassium levels increase. These changes occur at the same time as the marked preparturient rise in foetal plasma levels of corticosteroids, suggesting an association which is confirmed by the similar decrease in sodium and increase in potassium levels in allantoic fluid before ACTH-induced premature delivery.

During maternal hypoglycaemia these ionic changes occur at earlier gestational ages, suggesting an earlier increase in foetal plasma levels of corticosteroids, but not to sufficient levels to induce premature delivery. In fact, a reduction in the glucose levels of foetal plasma is associaterd with rising foetal corticosteroid levels; these ionic changes reflect the effects of maternal-induced foetal hypoglycaemia on foetal corticosteroid secretion.

(iii) Mild cold exposure and stress of the ewe.

During mild cold exposure, thyroxine secretion increases and, presumably through its glucogenic action, plasma glucose levels rise. Under these circumstances allantoic fluid sodium levels tend to rise and potassium levels to fall, suggesting a relative inhibition of foetal corticosteroid secretion by a maternal-induced rise in foetal plasma glucose levels. On the other hand, still using allantoic fluid as the index, a rise in foetal corticosteroid levels seems to occur during postoperative stress and during simulated stress (maternal ACTH infusion). This may be due to maternal corticosteroids crossing the placenta and acting once they reach the foetal circulation.

Although some of this remains to be confirmed by direct observation of foetal plasma, it indicates that we can no longer regard the foetus as a passive sponge, responding slightly, if at all, to changes in the ewe. It is now clear that the foetus responds actively to a whole variety of maternal stimuli. A most striking example of this foetal sensitivity to changes in the mother is the foetal panting which follows 5-10 minutes after the ewe has been disturbed by having its pen hosed out.

REFERENCES

Barcroft Centenary Symposium on Foetal and Neonatal Physiology (1973). Cambridge Univ. Press.

Mellor, D.J. and Slater, J.S. (1974) Some aspects of the physiology of sheep foetal fluids. Br.Vet.J., Vol. 130, No.3, 238-248.

British Medical Bulletin (1975) Vol.31, No.1.

UK, Ciba Foundation Symposium (1977) The Fetus and Birth. Amsterdam; Elsevier (Exerpta Medica) North Holland. viii + 481pp.

ENERGY METABOLISM

PROFESSOR A.J.F. WEBSTER

University of Bristol School of Veterinary Science, Langford, Bristol

The conversion by a sheep of the energy contained in the food it eats to that contained in meat or wool is a highly complex process. As a first step in its analysis one may identify three stages.

(i) The fermentation and digestion of food energy to yield a fuel for the processes of metabolism. The digestibility of energy is largely a function of the physical and chemical nature of the food.

(ii) The metabolism of the nutrients made available by digestion for the maintenance of body structure and function and the synthesis of new tissue. This depends both on the nature of the nutrients and the physiological use to which they are put.

(iii) The relationship between energy retention and the mass of the new material synthesised as meat, milk or wool. This depends on the proportions of the three major constituents of body tissues and secretions, namely protein, fat and water.

THE METABOLIZABLE ENERGY CONTENT OF FEEDS

A detailed description of the measurement of digestible energy (DE) or metabolizable energy (ME) and the properties of feeds that determine these values is outside the scope of this paper. This topic has been covered very thoroughly in a recent book by Schneider & Flatt (1975). Comprehensive tables of the DE and ME contents of feedstuffs are published by NRC (1975) and MAFF (1975); unfortunately many of the MAFF and most of the NRC values are not based on direct measurements of energy balance. The UK Feedingstuffs Evaluation Unit recently published its first report (1975) giving direct measurements of ME made at the maintenance level of nutrition with sheep for varieties of oats, barley, silages and dried grasses. This report also presented all the conventional and some unconventional measurements of the chemical composition of the same feeds in an attempt to establish what chemical measurement or combination of measurements would most accurately predict the metabolizable energy of the feed as determined in vivo. To date no singular characteristic or combination of characteristics has emerged which can predict metabolizable energy with a greater precision than that to be found in existing sets of tables of feed composition.

Two important points have however emerged from recent studies. Firstly, in most normal mixed diets, the ME values of the individual components are additive so that the ME value of the mixture can be estimated directly from the sum of its parts. Secondly, for most forages and mixtures of forages and cereals the ratio ME:DE is about 0.82. For lambs being fattened very rapidly on high energy diets containing 15% or less of forage, methane production is very low and the ratio ME:DE is about 0.9. This point is important when predicting ME values of diets from traditional digestibility measurements or from measurements made in vitro for DE value.

The DE and ME contents of feeds are affected by the amount of feed consumed. Measurements of ME are therefore conventionally made at about the maintenance level of nutrition. The Agricultural Research Council (1965) introduced a complicated calculation whereby the ME value of feeds was reduced according to the extent to which an animal consumed feed in excess of maintenance. In fact, this so-called 'plane of nutrition' effect is not only unnecessarily complicated but recent evidence has shown it to be also unjustifiable on biological grounds.

For most ruminant feeds, digestibility falls as intake relative to maintenance requirement increases, probably because the more an animal eats the faster is the rate of passage of food through the gut. Generally speaking, the less digestible the feed the greater the dependence of digestibility on level of intake. The proportion of DE lost as methane declines with decreasing intake, which tends to compensate for the decline in digestibility by increasing the ration ME:DE. In practice, the only feeds which show a marked decline in metabolizability with increased intake above maintenance are poorly digestible feeds which sheep are unwilling or unable to eat in large amounts. Even growing lambs seldom eat more than twice their maintenance requirement, unless given feeds of the highest quality for which ME content does not decline with increasing intake (Blaxter 1974). Thus there is in practice little point in applying a level of feeding correction to values for ME determined at the maintenance level in formulating rations for sheep.

THE HEAT PRODUCTION OF SHEEP

The four principle factors that determine the heat production of sheep are:

(i) Body size, a functon of body weight (W)

(ii) The heat increment of feeding, a function of ME intake

(iii) Activity

(iv) The work of thermoregulation.

In confined and thermally comfortable conditions similar to those existing in a calorimeter, or in a good intensive production unit, activity is minimal and the work of thermoregulation is nil. In these circumstances heat production (H) is determined by the amount of ME eaten and the amount of animal that eats it, a function of body weight (W). These two factors are not independent but there are certain effects on H which can be attributed chiefly to the quantity and quality of the food and others which can be attributed chiefly to the size and physiological state of the animal. These will be considered in order.

Body size and heat production

The conventional measurement of the effect of body size on heat production is fasting metabolism, which removes by definition the effect of food intake on heat production. It is not however a measurement of much practical significance and even its theoretical foundation is unsure (see Webster et al 1974). It is probably more useful to describe heat production (H) at maintenance (when ME intake = H). In these circumstances, H for sheep is about 380 kJ/kg$^{0.75}$d for yearlings, falling to about 280 at about five years of age. These values were determined in the strict confinement and seclusion of a respiration calorimeter but are probably close to the ME requirement for maintenance of housed adult ewes and lambs fattened intensively in confinement. The increased ME requirement for maintenance of animals on open range will be considered below.

Food intake and heat production

Heat production increases at an accelerating rate with increasing ME intake. Blaxter (1967) proposed that this curvilinear relationship between H and ME intake be described by two efficiency terms, k_m and and k_f, which relate increments of energy retention (RE = ME − H) to increments of ME, where k_m is the efficiency of utilisation of ME between fasting metabolism and maintenance and k_f is the efficiency of utilisation of ME between maintenance and twice maintenance. Blaxter also claimed that both k_m and k_f were simple functions of the ME content of dry matter (M/D, MJ/kg DM), itself inversely related to the amount of fibrous material in the diet.

Technical Bulletin 33 (MAFF 1975) sets out practical recommendations for the use of the metabolizable energy system for sheep. This suggests that k_m should be assumed constant at 0.70 and k_f (or k_g in their usage) should be calculated by

$$k_f \text{ (or } k_g\text{)} = 0.0435 \text{ M/D}$$

The explanation for this relationship given by Blaxter (1967) was that the amount of fibre in the diet (inversely related to M/D) determined the proportions of the volatile fatty acids (VFAs), chiefly acetate, propionate and butyrate, made available by rumen fermentation. Armstrong & Blaxter (see Blaxter 1967, Annison & Armstrong 1970) had infused different mixtures of VFAs into the rumen of sheep, obtained different heat increments and concluded that the heat increment of feeding was proportional to the molar proportion of acetate in the infusate, or in more normal circumstances in the end-products of digestion. Ørskov (1975) criticised this work on theoretical grounds and concluded that heat increment should not be much affected by quite wide differences in VFA proportions. This suggests that other explanations must be found to account for the variation in heat increment when sheep eat normal feeds. The contributors to heat increment may be divided into the following measurable components.

(i) The energy cost of eating

(ii) The energy cost of rumination

(iii) The heat of fermentation

(iv) The work of digestion

(v) The work of nutrient metabolism.

Only the last contributor can be considered analogous to the heat increment produced as a result of absorption of VFAs into the hepatic portal circulation.

The energy cost of eating is proportional not to the amount of food eaten but to the length of time spent eating. On average it amounts to about 30-40 J/kg body weight/min. The contribution of eating to heat increment therefore ranges from only about 3 kJ/MJ ME for concentrate feeds eaten very quickly to about 35 for grazed herbage (Osuji 1974). These costs do not include the energy costs of standing to eat or walking to graze. Rumination, curiously, is a much less expensive process and the total energy cost of rumination even for about 8 h/day is only about 20 kJ/MJ ME. The heat of fermentation is probably rather constant at about 65 kJ/MJ of fermented energy, irrespective of the nature of the other end-products of fermentation (Hungate 1966, Webster et al 1975). The energy costs of eating, rumination and fermentation together account for about 20-25% of total heat increment. Heat increment resulting from the infusion of VFAs would appear to be about 65-70% of that which occurs when a comparable amount of absorbed energy is presented in the form of real food. A residual 10% of total heat increment may therefore be attributed with some confidence to the 'work of digestion', i.e. the muscular and secretory processes arising from the presence of food in the gut. Table 1 presents an analysis of heat increment for an adult sheep either grazing chopped hay or grazing fresh herbage.

Table 1 Analysis of the heat increment of feeding for a 50 kg sheep getting 16.8 MJ/day (about twice maintenance)

	Chopped hay	Fresh grass
Total heat increment of feeding (kJ)	7610	7930
Work of ingestion and digestion (kJ)		
eating	223	543
ruminating	130	130
fermentation	1336	1336
digestion	353	353
% of total HIF	27	30
Work of nutrient metabolism (kJ)	5568	5568

This table and the evidence presented so far still do not account for the large differences in heat increment above maintenance (and thus k_f) claimed for different feeds. Current work by D.E. Beever, E.R. Ørskov and P.C. Thomas, much of it unpublished, indicates rather convincingly that differences in k_f cannot be attributed simply to differences in molar proportions of VFAs produced by ruminal fermentation. Moreover, Blaxter (1974) in a recent review showed that k_f and k_m do not vary as simple functions of M/D as he originally suggested (Figure 1). For most mixed feeds, k_m and k_f vary only slightly, to an extent in fact that can be accounted for by the differences summarised in Table 1. For grasses however there are large differences in k_f which must relate to the chemistry and structural conformation of the plant but which probably cannot simply be equated to molar yields of VFAs. This subject is still unresolved. For practical purposes however it would seem that the MAFF (1975) prediction of k_f = 0.0435 M/D is not much use. For mixed diets, k_f could be assumed constant at about 0.48 but for grasses k_f varies enormously with differences in plant structure and chemistry and does so in a way that is not simply related to M/D.

Activity

Table 2 lists the energy costs of major activities for sheep and compares the energy expenditure of a 50 kg sheep on range with that of a confined animal. Table 2 indicates that for sheep kept at a high stocking intensity on good pasture, heat production is less than 10% above that of the confined animal. On short grass ranges where animals have to work quite hard to harvest their food, the increment in heat production is nearer to 25% (Osuji 1974). Daily energy expenditure also depends on the state of vigilance of the animal and the length of time it spends asleep. Variations in the duration of the latter do not alter heat production by more than about 2% but variations in vigilance of nervousness while awake are probably more important (Toutain et al 1978).

Table 2 The daily energy expenditure of a 50 kg sheep on range compared with that of a similar animal confined in a calorimeter

Activity	Energy cost of activity	Amount of activity: Confined	Amount of activity: On range	Additional cost of activity on range (kJ/d)
Eating	36 J/kgW.min	1.5h	5-9h	370-790
Standing	5.5 J/kgW.min	5h	5-15h	0-160
Changing position	55 J/kgW	20 actions	20-30	0-30
Walking	2.1 J/kgW. horizontal m.			
	26.5 J/kgW. vertical m.	nil	1-6km	100-630

Extra energy cost of outdoor activity (MJ/d)	0.47 - 1.61
Heat production at maintenance in confinement (MJ/d)	7.15
Heat production on range (MJ/d)	7.62 - 8.76
Increase due to activity (%)	6.5 - 22.5

Figure 1

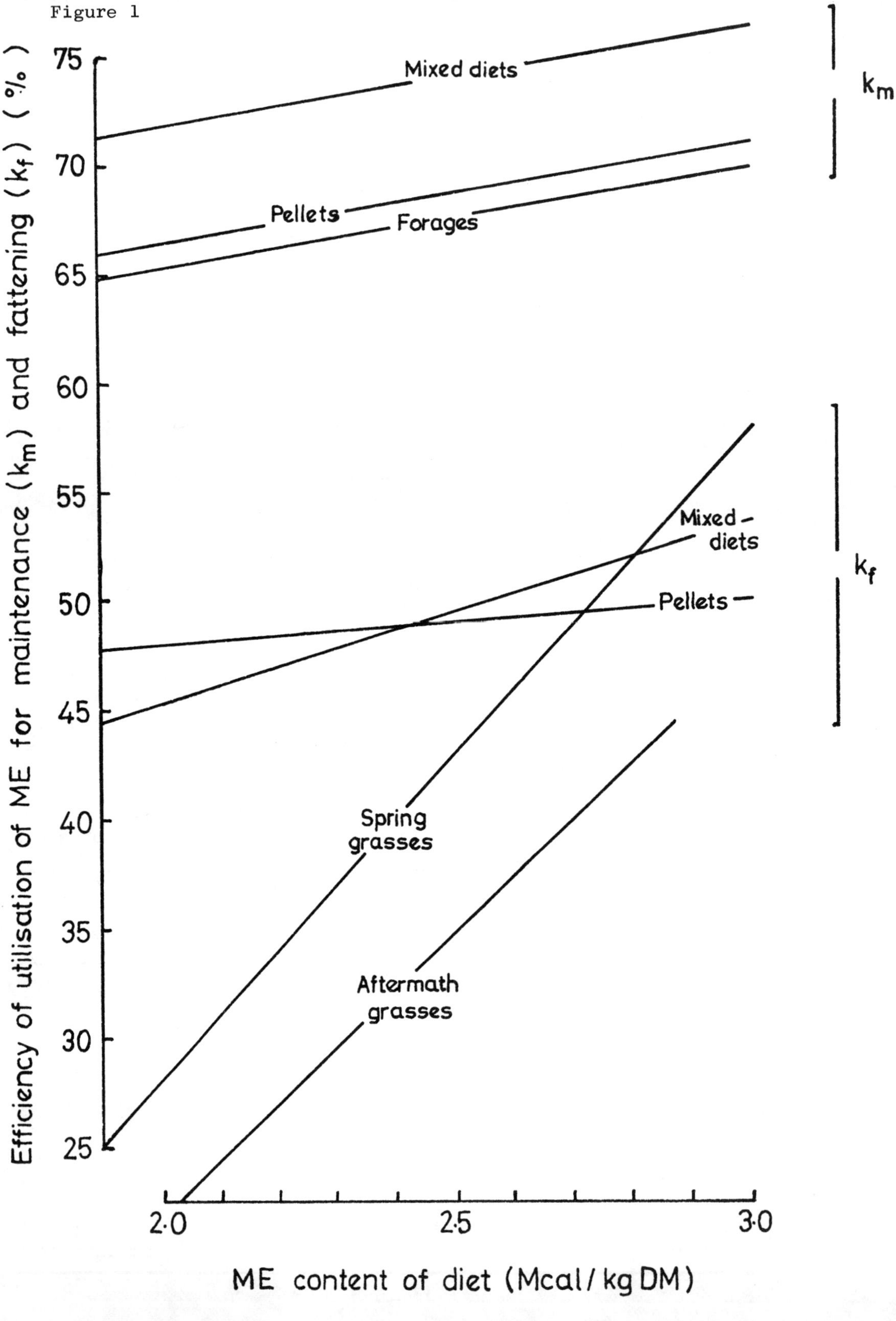
75
70
65
60
55
50
45
40
35
30
25
Efficiency of utilisation of ME for maintenance (km) and fattening (kf) (%)
Mixed diets
Pellets
Forages
km
Mixed-
diets
Pellets
kf
Spring
grasses
Aftermath
grasses
2·0
2·5
3·0
ME content of diet (Mcal/kg DM)

Thermoregulation

Sheep are well insulated animals and also have a very efficient mechanism for regulating evaporative heat loss by ventilating the mucous surfaces of the upper respiratory tract. Thus they are able to survive, produce and reproduce in a wide variety of thermal environments without that environment having any direct effect on energy metabolism. In these so-called thermoneutral environments, heat is produced only as an inevitable consequence of normal metabolic processes. The sheep does not produce heat to keep warm. Instead, it regulates heat loss so as to keep body temperature down to about 38-39°C. The thermoneutral zone for mature sheep in full fleece is very wide. Unless the fleece is saturated by prolonged rain, adult sheep are unlikely to experience cold stress in any of the livestock producing areas of the world (see Alexander 1974). Shearing of course markedly reduces cold tolerance but is done obviously at a time when the direct effects of weather on energy balance are likely to be minimal. The effect of cold on young lambs has also been reviewed by Alexander (1974) and is considered elsewhere by J. Slee. For practical purposes however there is no need to include in calculations of the energy requirements of adult ewes or intensively fattened lambs any term for the energy cost of maintaining homeothermy in cold environments. The effect of heat stress on energy intake and expenditure is outside the scope of this article.

Energy retention and body composition

In addition to the factors already discussed, the efficiency of conversion of ME into body tissue depends on the composition of the body gains. Body gains consist of protein at 23.5 kJ/g and fat at 39.3 kJ/g, which make up about 98% of the combustible material in the body. The other major constituents, water and ash, are not combustible. In muscle and soft tissues, protein is associated with about 4 times its own weight of water so that the energy content of 'wet' tissue protein is only about 4.7 kJ/g, only 12% that of fat. The energy content of 'dry' wool protein is assumed to be about 23.5 kJ/g. In theory therefore the energy retained in 1 kg LWG of an animal depositing energy entirely as fat is about 8 times that of an animal depositing energy only as 'wet' protein. In the USA the energy content of liveweight gains is assumed in practice to vary from about 10.2 MJ/kg at 20 kg liveweight to 23.3 MJ/kg at 60 kg liveweight (Rattray et al 1973). In the UK, it is assumed to vary from a minimum of 14.5 MJ/kg for lambs of 15 kg gaining 100 g/d to 22 MJ/kg for lambs of 45 kg gaining 400 g/d (MAFF 1975). These figures imply that the proportion of energy retained as protein in the young animal is about 80%, dropping to about 40% by slaughter weight. No allowances however are made for differences between breeds and strains of sheep in mature size and in related differences in the partition of retained energy between protein and fat. The information is at present simply not available but it is essential if the conversion of food energy to body gains is to be predicted with any precision in growing lambs.

It has also been demonstrated that the efficiency of retention of ME (k_f or kg) in the growing animal depends upon the partition of retained energy between protein and fat. For simple stomached species such as the rat and the pig eating a highly digestible diet, the efficiencies of utilisation of ME for protein and fat deposition are about 0.45 and 0.75 respectively (see Pullar & Webster 1977). Ørskov & McDonald (1970) observed similar relative efficiencies of protein and fat deposition for lambs raised intensively on cereal diets. If the relative efficiencies of protein and fat deposition are the same for sheep as for simple stomached species, one may estimate the extent to which differences in the partition of RE between protein and fat might influence k_f (kg) during growth. Assume that adult sheep used for conventional measurements of k_f retain 80% of energy as fat, and that for most mixed diets k_f, thus determined, is about 0.48. This would imply values for the apparent energetic efficiencies of protein and fat deposition of 0.32 and 0.52. During normal growth therefore variations in the proportion of energy retained as protein from 40 to 80% would cause the efficiency of utilisation of ME for gain (true k_g , as distinct from k_f) to vary from 0.44 to 0.36 respectively. This too needs to be substantiated by direct experiment.

REFERENCES

Alexander, G. (1974) Heat loss from sheep. In: Monteith, J.L.;Mount, L.E. (Eds) Heat loss from animals and man. Assessment and control. Proceedings of the Twentieth Easter School in Agricultural Science, University of Nottingham, 1973. London, UK; Butterworth & Co. Ltd., pp173-203. ISBN 0-408-70652-X.

Annison, E.F.;Armstrong, D.G. (1970) Volatile fatty acids and energy supply. In: Phillipson, A.T. (Ed) Physiology of digestion and metabolism in the ruminant. Proceedings of the Third International Symposium, Cambridge, August 1969. Newcastle upon Tyne; Oriel Press Ltd., pp422-437. ISBN 0-85362-053-9.

Blaxter, K.L. (1967) The energy metabolism of ruminants, 2nd edn. revised. London, UK; Hutchinson & Co. Ltd., 332pp.

Blaxter, K.L. (1974) Metabolizable energy and feeding systems for ruminants. In: Swan, H.; Lewis, D. (Eds) Nutrition conference for feed manufacturers: 7. London,UK; Butterworth & Co. Ltd., p3-25. ISBN 0-408-70704-6.

Hungate, R.E. (1966) The rumen and its microbes. New York: London; Academic Press, x +533pp.

Ørskov, E.R. (1975) Manipulation of rumen fermentation for maximum food utilization. World Review of Nutrition and Dietetics 22,152-182.

Ørskov, E.R.;McDonald, I. (1970) Utilisation of dietary energy for maintenance and for fat and protein deposition in young growing sheep. In: Schurch, A.; Wenk, C. (Eds) Energy metabolism of farm animals. Proceedings of the 5th Symposium, Vitznau, Switzerland, September 1970. European Association for Animal Production, EAAP Publication No.13. Zurich, Switzerland; Juris Verlag, p121-124.

Osuji, P.O. (1974) The physiology of eating and the energy expenditure of the ruminant at pasture. Journal of Range Management 27,437-443.

Pullar, J.D.;Webster, A.J.F. (1977) The energy cost of protein and fat deposition in the rat. British Journal of Nutrition 37,355-363.

Rattray, P.V.;Garrett, W.N.;Hinman, N.;Garcia, I.;Castillo, J. (1975) A system for expressing the net energy requirements and net energy content of feeds for young sheep. Journal of Animal Science 36, 115-122.

Schneider, B.H.;Flatt, W.P. (1975) The evaluation of feeds through digestibility experiments. Athens, Ga.,USA; University of Georgia Press, xxiii+423pp. ISBN 0-8203-0378-X.

Toutain, P.L.;Toutain, C.;Webster, A.J.F.;McDonald, J.D. (1977) Sleep and activity, age and fatness, and the energy expenditure of confined sheep. British Journal of Nutrition 38,445-454.

UK, Agricultural Research Council (1965) The nutrient requirements of farm livestock. No.2. Ruminants. Technical reviews and summaries. London, UK; Agricultural Research Council, xi+264pp.

UK, Ministry of Agriculture, Fisheries and Food (1975) Energy allowances and feeding systems for ruminants. Technical Bulletin No.33. London, UK; H.M.S.O. viii+79pp. ISBN 0-11-240894-X.

UK, Rowett Research Institute (1976) Feedingstuffs Evaluation Unit. First Report 1975. Aberdeen, UK, v+55pp.

USA, National Academy of Sciences, National Research Council, Committee on Animal Nutrition (1975) Nutrient requirements of domestic animals. 5. Nutrient requirements of sheep. 5th edn. revised, Washington, D.C., v+72pp. ISBN 0-309-02212-6.

Webster, A.J.F.;Brockway, J.M.;Smith, J.S. (1974) Prediction of the energy requirements for growth in beef cattle. 1. The irrelevance of fasting metabolism. Animal Production 19,127-139.

Webster, A.J.F.;Osuji, P.O.;White, F.;Ingram, J.F. (1975) The influence of food intake on portal blood flow and heat production in the digestive tract of sheep. British Journal of Nutrition 34,125-139.

MINERAL METABOLISM

DR A.C. FIELD

Animal Disease Research Association, Moredun Institute, Gilmerton Road, Edinburgh

Continued ingestion of diets that are deficient, imbalanced or excessively high in a particular mineral invariably leads to impaired growth, health, fertility and milk production. All minerals have a spectrum of activity, depending upon the intake of the element in question and the nutritional state of the animal. The relationship between dietary concentration and effect on the animal can be presented in the form of a curve. The first part of the curve, an increasing effect with increasing concentration until a plateau is reached, expresses the biological action of the element and the plateau expresses optimal supplementation and normal function. The width of the plateau is determined by the homeostatic capacity of the animal or system. With further increasing concentration the element enters a phase of irritation and stimulation of some function, expressing what may be called its pharmacological action. At still higher intakes this is followed by the appearance of signs of toxicity and perhaps by death. These phases of action differ widely between different minerals. The biological phase is a characteristic of 'essential' elements, those that play a vital role in the body. With some, such as copper in sheep, the margin between the biological and toxic actions of the elements can be small, and with others, the so-called toxic elements such as arsenic, lead, cadmium and mercury, the toxic phase occurs at very low concentrations.

The nutritionist is concerned with the essential nutrients in the deficiency state and then only with those which are associated with commonly occurring nutritional maladies or with poor animal performance. He must also avoid unsafe dietary levels of potentially toxic trace elements. His aim at all times is the plateau which represents optimal supplementation and normal function. The minerals most frequently considered in this context include calcium, phosphorus, magnesium, sodium, chlorine, iodine, iron, copper, manganese, zinc and cobalt. More recently very small amounts of selenium have been added to some sheep diets.

Essential elements play a static and dynamic role in the body. The static functions, mainly a property of the macro-elements, include increasing bone rigidity to protect the soft tissues and provide attachments for the muscles, maintenance of osmotic pressures in fluids and tissues, maintenance of an optimum hydrogen ion concentration to promote normal enzyme activity, and maintenance of irritability of the tissues and membrane permeability of the cells. The dynamic function concerns mainly the trace elements which act primarily as catalysts in enzyme systems in the cell. The roles that they play in enzymic reactions range from weak, ionic strength effects to highly specific, metallo-enzyme associations. In the latter the metal is firmly associated with the protein and there is a fixed number of atoms per molecule of protein which cannot be removed without loss of enzymatic activity. Two of the elements, iodine and cobalt, are unique because on present evidence their entire functional significance can be accounted for by their presence in single compounds, thyroxine and Vitamin B_{12} respectively.

The requirements for a mineral may be defined as the amount of a mineral that must be supplied in diet to meet the needs of a normal healthy sheep given a completely adequate diet. The United States National Research Council and the Agricultural Research Council of the United Kingdom have published a complete list of the mineral requirements of sheep. The precision of these estimates however varies greatly among the minerals and I wish to review the information on which these allowances are based, identify and discuss the validity of the assumptions that have been made and finally to assess their value within the general framework of sheep nutrition.

The estimation of a dietary allowance needs firstly an estimate of the net requirements of the animal for maintenance, growth, reproduction and lactation and, secondly, a factor, termed absorbability or efficiency of absorption for converting these requirements into dietary allowances.

NET REQUIREMENTS

Maintenance

The maintenance requirement for an element is defined as the inevitable loss in the faeces and urine which occurs when the element is absent from the diet. A more realistic definition equates maintenance requirement to the inevitable loss when absorption and requirements are in balance, but this is not applicable for the elements, Na, K and Cl, since there is substantial exchange of these elements across the gut wall. Inevitable faecal loss of these elements must therefore be determined in part by the loss of faecal water, a fact not generally recognised, probably because of the lack of interest in these elements which are rarely limiting in agricultural systems. Another factor will be dietary intake but only to the extent that intake determines the concentratin of K, Na and Cl in faecal water.

The difficulty of formulating diets adequate in all respects other than the complete absence of the mineral in question has proved to be insurmountable for most essential minerals. Quantitative information on endogenous secretion has been derived from studies involving the use of radioisotopes. In these studies the sheep were on normal diets and the isotopes were used to identify the fraction of the faeces which originated in the body. The inherent assumption is that faecal endogenous excretion is independent of the relationship between the amount absorbed and the animal's requirements. The assumption appears to be valid for Ca and Mg in so far as endogenous excretion of these elements appears to be independent of the dietary intake. On the other hand, recent evidence suggests that the faeces is a true route of excretion of P absorbed in excess of requirements. In this instance, the P content of the diet must be so chosen for radioisotope studies that the amount of P absorbed is equal to or less than requirements.

Inevitable losses in faeces have been predicted from the dietary intake of dry matter for K and from the dietary intake for Na and Cl.

The best estimates for endogenous excretion and hence maintenance of the macro elements are Ca 16 , P 14 and Mg 3 mg/kg liveweight per day. For K a figure of 1.0 g/kg dry matter intake is recommended for the faecal losses and one of 37.5 mg/kg liveweight for urinary losses. The maintenance requirements for Na and Cl are taken as 5.8 mg (plus 2 g) Na and 7.9 mg Cl/kg liveweight per day.

The maintenance requirements for suckling lambs are considered to be lower for P and Mg, at 4 and 2 mg/kg liveweight per day respectively.

There are few estimates of the maintenance requirements for trace elements, mainly because of the difficulty in identifying the transport form of these trace elements in the body, a pre-requisite for radioisotope dilution studies. The only ones available are for Cu and Zn and are 0.004 and 0.076 mg/kg liveweight per day respectively.

Mineral Accretion During Growth

The mineral content of the individual tissues of the body varies considerably and, unless the relative rates of growth in soft tissue, bone and fat are similar throughout somatic growth, the mineral content of each increment in weight during development cannot be constant. As an animal ages any gain in weight contains an increasing proportion of fat, and furthermore at a given age large weight gains contain proportionally more fat than do small ones. Although skeleton growth rate falls markedly with age, there is ample evidence that in young animals the skeleton becomes more mineralised with increasing age, until at maturity changes in body weight are not necessarily associated with changes in deposition of major mineral components of the skeleton.

For the sheep the ARC have ignored the above effects and have adopted a constant mineral composition for each increase in bodyweight throughout somatic growth. Their values for each kg gain in bodyweight are (in g):

Ca 11	K 1.8
P 6	Na 1.1
Mg 0.41	Cl 0.8

The importance of this over simplification as a source of error in total requirements will clearly depend upon the contribution of growth to total requirements. Growth plays an important role in determining total requirements for those minerals found chiefly in the skeleton.

The only values (mg/kg carcass gain) for the trace element composition of growing sheep refer to Fe 60, Zn 24, Mn 0.9 and Cu 0.8.

MINERAL DEPOSITION IN THE FOETUS AND CONCEPTUS

To calculate the quantities of minerals deposited in the uterus at successive stages of pregnancy we need to know the total quantities deposited in the whole product of conception (foetus, membranes and fluids, and uterine wall, minus the uterus of the non-pregnant ewe) and the time course in nutrient deposition. Table 1 gives the best estimates of mineral composition of the whole products of conception at various stages of pregnancy.

Table 1. Whole Products of Conception (representative values for a lamb weighing 4 kg at birth).

Component	Weight (g) at age (days)					daily gain	
	35	63	91	119	147	91-119	119-147
Total	138	1141	2971	4593	5600	58	36
Ca	0.01	0.48	4.7	20.3	53	0.56	1.17
P	0.05	0.81	4.8	14.3	29	0.34	0.52
Mg	-	-	0.19	0.94	2.2	0.027	0.045
K	0.06	0.92	3.7	7.2	10.0	0.12	0.10
Na	0.33	2.6	6.7	10.0	12.0	0.11	0.071

Information on the trace element deposition in the conceptus is confined to Cu and Zn. Cu accumulates at rates of 15, 85 and 186 μg/day during the first, second and third trimesters and Zn at rates of 0.28 and 1.5 mg/day during mid and late pregnancy respectively.

MINERAL EXCRETION IN LACTATION

The ARC gives the following average composition for ewe's milk:

Ca	1.6 g/kg	Cu	0.30 mg/kg
P	1.3	Zn	7.2
Mg	0.17	Fe	0.77
Na	0.4		
K	1.4		

The concentrations of the remaining trace elements, Se, Co, I, Mn and Mo, vary greatly with the dietary intake and it is difficult to decide which value to adopt. For instance, values for Mo can vary between 10 and 980 μg/kg and those for I from 30 to 1000 μg/kg.

The excretion in milk of a mineral is the product of the animal's milk yield and the concentration of the mineral in milk. Total milk yield of ewes varies between 50 and 180 kg, depending on the breed and the individual and the overall nutritional status of the animal. Unlike the cow, direct determination of milk yield in sheep is difficult and milk yields must be estimated, often with low precision.

There are insufficient data to identify which of the many factors known to influence chemical composition of milk are of importance for each mineral. For many elements, such as copper, the important factors are stage of lactation, individual variation and the nutritional status of the animal.

ABSORBABILITY OF MINERAL ELEMENTS IN THE DIET

The efficiency of absorption of a mineral in the diet is affected by the chemical nature of the element, the forms in which it is ingested, the dietary levels of several other elements and organic substances and the age of the animal. In addition the variation in the efficiency with which individuals can absorb certain minerals from the same diet can be large. The precision with which absorbability can be predicted is the main source of error in the estimates of dietary allowances, especially where absorbability is low. For example, an error of 0.05 in the efficiency of absorption will generate errors in net requirements of 10 and 100% for efficiencies of absorption of 0.5 and 0.05 respectively.

Most minerals have two levels of absorbability depending upon whether they are present in milk or dry diets. In general the efficiency of absorption of minerals in milk is of a very much higher order, ranging mainly between 0.8 and 1.00. For instance, the figure for Cu in milk is 0.90, falling to 0.22 in the first 6 weeks of life, and only 0.06 for ruminating animals.

For ruminating animals the minerals can be classified according to their absorbability into a number of broad groups.

Total Absorbability

Sodium, Chlorine, Potassium and Iodine

These elements are either highly electro-positive or electro-negative (Group I and VII, periodic table) and occur in tissues and fluids as ions. The exception is I which is also found as an integral part of thyroxine.

For this group of elements the only source of error in dietary allowances lies in the error attached to net requirements, which, although difficult to quantify, is usually taken as 15-20%.

Absorbability between 0.3 and 0.9

Calcium

This element is less electro-positive than those listed above and can exist both in the ionic and in combined forms in tissues and fluids.

The concept of absorbability does not apply to calcium; sheep absorb calcium according to their needs and no relationship has been found between amount absorbed and ingested. Thus for a given sheep, the "absorbability" will be inversely proportional to feed calcium. We must therefore arrange that the diet contains sufficient Ca in a form that can be absorbed. At present there is no information on what proportion of dietary Ca is in this form. An estimate can be made from balance trials by considering only those trials where dietary intake and requirements are of a similar order. Such an exercise suggests that an average 0.6-0.7% of dietary Ca is in an available form.

Phosphorus, Sulphur and Molybdenum

These elements are electro-negative and exist in tissues and fluids as highly oxidised ions and in organic complexes. Information from radioactive studies with ^{32}P suggest that lambs up to 1 year old absorb dietary P more efficiently (0.73) than older animals (0.60). Recent studies have shown that S intake has a marked effect on the absorption of Mo. At low intakes of S the efficiency of absorption of Mo can be as high as 0.80; when a S-rich diet results in a high intake of S, the efficiency of absorption of Mo falls to 0.30. There is limited information on the absorption of S, but it appears to be about 0.60.

Absorbability between 0.1 and 0.3

Magnesium

It is surprising that the absorbability of this element is not higher when one considers its chemical properties and the fact that it exists in biological systems partly in the ionic form. Figures for the absorbability of dietary Mg, obtained with ^{28}Mg, range from 0.10 to 0.32 with a mean of 0.24. In view of the frequently observed variation in absorbability of dietary Mg between individual sheep consuming the same diet, a lower figure of 0.17 is recommended for the calculation of dietary allowances.

Within this group there appears to be a small number of trace elements. Studies with roughage diets suggest that the absorbability of dietary Fe is 0.21, Zn 0.25-0.31, and Se 0.30.

Absorbability less than 0.10

In this group are found most of the trace elements. These rarely exist in the ionic form in tissues but form compounds by co-ordinate valency links; in fact the copper ion is extremely toxic to living tissues. Each element is absorbed as a complex, the so-called transport form, and any dietary constituent which limits the formation of the latter either by competing for the metal or for the other part of the complex, will depress absorption. Many such relationships have been recognised and are typified by the inter-relation between Cu, Mo and S.

The consequence is that for most trace elements the amount absorbed cannot be predicted from the amount ingested and dietary allowances can only be formulated in the most general terms.

REQUIREMENTS

It is possible to derive the dietary macro-element requirements for the various classes of sheep from the data given previously. This is not the case for the trace elements and it is the practice to state the dietary allowances in terms of the concentration in the dietary dry matter. The generally accepted estimates are:

Fe	20 mg/kg DM	Cu	7	
Zn	20-33	Se	0.05	
Mn	20-25	I	0.5	or 2 mg in the presence of substantial quantities of goitrogens.

DIETARY INTAKE OF MINERALS

Most sheep spend most of their life at pasture where they are frequently subjected to successive periods of abundance and shortage of herbage. Drought and cold temperature are the usual causes of shortages of herbage but, occasionally, high stocking rates have been held responsible. The estimation of the mineral intake of the grazing sheep during herbage shortage is almost impossible. Obtaining a sample, representative of the herbage actually consumed, particularly in extensive systems, is only possible with sheep fitted with oesophageal fistulae. The animal's potential for altering mineral intake by selective grazing is particularly great; the mineral composition of herbage varies greatly among plant species and between parts of the same plant. Determination of herbage intake is also difficult and often necessitates confining the sheep to plots much smaller than the area normally grazed.

Thus during the periods of greatest need for supplementation it is not possible to measure or even estimate mineral intake with an acceptable degree of precision.

CONCLUSIONS

The scientific approach to mineral nutrition of sheep has been shown to have limited applicability; net requirements can be estimated with a high degree of precision but neither the availability of minerals in the diet nor the actual amount ingested by the sheep can be predicted with sufficient accuracy to be of use in practical agriculture. This applies particularly to trace elements but is true in part for all elements. Assessment of the adequacy of a particular diet appears to be possible only by measuring the performance of the animals actually consuming the diet. For acute disorders associated with deficiencies of specific minerals, such as the various manifestations of copper deficiency, this approach is of immense value. For the more widespread and mild nutritional maladies, often associated with no discernible symptoms other than poor animal performance, the farmer knows the diet is deficient but has no knowledge of which mineral is lacking. Our hope for the future lies in using our knowledge to develop a series of biochemical tests based either on blood or tissue analysis for mineral deficiency.

PROTEIN METABOLISM

DR J.C. MacRAE

Rowett Research Institute, Bucksburn, Aberdeen

Modern intensive agriculture places ruminants in systems which are far removed from the rangeland-type environment in which they evolved. In some areas they are still expected to subsist on poor-quality, high-roughage diets, often containing mainly structural carbohydrates with very little nitrogen, for at least some part of the year; they are more often expected to adapt their digestive and metabolic processes to the efficient utilisation of high quality diets such as improved pasture with high soluble carbohydrate/high protein, or even cereals and other components perhaps more suited to monogastric digestive systems. It is important therefore that we know precisely how ruminants digest and metabolise this wide range of diets so that we can develop more productive feeding systems and ensure the correct balance of dietary constituents to optimise the efficient conversion of such feedstuffs into meat, wool or milk.

DIGESTION

Host-animal enzymic digestion

The main sites of mammalian protein digestion are the stomach (the abomasum in ruminant animals) and the small intestine (Figure 1). Here, proteolytic enzymes hydrolyse protein to its constituent peptides and amino acids, which are subsequently absorbed from the small intestine. The main enzymes involved in these reactions are detailed in Table 1.

All the enzymes other than the peptidases are secreted into the lumen of the gut as inactive percursors known as zymogens and are converted to the active forms by the catalysts indicated. Procarboxypeptidase is converted to its active form in the pancreas. Pepsin, trypsin and the chymotrypsins are able to attack a wide range of proteins whereas the peptidases can only attack peptides at specific points in their molecules.

The complications of pre-gastric fermentation

In the monogastric animal, food passes directly into the stomach where enzymes can initiate proteolytic action. In contrast, in the ruminant animal, protein digestion is complicated by the fact that before dietary protein reaches the site of host-animal proteolytic enzyme secretion it has to pass through the rumen or forestomach (Figure 1) where the microbes present have a considerable influence on its digestion. These microbes degrade the bulk of the dietary protein to its constituent peptides and amino acids, which are then largely deaminated to ammonia. This ammonia has then to be utilised along with amino acids and shorter chain peptides for microbial protein synthesis, if the nitrogen passing out of the rumen to the site of host-animal enzyme digestion in the abomasum and small intestine is to be of any use to the animal (see Lewis 1960).

Fig. 1. SCHEMATIC REPRESENTATION OF N DIGESTION IN RUMINANTS

$$\text{N digestibility (\%)} = \frac{\text{Feed N} - \text{Faecal N}}{\text{Feed N}} \times 100$$

$$\text{N balance (g/d)} = \text{Feed N} - (\text{Faecal N} + \text{Urinary N})$$

$$\text{Biological Value} = \frac{\text{Feed N} - (\text{Faecal N} - \text{MFN}) - (\text{Urinary N} - \text{EUN})}{\text{Feed N} - \text{Faecal N} - \text{MFN})} \times 100$$

Where MFN (metabolic faecal -N) and EUN(Edogenous urinary N) are components of faeces and urine which are derived from metabolic processes rather than directly from the diet (see any animal nutrition text book)

Figure 1 (contd)

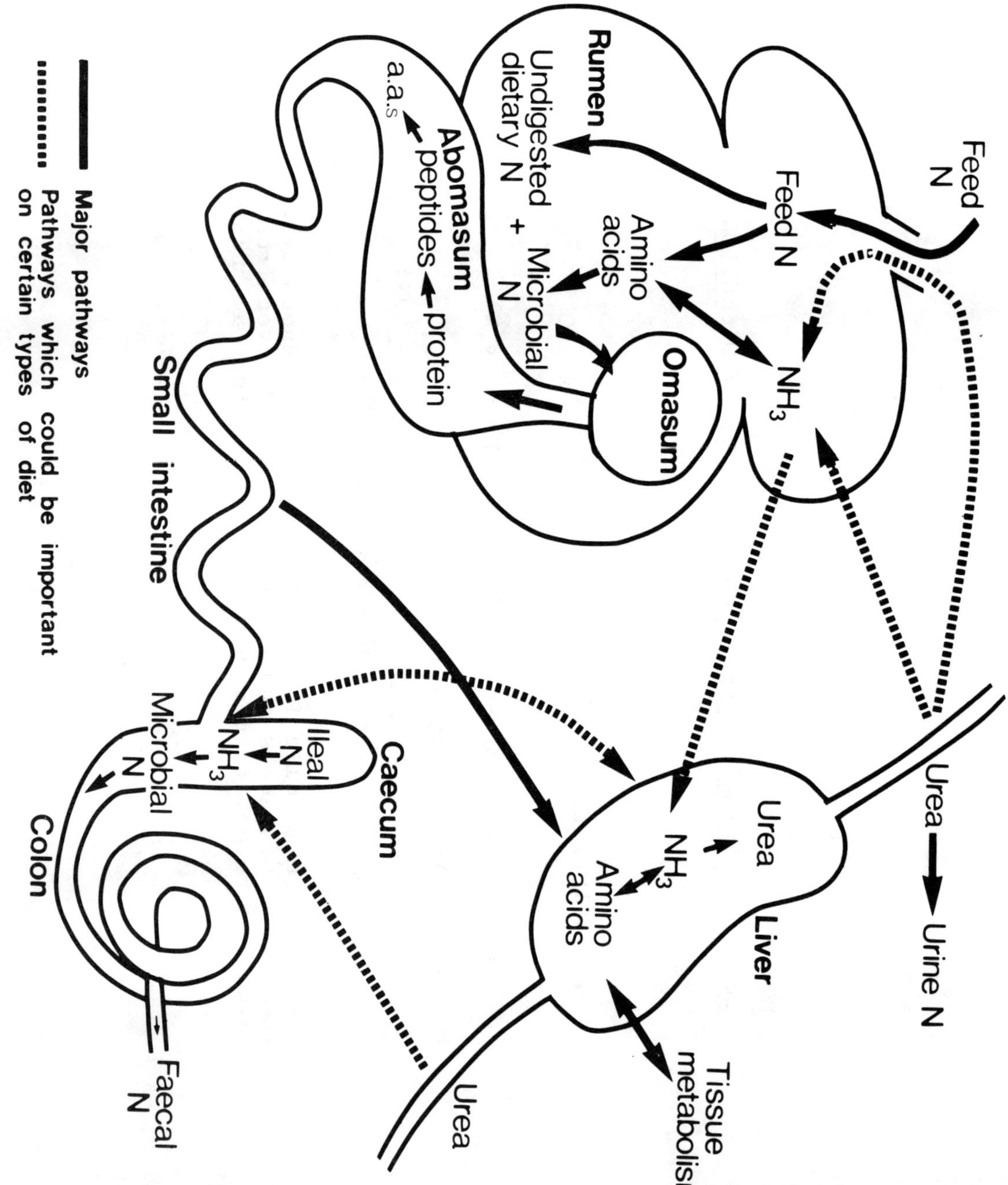

Table 1 Mammalian Proteolytic Enzymes

Source	Secreted zymogen	(catalysts) →	Proteolytic enzyme
Stomach	Pepsinogen	(HCl + pepsin)[1] → (pH 6, Opt.2)	Pepsin
Pancreas	Trypsinogen	(enterolinase + trypsin)[1] → (opt. pH 8)	Trypsin
Pancreas	Chymotrypsinogens	(trypsin + chymotrypsin)[1] →	Chymotrypsin (man)
Pancreas	Procarboxypeptidases[2]	? →	Carboxy-peptidases
Intestinal mucosa			Amino-peptidases
Intestinal mucosa			Dipeptidases

[1]reaction occurs in lumen of tract

[2]reaction occurs in pancreas

On poor quality, low nitrogen diets, this process is thought to be advantageous to the animal because, as well as degrading dietary protein the micro-organisms can convert non-protein nitrogen (NPN) from either dietary or endogenous sources such as recycled urea-N via saliva or diffusion, into ammonia which can then be used to produce microbial protein with a relatively high biological value ($\simeq$80, McNaught et al 1954). On a good quality, high nitrogen diet however the process can be disadvantageous because the ammonia tends to be produced more quickly than the microbes can use it and so considerable amounts can be absorbed from the rumen and passed to the liver for conversion to urea. Although some of this urea may be recycled back into the rumen via saliva, a large proportion is likely to be excreted in the urine. This excretion of urinary urea represents a net loss of nitrogen to the animal.

Qualitative influences

As pointed out above, amino acids absorbed from the small intestine are the main nitrogenous substrates available for mammalian protein metabolism.

Comparison of amino acid compositions of feed and duodenal digesta would suggest that there are some qualitative changes in the amino acid mixture between consumption and arrival at the small intestine. In Figure 2 the individual amino acids consumed and entering the small intestine of sheep, given freshly cut short rotation ryegrass at 500 g OM/24 h (MacRae & Ulyatt 1974) are presented as a graph. This diet has been selected because the amounts of total amino acid eaten and arriving at the duodenum were almost identical (107 g and 109 g respectively). The data are therefore useful in illustrating which amino acids appear to be degraded or synthesised in their passage through the stomach.

Net synthesis of lysine and methionine (with this diet 20% and 55% respectively) and net degradation of proline (with this diet 25%) are characteristic of most diets studied. Other changes are usually found to be much smaller and variable. In the case of this fresh ryegrass diet there was a 40% increase in tyrosine and a 15-20% degradation of arginine but other observations on dried diets (e.g. Clarke et al 1966, Coehlo da Silva et al 1972, MacRae et al 1972) show lesser changes in these two amino acids.

Quantitative influences

Table 2 presents experimental data to illustrate how the amounts of nitrogenous material reaching the small intestine can be quantitatively influenced by dietary characteristics.

In Table 2(a), data from two separate experiments, involving widely differing levels of fed N, produced duodenal non-ammonia-N (NAN) flow rates which were considerably less variable. Thus on the very low-N diet (5-6 g N/d) there was an apparent net synthesis of NAN anterior to the duodenum, whereas on the high-N diets (> 15-20 g N/d) there was considerable apparent degradation of N.

The data in Table 2(b) indicate that the soluble carbohydrate content of the diet may influence the amount of N reaching the duodenum, presumably by providing a readily available source of carbohydrate structures which permit microbes to utilise the large amounts of ammonia being produced concurrently.

Alteration of the physicochemical condition of a diet can also influence the fermentation processes. Until recently the evidence for this effect was mainly circumstantial. When fresh herbage is frozen and thawed, its content of readily soluble N falls by as much as 50% (MacRae 1970, MacRae et al 1975); it is reduced even further if the material is dried. This reduction would appear to allow the micro-organisms to utilise better the nitrogenous component of the diet, presumably because the rate of release of NH_3 is more closely matched to the microbial capacity for NH_3 utilisation. The between-experiment evidence on which this hypothesis was first formulated is given in Table 2(c). More recently however workers at the Grassland Research Institute, Hurley (Beever et al 1974) have demonstrated that when fresh and frozen perennial ryegrass is fed to sheep the animals receiving the frozen herbage have a 26% increase in microbial protein production and a 15% increase in the quantity of amino acids entering the small intestine.

Figure 2

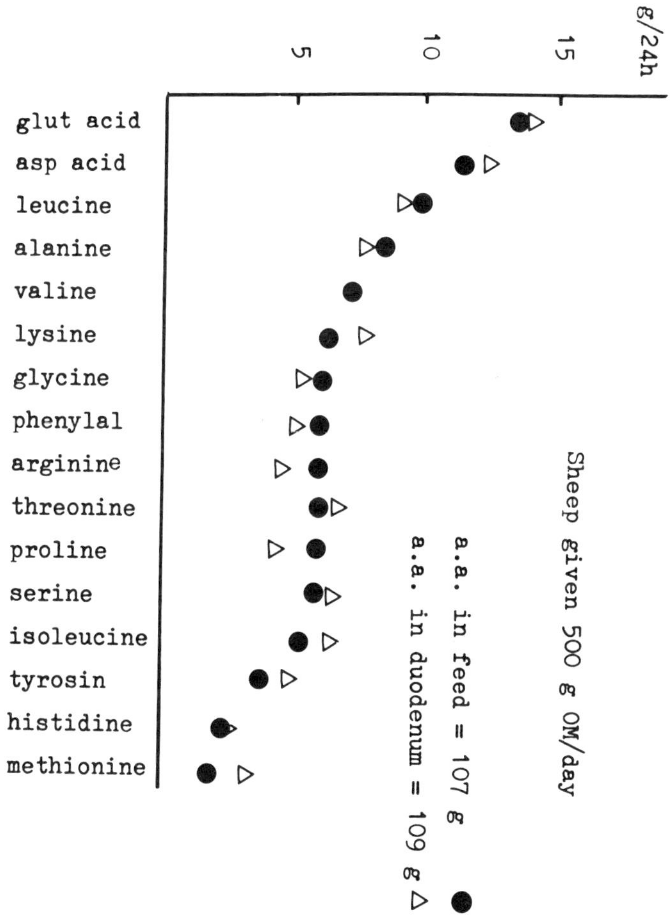
g/24h
5
10
15
glut acid
asp acid
leucine
alanine
valine
lysine
glycine
phenylal
arginine
threonine
proline
serine
isoleucine
tyrosin
histidine
methionine
Sheep given 500 g OM/day
a.a. in feed = 107 g
a.a. in duodenum = 109 g

Table 2 Dietary Factors Affecting the Amounts of Nitrogenous Constituents Reaching the Small Intestine

(a) N Intake

Diet	Hay	Hay + maize	Hay + maize + soya bean	Hay + soya bean	Forage oats Maturity III	II	I
N intake (g/24 h)	5.1	7.3	16.4	24.8	6.2	28.0	38.0
NAN[1] reaching SI	8.5	11.6	16.4	17.4	9.9	23.9	31.1
NAN app.absorpt. from SI	4.6	7.0	11.3	12.1	7.1	17.8	22.6
	Clarke et al (1966)				Hogan & Weston (1969)		

(b) Soluble Carbohydrate (CHO) Content of Ration

Diet	Forage Oats Maturity I		Chopped Lucerne	Chopped Dried
N intake (g/24 h)	37.3	38.0	30.9	20.6
Sol CHO intake	142	35	74	173
NAN reaching SI	38.3	31.1	23.8	21.4
NAN app. absorbed from SI	30.3	22.6	16.2	13.7
	Hogan & Weston (1969)		Hogan & Weston (1967)	MacRae et al (1972)

(c) Solubility of N in diet

Diet	Perennial ryegrass fresh	frozen	dried	Clover fresh (white)	frozen (red)	Dried & wafered (red)
N intake	26	26	27			
Amino acid intake				127	127	123
N reaching duodenum	20	25	32.5			
Amino acid reaching duodenum				80	133	148
N app. absorbed from SI	13	15.5	20			
Amino acid absorbed from SI				50	95	79
	MacRae & Ulyatt (1974)	Beever et al (1969)		MacRae & Ulyatt (1974)		Beever et al (1971)

[1]NAN denotes non-ammonia-N

Apparent absorption of protein from the small intestine

When digesta leaves the rumen it enters the abomasum and small intestine where the mixture of dietary and microbial protein comes under the influence of the digestive enzymes of the host animal.

Unfortunately, while the rumen processes and net changes in the stomach region have been thoroughly investigated by collecting either abomasal or duodenal digesta, digesta samples have much less often been taken from the ileal region. Thus in many experiments it is not possible to differentiate between small intestine enzymatic digestion and large intestine microbial digestion. Those studies in which the amounts of protein leaving as well as entering the small intestine have been measured indicate that, on most diets studied (mainly roughages), some 55-75% of the protein which enters the small intestine disappears there. Exceptions to this finding can occur either where the dietary proteins have been denatured in some way, perhaps during drying processes as possibly happened to the dried red clover diet in Table 2(c), or in diets such as heather (Calluna vulgaris) and Sainfoin, which contain appreciable amounts of tannin which bind onto the digesta protein and make it unavailable. Apparent digestibility of the NAN entering the small intestine of sheep fed heather diets is generally only in the region of 40-45% (MacRae & Milne, unpublished), while that in sheep given Sainfoin has been found to be 7-10% lower than with a ryegrass or clover of similar N intake (Ulyatt & Egan 1979).

There is also some suggestion of preferential absorption of "essential" as opposed to "non-essential" amino acids. Clarke et al (1966) and Coehlo da Silva et al (1972) pointed this out on dried diets, and MacRae & Ulyatt (1974) reported the same in sheep consuming fresh grass and clover diets. On the fresh herbage diets the apparent availabilities of the total amino acids entering the small intestine (i.e. duodenal-ileal x 100%) were 73%, 75% and 66% respectively for perennial ryegrass, short rotation ryegrass and white clover. When the amino acids were differentiated as "essential" and "non-essential", the availabilities were; essential 75%, 77%, 69%, and non-essential 70%, 73% and 63% PRG, SRR and WC respectively. Figure 3 illustrates these differences for the individual amino acids. In a situation where the apparent availabilities of all the amino acids were identical, the points on Figure 3 would all lie on the 45° line. It can be seen that nearly all the essential amino acids (O) lie above this line while the converse is true for the non-essential (●); proline and serine have particularly low availabilities. There are two noticeable exceptions to this general finding, the essential amino acid threonine had a very low availability relative to the others (65%, 68% and 60% respectively for the perennial, short rotation and clover herbages), while the non-essential amino acid tyrosine had a higher availability than the other non-essentials (76%, 80% and 72% respectively for the three herbages).

It must be noted that these values are apparent or net availabilities. Clarke et al (1966) point out one important qualification to such findings. Mucins, which are continuously secreted into the small intestine, have high concentrations of threonine, proline and alanine but low

Figure 3

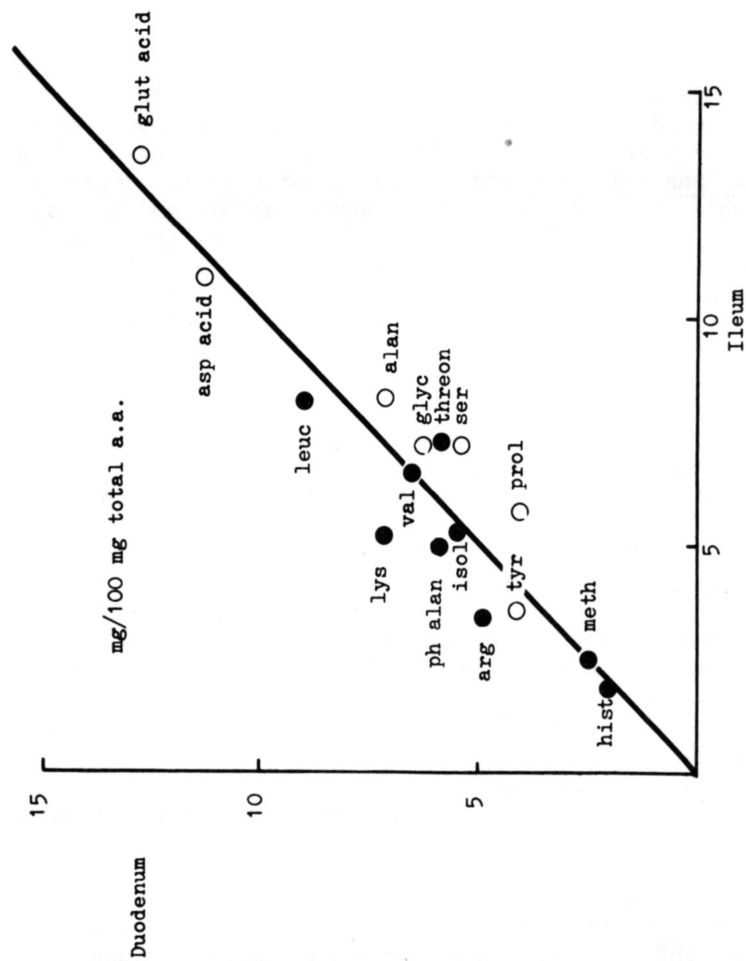
mg/100 mg total a.a.
Duodenum
15
10
5
Ileum
5
10
15
glut acid
asp acid
alan
glyc
threon
ser
leuc
val
prol
lys
isol
tyr
ph alan
meth
arg
hist

concentrations of leucine, tyrosine, phenylalanine and lysine and so these apparent preferential absorptions may to some extent be artifacts. On the other hand, Purser (1976) reviewing protein digestion and amino acid absorption presented evidence which suggested that the transport sites of the small intestine mucosa have a low affinity for threonine. It is perhaps interesting to note that, when Williams (1969) measured comparative rates of absorption of equimolar solutions of 16 amino acids in washed-out, isolated loops of sheep small intestine, he found that all the "essential" amino acids other than threonine and histidine were placed in the two fastest groups (six groups in all).

N digestion in the large intestine

The experiments referred to above have generally shown that digestion in the large intestine accounts for 5-10% of the overall N digestion of ruminants. The significance to the host-animal of this nitrogen digestion is not fully understood. It is not generally believed that amino acids can be absorbed as such from the large intestine and so it is thought that any losses of N between the ileum and the faeces are due mainly to absorption of ammonia-N, the bacterial protein produced in caecal fermentation being excreted in the faeces. Thus when Ørskov (1969) infused 250 g starch into the caeca of 2 sheep their faecal-N output increased by 4 g N/day.

It is probable that the so-called "associative effects" (Blaxter 1962), observed in N digestibility when ruminants are given different ratios of roughage concentrate in their diets, are due mainly to an enhanced hind-gut fermentation of structural carbohydrates (see Armstrong & Beever 1969) resulting in increased microbial protein excretion in the faeces. Whether this increased output of faecal-N is detrimental to the N economy of the animal is debatable. The extra N required by the microorganisms involved in this enhanced fermentation probably comes from blood urea-N sources. It was thought until recently that this might seriously alter the amount of urea-N being "recycled" into the rumen and subsequently converted into microbial amino acids which could be absorbed and used by the host animal. Recent observations have suggested that probably little of the circulating urea-N is utilised in this way (Nolan et al 1976, MacRae et al 1977) and, as the increased faecal N output is largely compensated for by a reduction in urinary-N excretion, there may be little penalty to the N economy of the animal. Thus the net result may only be a beneficial effect on the animal's energy economy, with extra VFA arising from the caecal fermentation.

METABOLISM OF ABSORBED NUTRIENTS

Up to now we have called the disappearance of nitrogenous material from the alimentary tract apparent absorption. What happens to these nutrients is not well understood because only recently have physiologists tackled the practical problems of quantifying the movement of metabolites in blood and the precursor/product relationship of N metabolites in the whole animal.

During absorptive processes

There is evidence to suggest that amino acid absorption is an active function. The data of Williams (1969) on relative rates of absorption for equimolar solutions supports this concept and work with preparations such as isolated loops of washed out small intestine has shown that certain amino acids can be absorbed against a concentration gradient. When amino acids are absorbed from the lumen of the small intestine, those which escape metabolism in intestinal cells drain into the portal blood stream and pass to the liver. Accurate measurement of portal blood flow is difficult and few data are available on the quantitative passage of metabolites in the blood. Human nutritionists have for some time talked about the probability that considerable amounts of amino acid may be metabolised during the absorptive processes (Texter et al 1968) but there is little direct evidence on the quantitative significance of absorptive metabolism in ruminants. Workers at Cornell University (Wolff et al 1972) and Kentucky (Hume et al 1972) have measured "net absorption" (portal blood flow x A-V difference) of amino acids in sheep given diets of alfalfa. Unfortunately neither group measured corresponding apparent absorptions (disappearance from the small intestine). To date, the best estimates of the magnitude of absorptive metabolism have to be derived from comparisons of the portal vein data with other independent reports on apparent absorptions made in sheep given similar diets. Table 3 gives two such comparisons and suggests that, with most of the amino acids, as much as 50% of that leaving the small intestine will be used in metabolic processes associated with absorption. Higher proportions of apparently absorbed valine, proline and perhaps tyrosine appear in the portal blood while aspartic acid, methionine and glutamic acid give very low recoveries. Indeed, in the ten experiments carried out by Wolff et al (1972), the portal flow of glutamine plus glutamate was in all cases less than the arterial flow; they reported net removal of 1.45 ± 0.29m moles glutamine and 0.18 ± 0.08m moles glutamate per h from this region. The authors compared their blood flow data with the apparent absorption data of Clark et al (1966) and concluded that the large discrepancies observed between portal flow and apparent digestion of glutamine and glutamate, asparagine and aspartate, arginine, alanine and cystine were probably attributable to their metabolism by gut tissues.

There was a very high portal flow of ammonia in both studies and some of this could have come from the deamination of amino acids during absorption. Portal blood also drains the reticulo-rumen but, without a precise knowledge of direct absorption of NH_3 from that organ, it is not possible to say how much of the amino-N was deaminated during absorption.

By the liver

Wolff et al (1972) showed that, in sheep fed the maintenance energy requirement, most of the amino acids absorbed and passed with the portal blood were removed by the liver. Only for asparagine, valine, isoleucine, leucine, citrulline, ornithine and glutamate was net production by the total splanchnic viscera readily detectable; the last two amino acids were almost entirely produced by the liver itself, the glutamate probably arising as a result of the large amounts of ammonia which had to be detoxified.

Table 3 Comparison of apparent absorption of amino acids from the small intestine and appearance of amino acids in portal drained viscera (m moles/24 h)

	Apparent absorption from small intestine		Uptake by portal drained Viscera		Percentage recovery in blood
diet (regimen)	lucerne cubes (2 x daily)	chopped dried grass (continuous)	alfalfa cubes (continuous)	chopped alfalfa (2 hourly)	
Arginine	35	26	8	19	23-73
Histidine	10	11	7	4	36-70
Isoleucine	50	34	16	17	32-50
Leucine	67	54	25	27	37-50
Lysine	46	40	17	24	37-60
Methionine	11	13	5	4	31-45
Phenylalanine	30	28	19	19	63-67
Threonine	37	39	17	26	43-70
Valine	48	44	20	35	42-79
Alanine	71	62	55	68	77-110
Aspartic acid	82	66	24	13	16-36
Glutamic acid	81	66	-39	5	-ve to 7
Glycine	75	53	37	26	35-68
Proline	40	25	18	47	45-188
Serine	40	32	26	25	62-81
Tyrosine	24	18	15	18	62-100
Total amino acids	759	611	294	376	39-61
Ammonia	15	89	696	222	
	Coehlo da Silva et al (1972)	MacRae et al (1972)	Wolff et al (1972)	Hume et al (1972)	

This finding helps to explain why, when Reis & Tunks (1970) infused casein into the abomasum of sheep, they found the greatest increases in jugular plasma amino acids were for valine, isoleucine and leucine, and also why few conclusions on the amino acid nutrition of ruminants have been obtained from measurements of peripheral plasma amino acids concentration.

Amino acids taken up by the liver, and those later taken up by the peripheral tissues, can be used either as precursors for protein synthetic pathways or as energy yielding substances. The relative contribution of any individual amino acid to these alternative processes will depend on a large number of complicated, and as yet not well understood, factors associated with the particular physiological state of the animal and the balance of available nutrients. With the development of tracer kinetics techniques, employing isotopically labelled metabolites ([^{14}C], [^{15}N], [^{3}H] and [^{35}S] labelled compounds), researchers have started to look at phenomena such as protein turnover rates in ruminants and the relative contribution of amino acids to anabolic and catabolic processes. For example it is now generally accepted that proteins laid down in tissues are not static substances but are being continuously synthesised and degraded; the current data suggests that tissue protein is probably completely turned over every 8-10 days.

When one considers that each and every protein in the body has a characteristic amino acid pattern, requiring the exact ratio of amino acids to be available at the site of synthesis and sufficient energy to "drive" the process, one begins to appreciate the complexities of protein metabolism; it will be a long time before the whole intricate picture is fully understood. Very valuable information is starting to accrue from the use of tracer kinetic techniques in studies of specific aspects of protein digestion and metabolism. For example, in the study of rumen microbial protein production, tracer amounts of [^{35}S]-sulphate (Beever et al 1974) or [^{15}N]-ammonia (Nolan et al 1976) infused into the rumen were taken up by the microorganisms during fermentative processes and incorporated into microbial protein. The dilution of the tracer isotope in the microbes and duodenal digesta can be used to give estimates of microbial protein production rates and the proportion of duodenal-N which is of microbial origin (see Figure 1). Similarly in studies of amino acid metabolism, incremental increases in the rate of urea synthesis, determined with [^{15}N]- or [^{14}C]-urea (Nolan & Leng 1968) or the incorporation of the [^{14}C] label of specific amino acids into glucose or products of energy metabolism (Black et al 1968, Egan & MacRae 1978), provide a means to follow changes in the role of plasma amino acids when the physiological state of the animal changes, e.g. during pregnancy and lactation or when the balance of N/energy in the diet is altered.

PROTEIN AVAILABILITY AND ANIMAL PERFORMANCE

Having discussed the digestion and metabolism of proteins in ruminants, we ought to attempt to relate the availability of protein to animal performance. Experiments are usually designed either to indicate production responses to dietary manipulations and supplements or to obtain detailed digestive and metabolic measurements on diets which have little direct relevance to any production situation. Unfortunately, experiments are seldom designed so that digestive and/or metabolic parameters can be closely related to production data, but occasionally it is possible to relate the two. In the fresh-herbage experiments of MacRae & Ulyatt (1974), digestive parameters were measured in sheep given a range of herbage intakes. They were found to be linearly correlated with intake. When these data were then related to the liveweight gains of lambs grazing the same herbages (Table 4), there appeared to be a very good relationship between absorbed protein and production (r = 0.79) but none between absorbed energy and production (r = 0.02). In these experiments the availability of protein apparently absorbed from the small intestine appeared to be a limiting factor to animal performance.

Hutton & Annison (1972), when reviewing the control of N metabolism in the ruminant, did calculations in which they compared the amounts of essential amino acids reaching the duodenum of a steer with the tissue concentrations of each acid in an attempt to establish whether any one amino acid might be limiting production. They found that the ratio of required:available amino acids was in most cases less than 0.50 but that for methionine it was 1.04. In sheep the requirement for methionine is likely to be even higher relative to the other amino acids, because of the high S-amino acid content of wool; infusion of L-cysteine or DL-methionine directly into the abomasum of sheep can markedly increase wool growth (Reis 1967). Armstrong & Annison (1973) concluded that the availability (apparent absorption from the small intestine) of methionine and threonine could well be inadequate to meet calculated tissue requirements on several forage and forage/grain diets.

PRESENT STATE OF KNOWLEDGE

The results of numerous studies over the last 20 years, on the digestive processes in ruminants, allow us to manipulate these processes to some extent. By comparison little is yet known about the quantitative aspects of absorption and subsequent metabolism. It is hoped that in the next few years, techniques of the types mentioned in this paper can be used to form dynamic models of N and energy metabolic transactions in the ruminant and lead to a better understanding of the quantitative aspects and controlling mechanisms of protein metabolism.

Table 4 Relationships between the observed live-weight gains of sheep-grazing perennial ryegrass (R), short-rotation ryegrass (M) and white clover (C) and the estimated N and energy digestion values pertaining to the grazing intakes.

	1967[1]			1968[1]			Correlation with live-weight gain
	R	M	C	R	M	C	
Live-weight gain[1] (g/day)	227	270	331	156	241	249	
Protein (NAN x 6.25) apparently absorbed from small intestine (g/day)[2]	119	175	188	113	188	163	0.79
Energy absorbed as VFA from stomach and large intestine (Kcal/day)[3]	2431	2057	2359	2237	2462	2176	0.02

[1]Experiments reported by Ulyatt (1971)

[2]Calculated from regression equations in Figure 1 MacRae & Ulyatt (1974)

[3]Calculated as 75% of apparent digestion of gross energy from stomach and large intestine using regression equations in Figure 2, Ulyatt & MacRae (1974)

REFERENCES

Armstrong, D.G.;Annison, E.F. (1973) Amino acid requirements and amino acid supply in the sheep. Proceedings of the Nutrition Society 32,107-113.

Armstrong, D.G.;Beever, D.E. (1969) Post-abomasal digestion of carbohydrate in the adult ruminant. Proceedings of the Nutrition Society 28,121-131.

Beever, D.E.;Cammell, S.B.;Wallace, A. (1974) The digestion of fresh, frozen and dried perennial ryegrass. Proceedings of the Nutrition Society 33,73A-74A.

Beever, D.E.;Thomson, D.J.;Harrison, D.G. (1971) The effects of drying and the comminution of red clover on its subsequent digestion by sheep. Proceedings of the Nutrition Society 30,86A.

Beever, D.E.;Thomson, D.J.;Pfeffer, E.;Armstrong, D.G. (1969) Effect of drying grass on sites of its digestion in sheep. Proceedings of the Nutrition Society 28,26A-27A.

Black, A.L.;Egan, A.R.;Anand, R.S.;Chapman, T.E. (1968) Role of amino acids in gluconeogenesis in lactating ruminants. In: Isotope studies on the nitrogen chain. Proceedings of the symposium on the use of isotopes in studies of nitrogen metabolism in the soil-plant-animal system jointly organised by the International Atomic Energy Agency and the Food and Agriculture Organization of the United Nations in cooperation with the Joint Commission on Applied Radio-activity (ICSU), Vienna, August-September 1967. Vienna; International Atomic Energy Agency, pp247-263.

Blaxter, K.L. (1962) The energy metabolism of ruminants. London, UK; Hutchinson & Co. Ltd., p195.

Clarke, E.M.W.;Ellinger, G.M.;Phillipson, A.T. (1966) The influence of diet on the nitrogenous components passing to the duodenum and through the lower ileum of sheep. Proceedings of the Royal Society of London, B 166,63-79.

Coehlo da Silva, J.F.;Seeley, R.C.;Thomson, D.J.;Beever, D.E.; Armstrong, D.G. (1972). The effect in sheep of physical form on the sites of digestion of a dried lucerne diet. 2. Sites of nitrogen digestion. British Journal of Nutrition 28,43-61.

Egan, A.R.;MacRae, J.C. (1978) The contribution of threonine to the glucose economy of hill sheep. Proceedings of the Nutrition Society 37,15A.

Hogan, J.P.;Weston, R.H. (1967) The digestion of chopped and ground roughages by sheep. 2. The digestion of nitrogen and some carbohydrate fractions in the stomach and intestines. Australian Journal of Agricultural Research 18,803-819.

Hogan, J.P.;Weston, R.H. (1969) The digestion of pasture plants by sheep. 3. The digestion of forage oats varying in maturity and in the content of protein and soluble carbohydrate. Australian Journal of Agricultural Research 20,347-363.

Hume, I.D.;Jacobson, D.R.;Mitchell, G.E. Jr. (1972) Quantitative studies on amino acid absorption in sheep. Journal of Nutrition 102,495-505.

Hutton, K.;Annison, E.F. (1972) Control of nitrogen metabolism in the ruminant. Proceedings of the Nutrition Society 31,151-158.

Lewis, D. (1960) Fate of nitrogenous compounds. In: Digestive physiology and nutrition of the ruminant. Proceedings of the University of Nottingham Seventh Easter School in Agricultural Science, 1960. London; Butterworth & Co. (Publishers) Ltd., pp127-139.

MacRae, J.C. (1970) Changes in chemical composition of freeze-stored herbage. New Zealand Journal of Agricultural Research 13,45-50.

MacRae, J.C.;Campbell, D.R.;Eadie, J. (1975) Changes in the biochemical composition of herbage upon freezing and thawing. Journal of Agricultural Science, UK 84,125-131.

MacRae, J.C.;Ulyatt, M.J. (1974) Quantitative digestion of fresh herbage by sheep. 2. The sites of digestion of some nitrogenous constituents. Journal of Agricultural Science, UK 82,309-319.

MacRae, J.C.;Ulyatt, M.J.;Pearce, P.D.;Hendtlass, J. (1972) Quantitative intestinal digestion of nitrogen in sheep given formaldehyde-treated and untreated casein supplements. British Journal of Nutrition 27,39-50.

MacRae, J.C.;Wilson, S.;Milne, J.A.;Spence, A.M. (1977) Urea-nitrogen recycling in sheep given low quality hill herbages. Proceedings of the Nutrition Society 36,77A.

McNaught, M.L.;Owen, E.C.;Henry, K.M.;Kon, S.K. (1954) The utilization of non-protein nitrogen in the bovine rumen. 8. The nutritive value of the proteins of preparations of dried rumen bacteria, rumen protozoa and brewer's yeast for rats. Biochemical Journal 56,151-156.

Nolan, J.V.;Leng, R.A. (1968) Proceedings, Australian Society of Animal Production 7,348.

Nolan, J.V.;Norton, B.W.;Leng, R.A. (1976) Further studies of the dynamics of nitrogen metabolism in sheep. British Journal of Nutrition 35,127-147.

Ørskov, E.R.;Foot, M.H. (1969) Influence of caecal starch infusion in sheep on faecal output of N, starch and dry matter. Proceedings of the Nutrition Society 28,31A-32A.

Purser, D.B. (1976) Protein digestion and amino acid absorption. In: Sutherland, T.M.;McWilliams, J.R.;Leng, R.A. (Eds) From Plant to Animal Protein - Reviews in Rural Science II. Armidale, New South Wales, Australia; University of New England Publishing Unit, pp39-45.

Reis, P.J. (1967) The growth and composition of wool. 4. The differential response of growth and of sulphur content of wool to the level of sulphur-containg amino acids given per abomasum. Australian Journal of Biological Sciences 20,809-825.

Reis, P.J.;Tunks, D.A. (1970) Changes in plasma amino acid patterns in sheep associated with supplements of casein and formaldehyde-treated casein. Australian Journal of Biological Sciences 23,673-680.

Texter, E.C. Jr.;Chou, C-C.;Laureta, H.C.;Trappen, G.R. van (1968) Ch.25. In: Physiology of the Gastro-intestinal Tract. St. Louis, Mo.;Mosby, C.V.

Ulyatt, M.J.;Egan, A.R. (1979) Quantitative digestion of fresh herbage by sheep. 5. The digestion of 4 herbages and prediction of sites of digestion. Journal of Agricultural Science, UK 92,605-616.

Williams, V.J. (1969) Comparative Biochemistry and Physiology 29,865.

Wolff, J.E.;Bergman, E.N.;Williams, H.H. (1972) Net metabolism of plasma amino acids by liver and portal-drained viscera fed sheep. American Journal of Physiology 223,438-446.

UTILISATION OF ROUGHAGES

PROFESSOR J.F.D. GREENHALGH

School of Agriculture, 581 King Street, Aberdeen

INTRODUCTION

In Britain and elsewhere the intensification of meat and milk production has led to an increase in the proportion of concentrates in the diets of ruminants. In general, the increase has been much greater for cattle than for sheep, and recent calculations (Greenhalgh 1977) show that for each 1 kg of carcass meat produced in Britain, cattle consume 4 kg concentrates and sheep only about 2 kg. In the United States (Wedin et al 1975) beef cattle obtained 27% of their energy requirements from concentrates and sheep only 11%. The rest of the sheep's food requirements are met by bulkier feeds, including succulents such as Brassica napus but for the most part consisting of roughages. There is no strict definition of the term roughage and it will be taken here to mean feeds containing more than 20% crude fibre in dry matter. The roughages therefore include most pasture herbage (fresh or conserved), forage crops such as kale (B. oleracea) and many fibrous by-products such as straw, chaff and sugar beet pulp. Grazed and conserved forages will be discussed in detail in later papers; my objective is to discuss the general principles governing roughage utilization by sheep.

STRATEGY OF ROUGHAGE UTILISATION BY SHEEP

The sheep is often considered to be an animal capable of meeting its energy requirements largely from poor quality roughages. An examination of the metabolizable energy (ME) requirements of the twin-bearing ewe and of her potential intake of dry matter (DM) shows that for 8 months of the year she can be fed satisfactorily on diets containing less than 9 MJ ME/kg DM (equivalent to moderate quality hay or poorer). For the remaining 4 months, the final month of pregnancy and the three months of lactation, the ME concentration of the diet must be considerably higher (10.5-12.0 MJ/kg DM) if energy requirements are to be met. In these 4 months the ewe will consume 50% or more of her annual feed requirement. Furthermore, if the lambs are to grow at their maximum rate, they also will require a diet containing a high concentration of ME. Of the total annual feed requirement of the ewe and her twin lambs, about two-thirds must be supplied as diets of high energy concentration (10.5-12.0 MJ ME/kg DM). Roughages which meet this specification are derived mainly from immature pasture herbage, either grazed in situ, ensiled or artificially dried. Silage, for reasons discussed later, is less suitable for sheep than for cattle. Artificially-dried herbage is expensive in terms of both money and fossil fuel energy. The sheep is therefore heavily dependent on high quality grazed pasture at the more critical stages of its production cycle. In practice its dependence on such herbage may be reduced by the use of concentrates. Peak demands for high quality roughage may be also reduced by using the ewe's body reserves, which are later replaced.

FEED INTAKE

Published data on the intake of UK feeds by ruminants have recently been surveyed for the Agricultural Research Council (Greenhalgh & McDonald 1978). When given 'coarse' diets, mainly unprocessed roughages with or without concentrates, sheep ate considerably less per unit of metabolic weight ($W^{0.75}$) than did cattle (Table 1). The difference is presumed to reflect the need of sheep to comminute roughages to a greater degree than cattle before moving them from the rumen to the intestine. Comminution takes time, and delayed passage from the rumen can reduce intake. It is significant that sheep, given coarse diets, increase their intake per unit $W^{0.75}$ as they grow. Thus a 30-kg sheep fed on poor hay (8.5 MJ ME/kg DM) is predicted to eat 42 g DM/kg $W^{0.75}$ per day, whereas a 60 kg sheep should eat 51 g DM/kg $W^{0.75}$. If roughages are ground and pelleted to bring them into the category of 'fine' diets the quantities in which they are eaten per unit $W^{0.75}$ are similar for sheep and cattle (Table 1).

Table 1 Mean values for the dry matter intake of growing sheep and cattle, as estimated from data reported in the literature (Greenhalgh & McDonald 1978)

Species	Sheep		Cattle	
Diet type[1]	Coarse[2]	Fine	Coarse	Fine
No. of diets	107	85	151	159
Metabolizability	0.51	0.53	0.54	0.58
Dry matter intake (g/kg $W^{0.75}$/d)	57.0	90.5	87.2	89.8

[1]Coarse diets: long roughages, with or without concentrates.
Fine diets: all concentrate diets, pelleted roughages, roots.

[2]Silages excluded (see text).

Sheep have particularly low intakes when fed on grass silages. From the ARC survey they were found to eat on average only 46 g DM/kg $W^{0.75}$ per day from silages, or 20% less than from other roughages of comparable ME value. Discrepancies of the same order were found in surveys of Canadian feeds (Heaney et al 1968) and French feeds (Demarquilly 1973). Reducing the particle size of silages by chopping before or after fermentation increased intake by sheep by as much as 90% (Dulphy & Michalet 1975).

The survey data confirmed the now well-known relationship between intake and either the digestibility or the metabolizability (ME/GE) of the diet. For sheep, intake increased by 10 g DM/kg $W^{0.75}$ per day for an increase in metabolizability of 0.1, but the variation around the regression line was considerable (SE of regression coefficient = 1.6). Metabolizability can be improved by raising the metabolizable energy content of the roughage component of diets, e.g. by harvesting at an earlier stage of growth, or by adding concentrates to the diet.

The addition of concentrates to a roughage generally increases total intake but may reduce roughage intake. The effect of concentrate supplements on intake is expressed as the replacement rate, i.e. the reduction in roughage intake per unit intake of concentrate. The replacement rate was found in the ARC survey data to be greater for high-quality than for low-quality basal, i.e. unsupplemented, feeds. For example, for each 1 kg of concentrate DM added to a basal diet containing 7.5 MJ ME/kg DM, the intake of the basal diet fell by 0.33 kg DM, whereas for a basal diet containing 11 MJ ME/kg DM the replacement rate was 0.87 kg/kg. Incidentally, both of these replacement rates were higher than those found for cattle. The effect of concentrate supplements on roughage intake is believed to be due to their effect on the digestion of cell wall components, the supposed chain of events being as follows. The rapid fermentation of starch and sugars leads to a fall in the pH of rumen liquor, the growth of cellulose-fermenting bacteria is reduced, cell wall components are less rapidly and completely digested in the rumen, rate of passage is reduced, and intake falls. Ørskov (1976) has shown that supplements of whole grain have less effect on roughage intake than do rolled or ground grain, presumably because whole grain is fermented less rapidly.

IMPROVEMENT OF THE DIGESTIBITY AND ACCEPTABILITY OF ROUGHAGES FOR SHEEP

The most fundamental means of improving roughages is to select among plant species and cultivars for those that will give the highest intakes. In the case of natural pastures, the sheep itself must exercise selection, but with sown pastures the choice is the farmer's. There is evidence that within the Gramineae some species promote higher intakes than others at the same level of digestibility (e.g. Miles et al 1969). Within a species, some cultivars may prove superior to others although the evidence for this is scanty, with the possible exception of that provided by comparisons of diploid and tetraploid varieties (Thomson 1971). There is more definite evidence that leguminous forages such as clovers, lucerne and sainfoin promote higher intakes and hence higher growth rates in sheep; data from the Grassland Research Institute, Hurley, are summarized in Figure 1. The higher intakes obtained with legumes appear to be associated with a faster rate of digestion (Ulyatt 1970) and they lead to higher growth rates (Figure 2).

The harvesting of roughages at an earlier stage of growth is a well-tried method of improving their quality. Better quality is obtained at the expense of yield per unit area (Table 2), but animal production per unit area is still likely to be higher with better quality roughage. The advantages of using immature herbage are diminished if concentrates make

Figure 1

Fig 1 DRY MATTER INTAKE OF A 40-kg LAMB OFFERED VARIOUS LEGUMES OR RYEGRASS (DATA FROM GRASSLAND RESEARCH INSTITUTE)

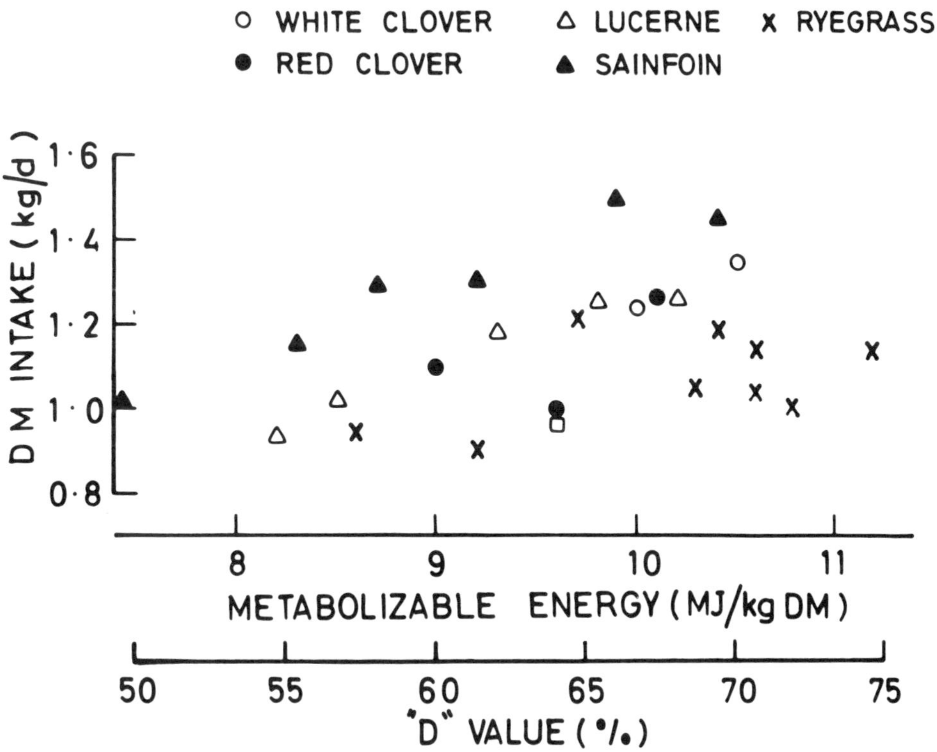

up a large proportion of the diet (Blaxter & Wilson 1963). Systems of sheep production often provide immature herbage for the immature animal (the young lamb, which is most sensitive to roughage quality) and mature herbage for the mature animal (the dry ewe).

Table 2 Nutritive value to sheep of perennial ryegrass (Lolium perenne) harvested three, four or five times per year and artificially dried (Greenhalgh et al 1971)

	Cuts/year			
	3	4	5	SE
Dry matter yield (t/ha)	13.0	11.7	10.7	0.27
Dry matter digestibility (%)	68.4	74.1	74.6	0.37
Dry matter intake (g/kg $W^{0.75}$/d)	51.2	57.6	62.6	1.4
Estimated energy retention				
(MJ/sheep/d)	1.26	2.18	2.75	
(GJ/ha/yr)	16.9	23.6	24.9	

Improved methods of conservation will be covered later but from results discussed earlier it seems that silage intended for sheep should be made from well-comminuted forages.

The effects of the mechanical processes of grinding and pelleting on the utilization of roughages are well known and have been reviewed by Greenhalgh & Wainman (1972) and Osbourn et al (1976). Grinding and pelleting increase intake by about 40% in sheep and reduce digestibility but either have no effect on net energy value or else increase it (Osbourn et al 1976). The improvements are greatest with the poorest parent materials and are again associated with increased rates of passage (e.g. Thomson et al 1969). Pelleted roughages can damage the rumen wall and tend to suppress rumen development (Greenhalgh & Reid 1974). Nevertheless, lambs fed on pelleted artificially-dried herbages will grow as fast as with all-concentrate diets (Greenhalgh et al 1976). The major disadvantage of pelleted roughages is their cost, which is often no less than that of the cereal concentrates they might replace. There are other, less extreme types of mechanical processing which may be beneficial to sheep. Fine chopping of silage has already been mentioned and the screw pressing used to extract protein-rich juice from forages (Houseman & Connell 1976) leaves a "shredded" residue which, despite its diminished

Figure 2

Fig 2 LIVEWEIGHT GAINS OF LAMBS GRAZED OR FED ON LEGUMES OR GRASSES (SUMMARY OF RESULTS FROM BRITAIN, NEW ZEALAND AND AUSTRALIA)

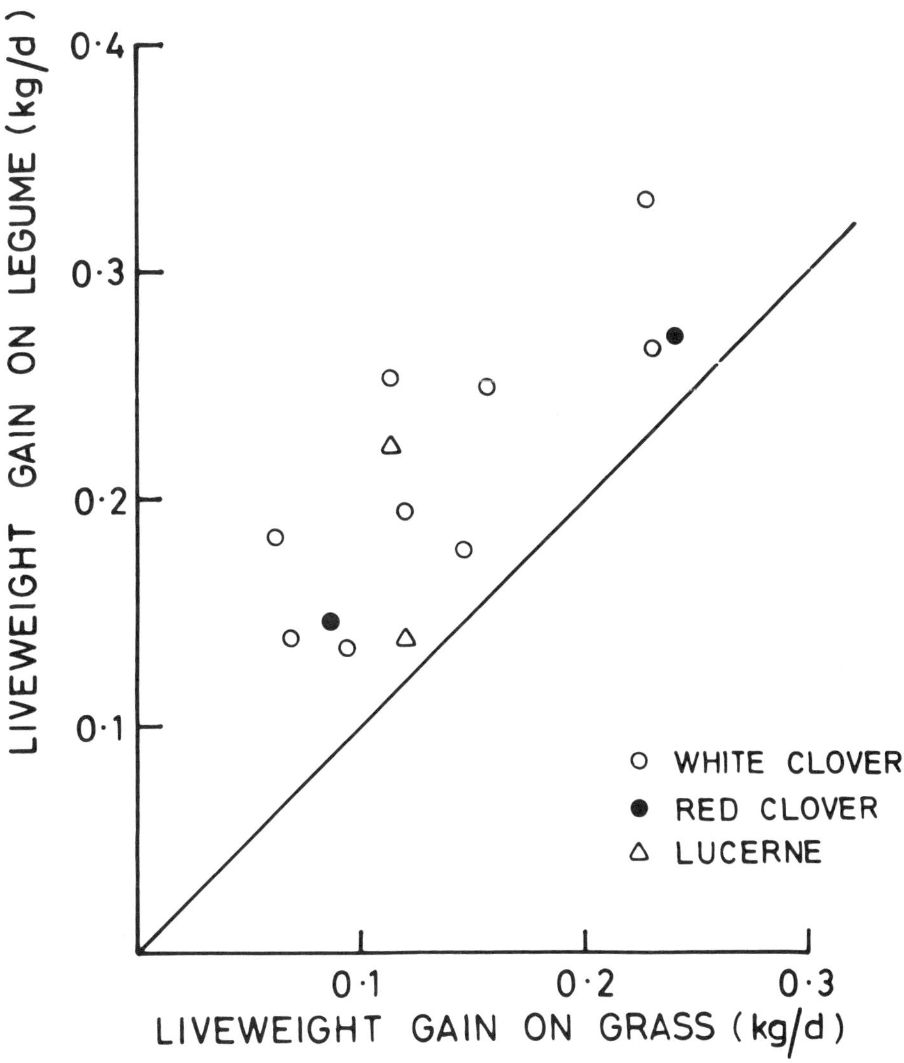

protein and soluble carbohydrate contents, promotes a high intake in sheep. When the juice was re-added to screw-press pulp, the intake of the resulting "reconstituted" grass was 25-45% higher than that of unprocessed grass (Greenhalgh & Reid 1975). In this case, processing did not reduce digestibility, but Thomas et al (1976) found that mincing silage for sheep reduced digestibility but increased intake.

For poorer quality roughages such as cereal straws, chemical processing is considerably more valuable than mechanical treatment because it increases digestibility as well as intake. The alkali treatment of straws has been reviewed by Jackson (1977). The original process involved the soaking of long straw in dilute NaOH solutions (1.5-3.0%) and removal of excess alkali by washing. Modern versions of the process involve the spraying of milled or chopped straw with concentrated NaOH (12-30%), without washing, although excess alkali may be neutralized by mineral or organic acids. Both the soaking and the spraying processes improve the digestibility and intake of straw (Table 3). The spraying process is preferred because it can be mechanised and because it gives rise to a reasonably dry product without effluent.

Table 3 Effects of mechanical and chemical treatment on intake and digestibility of barley straw by sheep (Fernandez Carmona & Greenhalgh 1972)

Mechanical treatment	Chemical treatment	Organic matter: Consumed (g/d)	Organic matter: Digested (%)
Chopping	None	471	45
	Soak in 3% NaOH	870	70
	Spray with 16% NaOH	950	61
Milling	None	696	45
	Spray with 16% NaOH	1050	64

Alkali treatment may also be used for rice hulls (Choung & McManus 1976), maize husks (Rounds et al 1976) and even good quality roughages (Mwakatundu & Owen 1974). Feeds so treated are commonly mixed with concentrates to make complete diets for fattening lambs. In general, alkali applied at 40-80 kg NaOH per tonne of straw increases straw intake in sheep by about 50% and increases digestibility by 10-20%; there is some evidence that higher rates of alkali addition may depress intake (Shin et al 1975). The increases in intake and digestibility lead to substantial improvements in animal growth rate, as illustrated in Table 4.

Figure 3

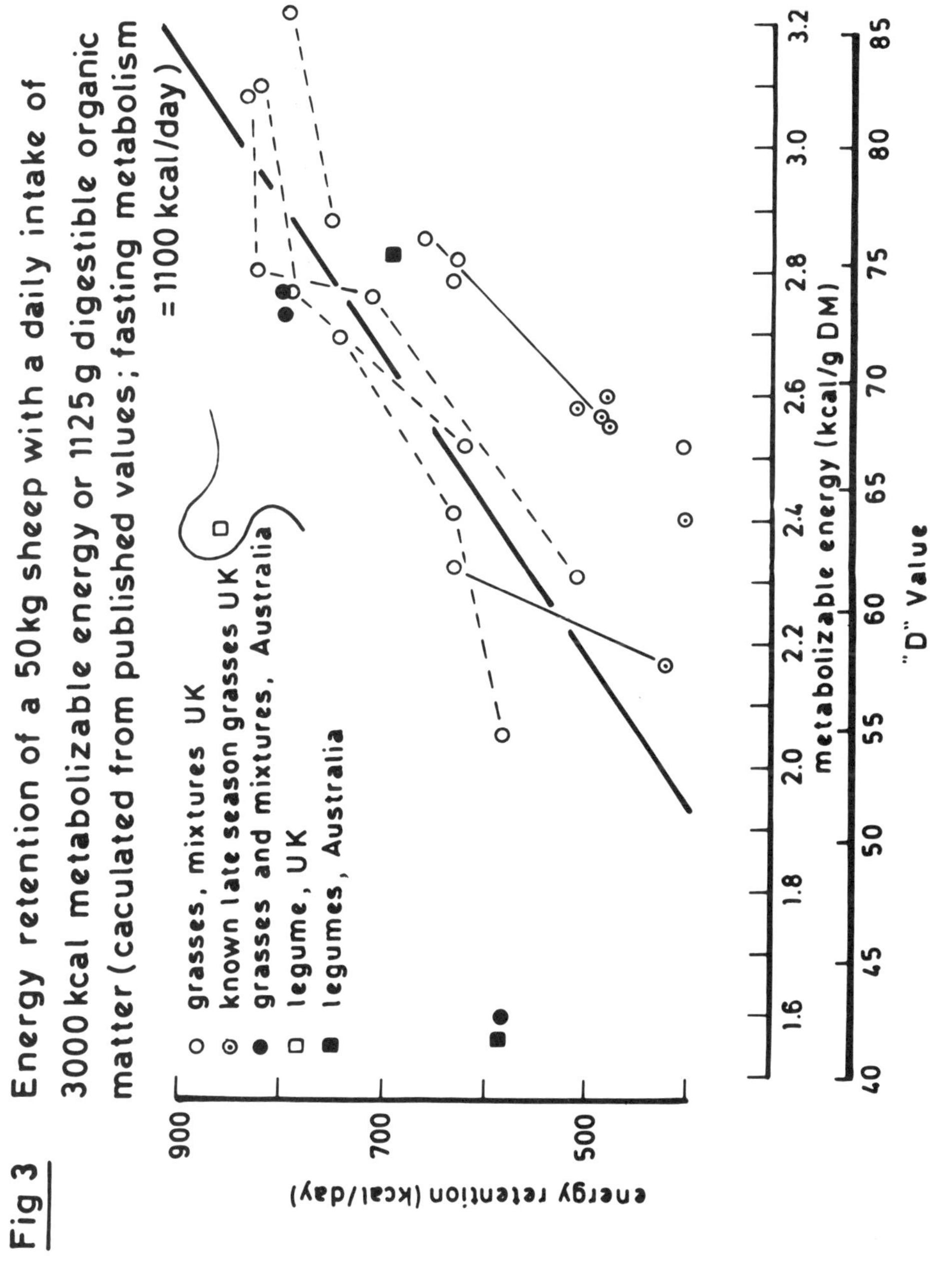
Fig 3
Energy retention of a 50kg sheep with a daily intake of 3000 kcal metabolizable energy or 1125 g digestible organic matter (caculated from published values; fasting metabolism = 1100 kcal/day)
grasses, mixtures UK
known late season grasses UK
grasses and mixtures, Australia
legume, UK
legumes, Australia
energy retention (kcal/day)
900
700
500
1.6
1.8
2.0
2.2
2.4
2.6
2.8
3.0
3.2
metabolizable energy (kcal/g DM)
40
45
50
55
60
65
70
75
80
85
"D" Value

Table 4 Intake and growth rate of lambs fed on an all-concentrate diet or diets containing alkali-treated or untreated straw (from Greenhalgh et al 1976)

Diet	DM intake (g/d)	Empty body gain (g/d)
All concentrate	910	235
Concentrate and alkali-treated straw (1:1)	908	140
Concentrate and untreated straw (1:1)	680	74

Bases other than NaOH may be used for alkali treatment. Calcium hydroxide may partially replace NaOH, if the feed is stored for long enough for this insoluble compound to react (Rounds et al 1976). Anhydrous ammonia is another slow-acting base which has been used successfully to improve digestibility; Coxworth et al (1976) report an average improvement from 50 kg NH_4OH/tonne straw of about 10 percentage units.

Chemical treatment may have a place when sheep are used as scavengers and fed on crop by-products and treated straw has been used successfully as a feed for pregnant ewes (Young & Orr 1976). For such an application, a cheap and simple on-farm method of treatment seems preferable to the more elaborate type of factory operation.

UTILIZATION OF THE ENERGY OF ROUGHAGES

The ME concentration of roughages can vary over a very wide range, from 6 MJ/kg DM for unimproved cereal straws to 11 MJ/kg DM for immature pasture herbage. The efficiency with which ME is used for maintenance and/or production is positively related to its concentration in the feed. In Figure 3, published values for the efficiency of utilization of various herbages have been brought to a common basis by using them to predict the energy retention of a standard sheep given a fixed quantity of ME per day. Figure 3 includes a general prediction line which was derived from the points joined by broken lines. The additional points do not fit this line so well. In particular, grasses harvested late in the grazing season are apparently less well utilized than those harvested earlier and equal in ME concentration. Conversely, two out of three points for legumes suggest unusually efficient utilization of ME. In general, the variation in the relationship between ME concentration and utilization is large enough to make accurate prediction difficult, but it also indicates the scope for improving utilization.

While the factor k_f , efficiency of utilization of ME for growth, is regarded as a particularly important attribute of a feed or diet, only about 20% of the total feed required by ewes and lambs is used directly for growth in conventional systems of sheep production. The greatest requirement is for maintenance of both the ewe and the lamb for which the efficiency of utilization of ME varies relatively little with ME concentration. The remaining requirements are for lactation and for foetal growth; little is known of the effect of dietary energy concentration on the utilization of ME for these purposes.

CONCLUSIONS

The subject of roughage utilization by sheep presents a number of paradoxes. The first is that although sheep are more dependent on roughages than are cattle, they appear to be less well adapted to such feeds. Intake relative to metabolic body weight is lower for sheep, especially with silages, and concentrates depress roughage intake more in sheep than in cattle. There is evidence, much of it circumstantial, that the problems encountered by sheep in ingesting roughages are attributable to the rate at which these animals can comminute feed by physical and chemical means to a particle size appropriate to the lower gut.

The second paradox is that, although sheep benefit more than cattle from physical and chemical processing of the roughages in their diet, their dependence on grazed pasture herbage gives them little opportunity to profit from such methods of processing. For sheep, improvements in the nutritive value of the roughages they consume will generally need to come from the less spectacular procedures of modifying the sward and harvesting it at an earlier stage of growth.

Finally, it is important to remember that, although the ewe can subsist for two-thirds of the year on low quality roughages, for the remaining one-third she and her offspring require roughages of the highest quality. Although the feed used directly for lamb growth represents a relatively small proportion of the total requirements of the ewe and her lambs, the quality of that feed is often a crucial factor in fat lamb production.

REFERENCES

Blaxter, K.L.;Wilson, R.S. (1963) The assessment of a crop husbandry technique in terms of animal production. Animal Production 5,27-42.

Choung, C.C.;McManus, W.R. (1976) Studies on forage cell walls. 3. Effects of feeding alkali-treated rice hulls to sheep. Journal of Agricultural Science, UK 86,517-530.

Coxworth, E.;Kernan, J.;Nicholson, H.;Chaplin,R. (1976) Improving the feeding value of straw for ruminant animals. Unpublished mimeograph. Saskatoon, Canada; Saskatchewan Research Council, 23pp.

Demarquilly, C. (1973) Composition chimique, cáracteristiques fermentaires, digestibilité et quantité ingérée des ensilages de fourrages: modifications par rapport au fourrage vert initial. Annales de Zootechnie 22,1-35.

Dulphy, J.P.;Michalet, B. (1975) Influence compáree de la machine de récolte sur les quantités d'ensilage ingérées par des génisses et des moutons. Annales de Zootechnie 24,757-763.

Fernández Carmona, J.;Greenhalgh, J.F.D. (1972) The digestibility and acceptability to sheep of chopped or milled barley straw soaked or sprayed with alkali. Journal of Agricultural Science, UK 78,477-485.

Greenhalgh, J.F.D. (1975) Factors limiting animal production from grazed pasture. Journal of the British Grassland Society 30,153-160.

Greenhalgh, J.F.D. (1977) Present and future use of cereals as feeds for livestock. Proceedings of the Nutrition Society 36,159-167.

Greenhalgh, J.F.D.;McDonald, I. (1978) The metabolizable energy system in practice: predicting feed intake. Animal Production 26,350.

Greenhalgh, J.F.D.;Orskov, E.R.;Fraser, C. (1976) Pelleted herbages for intensive lamb production. Animal Production 22,148-149.

Greenhalgh, J.F.D.;Pirie, R.;Reid, G.W. (1976) Alkali-treated barley straw in complete diets for lambs and dairy cows. Animal Production 22,159.

Greenhalgh, J.F.D.;Reid, G.W. (1974) Long- and short-term effects on intake of pelleting a roughage for sheep. Animal Production 19,77-86.

Greenhalgh, J.F.D.;Reid, G.W. (1975) Mechanical processing of wet roughage. Proceedings of the Nutrition Society 34,74A.

Greenhalgh, J.F.D.;Reid, G.W.;Aitken, J.N. (1971) Production and consumption by sheep of dried grass prepared by various cutting sequences. In: Proceedings of the 10th International Congress on Animal Production, Versailles (Section VII).

Greenhalgh, J.F.D.;Wainman, F.W. (1972) The nutritive value of processed roughages for fattening cattle and sheep. In: Proceedings of the British Society of Animal Production. Edinburgh, UK; Longman Group Ltd., pp61-72.

Heaney, D.P.;Pritchard, G.I.;Pigden, W.J. (1968) Variability in ad libitum forage intakes by sheep. Journal of Animal Science 27,159-164.

Houseman, R.A.;Connell, J. (1976) The utilization of the products of green-crop fractionation by pigs and ruminants. Proceedings of the Nutrition Society 35, 213-220.

Jackson, M.G. (1977) The alkali treatment of straws. Animal Feed Science and Technology 2, 105-130.

Miles, D.G.;Walters, R.J.K.;Evans, E.M. (1969) Dry-matter intake and live-weight gain of cattle and sheep offered different grass varieties with and without clover. Animal Production 11,19-28.

Mwakatundu, A.G.K.;Owen, E. (1974) In vitro digestibility of sodium hydroxide-treated grass harvested at different stages of growth. East African Agricultural and Forestry Journal 40,1-10.

Ørskov, E.R. (1976) The effect of processing on digestion and utilization of cereals by ruminants. Proceedings of the Nutrition Society 35,245-252.

Osbourn, D.F.;Beever, D.E.;Thomson, D.J. (1976) The influence of physical processing on the intake, digestion and utilization of dried herbage. Proceedings of the Nutrition Society 35,191-200.

Rounds, W.;Klopfenstein, T.;Waller, J.;Messersmith, T. (1976) Influence of alkali treatments of corn cobs on in vitro dry matter disappearance and lamb performance. Journal of Animal Science 43,478-482.

Shin, H.T.;Garrigus, U.S.;Owens, F.N. (1975) NaOH treated wheat straw rations for sheep. Journal of Animal Science 41,417.

Thomas, P.C.;Kelly, N.C.;Wait, M.K. (1976) The effect of physical form of a silage on its voluntary consumption and digestibility by sheep. Journal of the British Grassland Society 31,19-22.

Thomson, D.J. (1971) The voluntary intake of diploid (S22) and tetraploid (Tetila tetrone) Italian ryegrass, and white clover by sheep. Journal of the British Grassland Society 26,149-155.

Thomson, D.J.;Beever, D.E.;Silva, J.F.C. da;Armstrong, D.G. (1969) Sites of digestion in sheep of a dried lucerne fed in 3 different physical forms. Proceedings of the Nutrition Society 28,24A-25A.

Ulyatt, M.J. (1970) Factors contributing to differences in the quality of short-rotation ryegrass, perennial ryegrass, and white clover. In: Proceedings of the 11th International Grassland Congress, Surfers' Paradise, p709-713.

Wedin, W.F.;Hodgson, H.J.;Jacobson, N.L. (1975) Utilizing plant and animal resources in producing human food. Journal of Animal Science 41,667-686.

Young, N.E.;Orr, R.J. (1976) The effect of treated straw on the performance of pregnant ewes. Annual Report, Grassland Research Institute, Hurley p131.

UTILIZATION OF CEREALS FOR SHEEP AND IMPLICATIONS OF PROCESSING

DR E.R. ØRSKOV

Rowett Research Institute, Bucksburn, Aberdeen

Pigs and other monogastric animals utilize cereal based diets theoretically better than ruminants because they do not incur fermentation losses of energy. The food conversion ratio however depends largely on the level of feeding obtained and on the composition of the live weight gain. The effect of level of feeding is illustrated in Table 1 calculated from data of Andrews & Ørskov (1970).

Table 1 Effect of growth rate on the dry matter required per kg gain during the growth period from 15 to 40 kg weight of Suffolk cross lambs

Growth rate g/d	Males	Females
50	10.1	8.4
100	5.9	5.5
200	3.8	4.0
300	3.1	3.5
400	2.7	3.3
500	2.5	3.1

It can be seen that the higher the level of feeding, expressed here as growth rate, the greater is the efficiency of conversion of cereal dry matter to live weight gain. It can also be seen that male lambs convert food better than female lambs. The food conversion ratios obtained during the fattening period of lambs is very similar to that normally obtained with pigs.

With increasing live weight the composition of the gain will alter to contain more fat and less protein and water and as a result the efficiency of food conversion will decrease. It was found by Ørskov et al (1973) that for each 10 kg increase in life weight the amount of cereal dry matter required per kg gain increased by 0.35 kg.

For the early weaned lamb, where the rumen has not reached mature proportions, the rate of digestion and the rate of removal of ingested residues is important, as was illustrated by results of Andrews et al (1969) when they compared a barley with an oats based diet. The fibre content and the particle size of the oat husk was such that lambs at 15 to 20 kg live weight consumed less oats than barley. At 40 kg live weight however they were able to compensate for the higher fibre content in oats by eating more, thus achieving equal intakes of digestible dry matter.

While it was observed that early weaned lambs (Andrews & Ørskov 1970, Ørskov et al 1971) utilized cereal based diets extremely well, there were two problems; (i) the carcasses had very soft subcutaneous fat which was undesirable for the meat trade, and (ii) all animals had some degree of rumenitis (see Fell et al 1972). The latter problem was not very serious as it was difficult to demonstrate an influence on food utilization but, in any development of new systems of feeding, it is of course desirable to avoid any pathological changes which can be demonstrated to be directly caused by it.

The cause of the soft fat was found to be an unusually large proportion of mono-methyl, branched chain fatty acid in the subcutaneous fat while there was little or no difference in the content of unsaturated fatty acids (Duncan et al 1972). The physical characteristics of the branched chain acids were responsible for the soft fat. There was also an unusually large proportion of odd-numbered fatty acids, which indicated the cause of the branched chain fatty acids, namely the very high proportion of propionic acid in the rumen fluid. It appeared that both propionic acid and one of its intermediary metabolic products, methylmalonic acid, were participating in fatty acid synthesis in the subcutaneous adipose tissue. Normally propionic acid is metabolized via succinc acid in the liver and only trace amounts appear in the peripheral circulation (see Garton 1975). It seemed that the quantity of propionic acid produced exceeded the capacity of the liver to metabolize it. It was first considered that the problem might be due to a deficiency of vitamin B_{12}, which is the co-factor in the conversion of methylmalnic acid to succinc acid but neither addition of cobalt nor injections of vitamin B_{12} had any effect on the composition of the fat (Duncan et al 1974).

EFFECT OF CEREAL PROCESSING ON CARCASS FAT AND RUMENITIS

It seemed that the only solution to the problem of soft fat was a system which gave rise to a lower proportion of propionic acid in the rumen. The effect of cereal processing was investigated as a possible means of changing the type of rumen fermentation. This proved a very fruitful avenue to pursue as not only did processing of cereals seem superfluous as far as digestibility was concerned but neither did it reduce food utilization (Table 2).

It did however cause a very substantial change in the type of rumen fermentation (Table 3), sufficient to overcome the problem of soft subcutaneous fat (Ørskov et al 1974a), and it also completely eliminated any problems of rumenitis (Ørskov 1973), which was probably due mainly to the changes in rumen pH from about 5.2 to 6.2. This whole development was a great encouragement to farmers involved in intensive lamb fattening systems as it meant that diets could be made cheaper and at the same time problems of carcass quality and health were eliminated.

Table 2 Effect of processing of different cereals on food utilization by lambs

Treatment	Initial weight kg	Final weight kg	Live weight gain g/day	Killing-out percentage %	Food consumed per kg liveweight gain kg	Organic matter digestibility %
Ground barley pelleted	15.0	35.3	347	49.9	2.79	77.2
Whole barley loose	15.3	35.2	340	47.8	2.75	81.1
Ground maize pelleted	14.9	35.9	346	50.9	2.62	82.1
Whole maize loose	15.0	36.1	345	48.6	2.52	84.3
Ground oats pelleted	14.8	35.1	238	42.2	3.33	67.5
Whole oats loose	14.9	34.9	241	42.0	3.07	69.9
Ground wheat pelleted	15.0	35.2	323	50.1	2.56	86.6
Whole wheat loose	15.0	35.2	303	50.0	2.97	82.7
SE of means	-	-	15	0.8	0.11	1.2

(from Ørskov et al 1974b)

Table 3 Effect of processing of different cereals in lamb diets on rumen pH and proportion of acetic and propionic acid

Cereal	Form	Rumen pH	Molar proportion of: Acetic acid	Propionic acid
Barley	Whole loose	6.4	52.5	30.1
Barley	Ground pelleted	5.4	45.0	45.3
Maize	Whole loose	6.1	47.2	38.7
Maize	Ground pelleted	5.2	41.3	43.2
Oats	Whole loose	6.7	65.0	18.6
Oats	Ground pelleted	6.1	53.2	37.5
Wheat	Whole loose	5.9	52.3	32.2
Wheat	Ground pelleted	5.0	34.2	42.6
SE of mean		0.14	2.4	3.2

(from Ørskov et al 1974a)

CEREAL PROCESSING AND DIGESTION OF ROUGHAGE

The significant alterations in rumen pH as a result of feeding whole rather than processed grain led us to investigate whether whole barley was also a more suitable supplement to roughage based diets, as one of the problems of supplementary roughage diets with cereals is the inhibition of the rate of cellulose digestion. It was found that even with restricted feeding the rumen pH remained quite stable when whole rather than processed grain was fed (Ørskov & Fraser 1975). Secondly, it was demonstrated that dried grass incubated in nylon bags in the rumen disappeared much more rapidly when the sheep were fed on whole rather than processed grains. It was finally demonstrated that the voluntary consumption of dried grass was considerably greater when whole rather than processed barley was used as a supplement at a high level of feeding. It can be seen in Table 4 that increasing the supplement of pelleted barley from 309 to 608 led to an equivalent decrease in intake of hay so that a substitution effect was the result of the intended supplementation. With whole grains there was also a decreased intake, but much less than with processed barley.

It has to be pointed out that when whole grains are fed some of the grain will appear in the faeces particularly when it is supplementing roughage diets. The reduction in digestibility of the grain when used as supplement is very low, in the region of 5%, and can be tolerated due to the benefits to roughage digestion and savings in cost of processing. It is however serious insofar as farmers everywhere appear to conclude that if a few whole grains can be found in the faeces then all the grain has passed through undigested. It is up to the advisor to convince the farmer that this is not important or to demonstrate by experiment that it is of little or no consequence.

Table 4 The effect of giving supplements based on whole or pelleted barley on voluntary intake of dried grass and on total intake of dry matter (DM)

Supplement		Intake of dried grass	Total daily intake (including supplement)
Form	g DM/day	g DM/day	g DM/day
	0	991	991
Whole	310	790	1100
barley	616	642	1258
Pelleted	309	820	1129
barley	608	472	1080
SE of treatment differences		45	44

(from Ørskov & Fraser 1975)

PROCESSING OF GRAIN AND FOOD PREPARATION

In general cereals are deficient in some nutrients required by the animal. For early weaned lambs and for lactating ewes, addition of dietary protein is required and for other types of production, feeding of store lambs, pregnant ewes, etc, it may be desirable to use urea. Also minerals and vitamins are required in addition to that contained in cereals.

ADDITION OF PROTEIN SUPPLEMENTS

When protein supplements have to be included in diets of whole grain we have generally used a small pellet, about 4.4 mm, which mixes easily with whole grain. This pellet also contains the required minerals and vitamins. The inclusion rate of the pellet depends upon the type of protein and the requirements of the animals.

Two examples are given below in Table 5, intended for early weaned lambs. Pelleting of the supplement is not required but, if it is added as a loose mix, the minerals should be mixed with the protein supplement separately before it is mixed with the grain.

Table 5 Protein, mineral and vitamin supplements for lambs for inclusion in whole grain diets

For 10% inclusion		For 15% inclusion	
Fish meal	80	Soya bean meal	85
Ground limestone	15	Ground limestone	10
Molasses	4	Sodium chloride	4
[1]Trace mineral/ vitamin	1	[1]Trace mineral/ vitamin	1

[1]to give per kg of total diet: 150 mg $ZnSO_4 7H_2O$, 80 mg $MnSO_4 H_2O$, 200 mg MgO, 5 mg $CoSO_4 7H_2O$, 1 mg KIO_3, 5000 i.u. Vitamin A, 1000 i.u. Vitamin D_3 and 20 i.u. Vitamin E.

As mentioned earlier, for some systems of production it is advisable to use urea, namely in circumstances where the protein needs of the animals are being met by microbial protein and where the basal feed itself contains insufficient N to meet the microbial need. This is the case for almost all cereals. Addition of urea to whole grain was thought to be a problem but it was solved when it was observed that a solution of urea could be completely absorbed into the grain. Urea is highly soluble; it will dissolve to give a urea: water ratio of up to 60:40, depending on the temperature of the water used for dissolution. Not only is this a cheap and easy method of incorporating urea in grain diets but it also liberates ammonia at a slower rate (Ørskov et al 1974c).

Following this observation minerals and vitamins have been added using the same method of incorporation. It is now possible to use soluble forms of minerals to make up for the deficiencies of a diet including roughages to match the requirements of the animal. One mixture can only be used in a few instances. For complete cereal feeding to sheep, the mixture given in Table 6 is adequate.

Table 6 Appropriate supplements for whole grain diets for late weaned and store lambs. (Added as solutions)

Urea	1.2%
CaCl	1.2%
NaCl	0.3%
Na_2SO_4	0.4%

Trace minerals to be added per kg of diet but dissolved before mixing are: 1000 mg $MgSO_4 7H_2O$, 150 mg $ZnSO_4 7H_2O$, 50 mg $MnSO_4 H_2O$, 0.4 mg $CoSO_4 7H_2O$, 1 mg $K1O_3$.

Vitamins to be added per kg of diet: 5000 i.u. Vitamin A, 1000 i.u. Vitamin D and 20 i.u. Vitamin E.

It can be seen from the above discussion that feeding of cereals to sheep can and should be a simple operation. Money spent on processing or even on methods of urea incorporation is not returned by a greater food utilization. In fact the reverse occurs, since processing of cereals in intensive systems can give rise to problems of carcass quality and rumenitis and, when given at a high level as a supplement to roughage based diets, it can inhibit the rate of cellulose digestion, which in turn results in inefficient digestion and reduced consumption of the roughage.

REFERENCES

Andrews, R.P.;Kay, M.;Ørskov, E.R. (1969) The effect of different dietary energy concentrations on the voluntary intake and growth of intensively-fed lambs. Animal Production 11,173-185.

Andrews, R.P.;Ørskov, E.R. (1970) The nutrition of the early weaned lamb. 1. The influence of protein concentration and feeding level on rate of gain in body weight. Journal of Agricultural Science, UK 75,11-18.

Duncan, W.R.H.;Ørskov, E.R.;Fraser, C.;Garton, G.A. (1974) Effect of processing of dietary barley and of supplementary cobalt and cyanocobalamin on the fatty acid composition of lamb triglycerides, with special reference to branched-chain components. British Journal of Nutrition 32,71-75.

Duncan, W.R.H.;Ørskov, E.R.;Garton, G.A. (1972) Fatty acid composition of triglycerides of lambs fed on barley-based diets. Proceedings of the Nutrition Society 31,19A-20A.

Fell, B.F.;Kay, M.;Ørskov, E.R.;Boyne, R.;Walker, J.T. (1972) The role of ingested animal hairs and plant spicules in the pathogenesis of rumenitis. Research in Veterinary Science 13,30-36.

Garton, G.A. (1976) The occurrence and origin of branched-chain fatty acids in bacterial, avian and mammalian lipids. Annual Report, Rowett Research Institute 31,124-135.

Ørskov, E.R. (1973) The effect of not processing barley on rumenitis in sheep. Research in Veterinary Science 14,110-112.

Ørskov, E.R. (1976) The effect of processing on digestion and utilization of cereals by ruminants. Proceedings of the Nutrition Society 35,245-252.

Ørskov, E.R. (1977) Nitrogen digestion and utilization by young and lactating ruminants. World Review of Nutrition and Dietetics 26,225-257.

Ørskov, E.R.;Fraser, C. (1975) The effects of processing of barley-based supplements on rumen pH, rate of digestion and voluntary intake of dried grass in sheep. British Journal of Nutrition 34,493-500.

Ørskov, E.R.;Fraser, C.;Gill, J.C. (1973) A note on the effect of time of weaning and weight at slaughter on feed utilization of intensively fed lambs. Animal Production 16,311-314.

Ørskov, E.R.;Fraser, C.;Gill, J.C.;Corse, E.L. (1971) The effect in an intensive production system of type of cereal and time of weaning on the performance of lambs. Animal Production 31,485-492.

Ørskov, E.R.;Fraser, C.;Gordon, J.G. (1974) Effect of processing of cereals on rumen fermentation, digestibility, rumination time, and firmness of subcutaneous fat in lambs. British Journal of Nutrition 32,59-69.

Ørskov, E.R.;Fraser, C.;McHattie, I. (1974) Cereal processing and food utilization by sheep. 2. A note on the effect of feeding unprocessed barley, maize, oats and wheat on food utilization by early-weaned lambs. Animal Production 18,85-88.

Ørskov, E.R.;Smart, R.I.;Mehrez, A.Z. (1974) A method of including urea in whole grains. Journal of Agricultural Science, UK 83,299-302.

THE NUTRITION OF THE PREGNANT EWE

DR A.J.F. RUSSEL

Hill Farming Research Organisation, Bush Estate, Penicuik, Midlothian

The principal reason for wishing to exercise control over the nutrition of the pregnant ewe is an economic one. In some instances it may be possible to improve production by increasing the level of feeding at certain stages of pregnancy but, if the cost of the additional food input is greater than the return from the increase in output, the change in nutrition will not be economically justified. Thus a consideration of the nutrition of the pregnant ewe must be concerned not only with the definition of nutrient requirements but also with an assessment of the effects on production of levels of feeding which are lower than these full requirements. I wish to explore what constitutes "acceptable levels of feeding", which in most cases will be less than requirements in the true sense of the term but which will nonetheless ensure a satisfactory level of production, bearing in mind that it is impossible to state precisely levels of nutrition which will apply to all sheep in all situations.

The increases in weight of the foetus, the uterus, its membranes and the intra-uterine fluids during pregnancy are shown in Figure 1. The patterns of increase in weight of each of the components are quite distinct; the placenta develops rapidly in the earlier stages of pregnancy, whereas the foetus grows most quickly during the final few weeks. It is convenient from a nutritional standpoint to divide the period of gestation as follows; (i) the first month (early pregnancy), (ii) the second and third months (mid-pregnancy) and (iii) the fourth and fifth months (late pregnancy).

THE FIRST MONTH (EARLY PREGNANCY)

In an earlier paper, which reviewed factors affecting fertility and reproduction and dealt with the role of nutrition as it affects embryonic and foetal loss, the need for ewes to be in good body condition at mating was stressed. A description of the system of body condition scoring is given in Appendix 1. Total losses are estimated to vary from about 10 to 40% of fertilised ova and it is generally accepted that the highest proportion of these losses occurs during the first month of pregnancy.

Evidence indicates that losses at this stage are likely to be affected by both body condition and current nutrition. In general, ewes in good body condition (condition score 3) at mating shed more ova and have lower rates of embryo mortality than ewes in poor condition (e.g. condition score 1½). There is thus a dual benefit to be gained by ensuring that ewes are in a proper body condition at mating.

The effects of current nutrition on embryonic mortality are less clear. There is limited evidence that high levels of feeding during the first month after mating can lead to increased losses. It has also been shown that embryo survival can be adversely affected by severe undernourishment even of very brief duration. The consensus of evidence suggests however that a sustained moderate degree of undernourishment resulting in a 3 to 4% loss of live weight over the first 30 days of pregnancy is unlikely to have any significantly harmful effect in ewes which were in good body condition at mating.

In most flocks the majority of matings take place over 2 to 3 weeks and it is generally considered to be good practice to avoid under-nourishment during this time and also in the subsequent 2 to 3 weeks. The levels of energy and protein intake required for this maintenance level of feeding are shown in Figure 2. In the lowground situation, where mating usually takes place in October, these levels of energy and protein intake should be readily available from pasture and even in hill flocks, where mating is later and herbage quality poorer, it should be possible to provide these levels of intake by using autumn-saved pasture i.e. pasture from which all stock have been excluded during the late summer and early autumn and to which the ewes are allowed access before mating. If, for any reason, these levels of intake cannot be supplied from pasture it will probably be uneconomic to provide supplementary feeding other than hay to hill sheep in the event of snow cover, which can occur at that time of year.

THE SECOND AND THIRD MONTHS (MID-PREGNANCY)

During the second and third months of gestation the various tissues and organs of the foetus are undergoing rapid development, although the absolute weight of the foetus is still relatively low. By the end of the third month each foetus will weigh about 600 g, or approximately 14% of its birth weight. By contrast, the placenta, through which the foetus receives nutrients from the ewe, makes its most rapid growth during mid-pregnancy, generally attaining at least 95% of its final weight of about 500 g per foetus by the 90th day of gestation.

The literature does not present a clear picture of what constitutes an optimum level of nutrition at this time. The majority of experimental results on the effects of mid-pregnancy nutrition on lamb birth weight indicate that nutrition during this stage is relatively unimportant. Others have demonstrated reductions in lamb birth weight due to low levels of mid-pregnancy feeding and recently there have been reports of reductions in lamb birth weight attributable to high levels of nutrition.

In attempting to resolve these apparently anomalous results it is necessary to take into account a number of factors, including the age and condition of the ewe and the level of late pregnancy nutrition. It is clear that severe undernourishment during mid-pregnancy can reduce the weights of foetuses and of the placental material at 90 days. Except in situations where the undernourishment is exceptionally severe, adult ewes are generally able to compensate during late pregnancy for the effects of low levels of mid-pregnancy nutrition. Compensatory growth of cotyledonary tissue, and particularly of that part associated with the foetus, is known to occur during late pregnancy, but this will of course require a relatively high level of nutrition at that later stage.

Figure 1

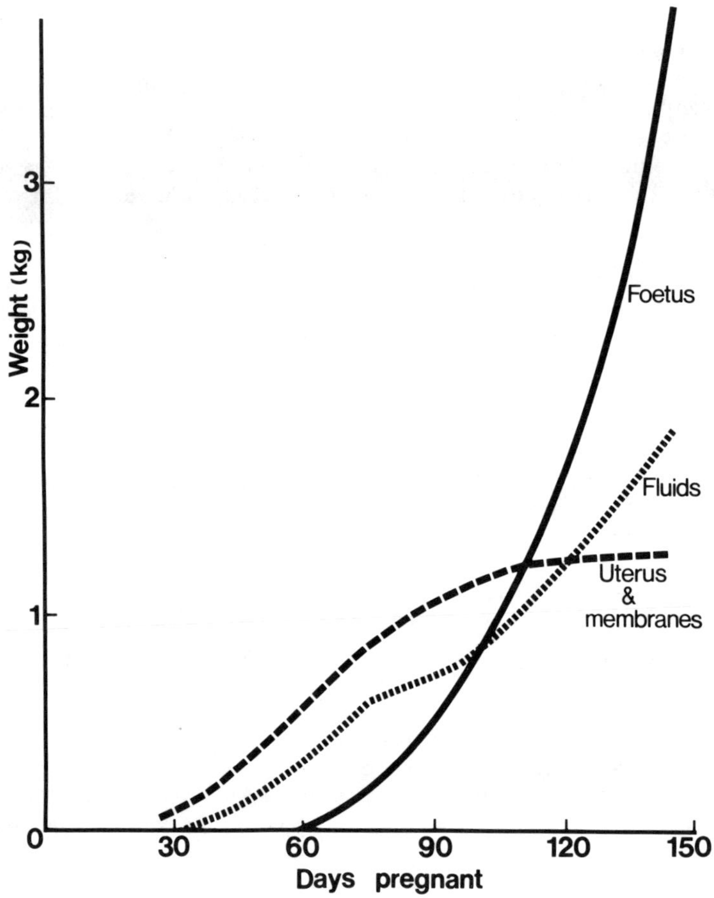

Weight (kg)
3
2
1
0
30
60
90
120
150
Days pregnant
Foetus
Fluids
Uterus
&
membranes

Figure 2

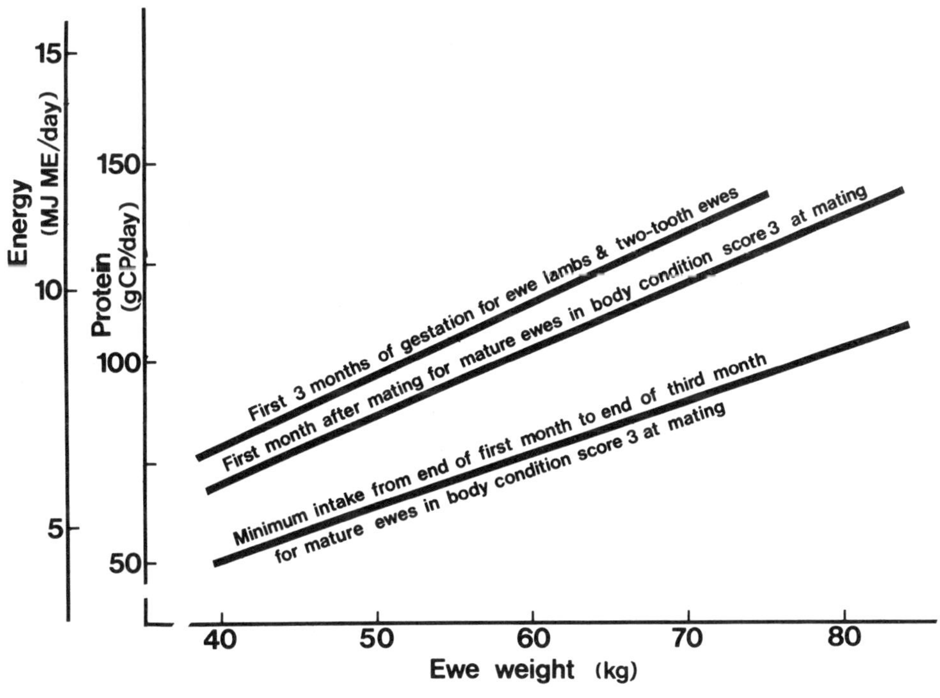

Energy (MJ ME/day)
15
10
5
Protein (gCP/day)
150
100
50
First 3 months of gestation for ewe lambs & two-tooth ewes
First month after mating for mature ewes in body condition score 3 at mating
Minimum intake from end of first month to end of third month for mature ewes in body condition score 3 at mating
40
50
60
70
80
Ewe weight (kg)

Although the evidence is not conclusive, young sheep, lambing for the first time at either one or two years of age, may be less well able to compensate than older ewes for a low level of nutrition during mid-pregnancy. Reductions in the birth weight of lambs from young ewes on a low level of mid-pregnancy nutrition are most likely to occur in animals underweight in relation to their age and in a relatively low body condition at mating. In young ewes which are well grown and in good body condition at first mating, reductions in lamb birth weights resulting from a high level of mid-pregnancy nutrition have been demonstrated. In work at the Rowett Research Institute, reductions in lamb birth weight of over 40% attributable to gross overnutrition of pregnant Finn x Dorset ewe lambs have been noted.

There is clearly an interaction between condition at mating and level of mid-pregnancy nutrition in young ewes and its effects are illustrated by the results of a recent experiment in the Hill Farming Research Organisation. Two groups of Scottish Blackface ewes, mated at approximately 18 months of age, were selected from two flocks in which the ewes, and particularly the younger ones, were of contrasting live weight, size and condition. One half of each group was fed to produce a weight loss of some 3 kg during mid-pregnancy (the loss of net maternal weight was calculated to be between 5 and 6 kg); the other half of each group was fed to gain 3 kg during the same period i.e. to maintain net maternal weight. In the ewes which were smaller and in poorer condition at mating the higher level of nutrition resulted in a small increase in lamb birth weight (7.3%), whereas the same level of nutrition (adjusted for differences in initial weight) caused a reduction (15.5%) in the birth weights of the lambs from the better grown and fatter ewes. It is unlikely that this interaction between body condition and level of mid-pregnancy nutrition is unique to ewes in their first pregnancy but it appears probable that animals of this age are more sensitive to the effects of both undernutrition and overnutrition during mid-pregnancy.

One must therefore take account of the age and body condition of the ewe at mating in making any recommendation regarding nutrition during mid-pregnancy. Adult ewes which are in good body condition (condition score 3) at mating are likely to be able to withstand a moderate degree of undernourishment during the second and third months of pregnancy, without this unduly affecting their subsequent performance. Indeed undernourishment at this period would appear not only to be acceptable but also desirable for this class of animal. In attempting to express the appropriate level of nutrition in terms of live-weight change, it must be remembered that by the end of the third month of pregnancy the uterus and its contents weigh between 3 and 5 kg depending on the size of the ewe and the number of foetuses. A lowground ewe, carrying twin lambs and apparently maintaining body weight during the first three months of pregnancy, will have drawn on her own body tissue to the extent of more than 4 kg. An acceptable apparent body-weight loss during mid-pregnancy in an adult lowground ewe in good condition at mating is about 5% of mating weight and should not be more than 8%. This latter figure would be an acceptable norm for most hill ewes where the incidence of twinning is lower than in lowground flocks. The lower (sub-maintenance) level of feeding illustrated in Figure 2 is calculated on the basis of an apparent 5% weight loss.

In hill flocks and October-mated lowground flocks, intakes from grazing will in most cases result in weight changes which are within the limits suggested above and supplementary feeding will be required only where stocking rates are exceptionally high or in snow. As already suggested, the nutrition during mid-pregnancy of animals lambing for the first time, either as ewe lambs or two-tooths, requires special consideration. These animals generally have not attained their mature size or weight and there is consequently a temptation to treat them more generously. This would appear to be justified in conditions where the juvenile nutrition is such that the ewe enters her first pregnancy in a relatively poor body condition and at a low weight for her age. The level of feeding in Figure 2 which allows animals to gain about 5% of their mating weight during the first three months of pregnancy would be indicated for this class of animal. With well grown animals in good body condition at first mating, a level of mid-pregnancy nutrition which does not result in a loss of weight or condition is likely to be detrimental and will in all probability lead to a reduction in lamb birth weight. Such animals should be fed according to the recommendations made for adult ewes and should lose some 5% of their mating weight during the mid-pregnancy period.

THE FOURTH AND FIFTH MONTHS (LATE PREGNANCY)

By the end of the third month of pregnancy, the uterus and its contents weigh between 3 and 5 kg, depending on the size of the ewe and the number of foetuses. This is equivalent to about one-third to one-half of the final uterine weight immediately before lambing. By this stage, some eight weeks before lambing, the placenta is fully developed, the major foetal tissues and organs have been formed and all that remains is an increase in size and weight. Approximately 70% of foetal growth takes place during the final six weeks of pregnancy.

The growth and development of the highly specialised tissues of the foetus are expensive in terms of nutrients and require considerably more food per unit of weight gain than is the case in an adult animal. Estimates of the efficiency of utilisation of metabolizable energy for foetal growth range from about 5 to 22%, compared with values of 40 to 60% in the growing or fattening animal. This low efficiency of conversion of nutrients to foetal tissues and the very rapid increase in foetal weight in late pregnancy result in a considerable increase in the ewe's requirements for nutrients and particularly for energy. In addition to the nutrients needed for foetal growth and develpment plus those required to support the other products of conception, nutrients are needed for the satisfactory development of mammary tissue and of course to meet the maintenance requirements of the ewe herself.

The energy requirements of the foetus in late pregnancy are of the order of 1.5 MJ ME/kg foetus/day. This means that by the end of pregnancy a ewe carrying a single foetus requires approximately twice as much food as is needed to prevent loss of weight in a non-pregnant ewe. The requirements of a ewe carrying twin foetuses are 2.5 to 3 times that of a non-pregnant animal (Figure 3). It must be emphasised that these very high nutrient requirements exist for only a relatively short period of

Figure 3

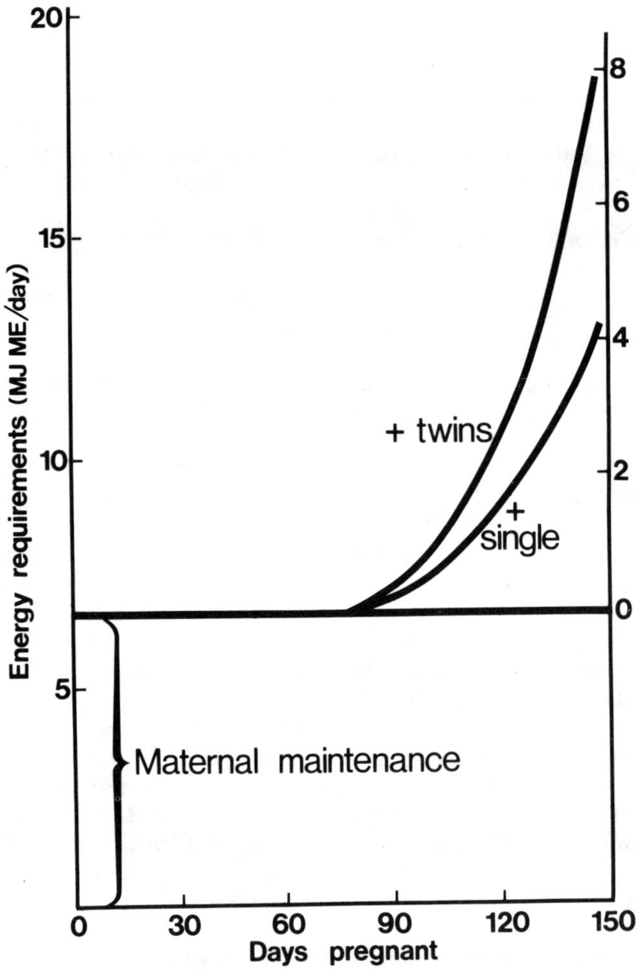

Energy requirements (MJ ME/day)
20
15
10
5
Foetal weight (kg)
8
6
4
2
0
+ twins
+
single
Maternal maintenance
0
30
60
90
120
150
Days pregnant

time and refer to the amounts of food needed to prevent the ewe from drawing on her own body reserves.

The effect on lamb birth weight of levels of late pregnancy energy intake which are less than the full requirements is given by the equation

$$y = -356.39e^{-0.0473x} + 103.37,$$

where y = lamb birth weight as a percentage of potential birth weight (i.e.the weight the lamb would have been if the ewe's energy requirements in late pregnancy had been met in full) and x = energy intake as a percentage of full energy requirements. This relationship, which is illustrated in Figure 4, is based on results obtained from experiments on Scottish Blackface and Greyface ewes in which different nutritional states were maintained in late pregnancy. Results from subsequent experiments may lead to small changes in the values of the coefficients, the constant term in the equation and hence to the precise shape of the curve. It is clear however that as energy intake is reduced from 100% of requirements the initial effect on lamb birth weight is relatively small but that as further reductions in intake are imposed the effect on birth weight becomes progressively greater. We estimate that with hill sheep in this country the level of undernourishment typically found in ewes carrying average-sized single foetuses will reduce the birth weight of these single lambs by about 10%. This is generally considered to be an acceptable reduction in birth weight and in most cases the input of feed required to prevent it would not be economically justified. Twin-bearing hill ewes are however more severely undernourished, and this greater degree of undernourishment reduces twin birth weights by about 25%, which must certainly prejudice the lambs' survival and perhaps also adversely affect the subsequent growth rate of those which do survive. Excessive undernourishment in late pregnancy may also lead to less vigorous lambs, and possibly to a delayed onset of lactation and a poorly developed maternal instinct in the ewe.

Although in practice it is neither necessary nor economic to attempt to meet in full the very high energy requirements of the ewe during late pregnancy, it is essential to avoid severe levels of undernourishment which will seriously affect production. The objective of a practical feeding system in late pregnancy can thus be regarded as the achievement and maintenance of a moderate degree of undernourishment in which the ewe is using body tissue to a limited and controlled extent to meet the deficit between intake and requirements and which will result in only a small and quite acceptable production penalty.

It is difficult to make general recommendations on practicable levels of feeding during late pregnancy which will apply to all ewes, but acceptable levels of energy and protein intake are suggested in Figure 5. These are based on a ewe allowance of 271 kJ ME/kg$^{0.75}$, plus a foetal allowance of 2.5 MJ ME/kg foetus at 8 weeks prepartum reducing to 1.5 MJ ME/kg foetus at term. By about six weeks before lambing, the quality and quantity of the diet should be such as to supply the levels of energy and protein suggested in Figure 5. This will mean either introducing

Figure 4

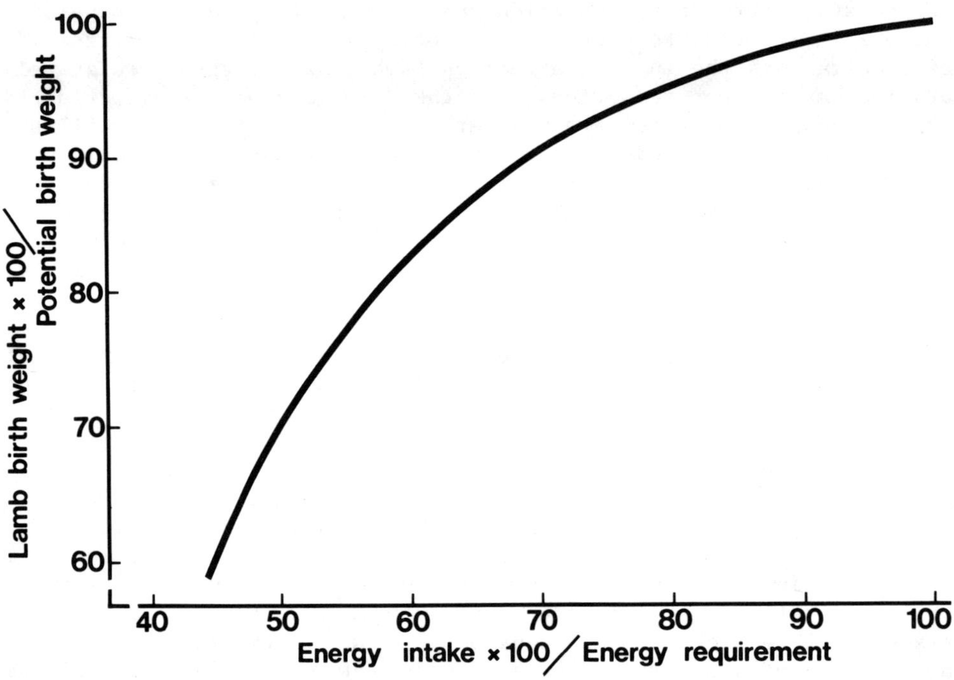

100
90
80
70
60
Lamb birth weight × 100/Potential birth weight
40
50
60
70
80
90
100
Energy intake × 100/Energy requirement

supplementary feeding at or before this time or changing the formulation of the diet of ewes which are already receiving feeding.

The pattern of feeding during the final few weeks of pregnancy should be matched to the growth of the developing foetus or foetuses, i.e. the level of feeling should be increased progressively, changes being made at intervals of one to two weeks. This will make more efficient use of a limited amount of food than supplying the same total quantity at a constant rate over the last six or eight weeks of pregnancy.

Examples of amounts of food, which will supply the quantities of energy and protein suggested in Figure 5, are given in Table 1. These figures are based on medium quality (i.e. early flowering stage) hay, supplying about 7.5 MJ ME/kg fresh weight, with a CP of 10%, and a concentrate mixture containing e.g. 17.5% ground nut meal, 82.5% mixed cereal grains and mineral and vitamin supplements. This concentrate diet would contain about 11.3 MJ ME/kg fresh weight and have a CP of about 17.5%. Lesser amounts of a lower protein concentrate could equally well be used with a higher quality hay. ME and CP values of some common feedstuffs are given in Appendix 2.

In practice most ewes will be at pasture and some allowance must be made for the contribution obtained from grazing. The nutrient intake from grazed herbage is low at the time of year which usually coincides with the later stages of pregnancy and the quality no better than that of a very moderate hay. Allowance must be made for this and for changes in pasture quality which may result from the initiation of grass growth about this time.

Body condition of the ewe is also important in deciding on the optimum level of feeding during late pregnancy and ewes in poor condition (condition score 1½ or less) should receive more liberal feeding.

The form in which feeding is given, and particularly the optimum ratio of roughage to concentrates, is difficult to decide arbitrarily. In many but by no means all situations, concentrates are a more expensive way of supplying nutrients than are roughages and a case for making maximum use of good quality hay (7.5 MJ/kg fresh material or better) can be argued. It is important however to ensure that the total quantity of food offered is not more than the ewe can reasonably be expected to eat, i.e. it must not exceed the ewe's maximum voluntary intake. As a general rule, the total weight of dry matter intake per day should not exceed 3% of the animal's body weight. Over-fat ewes (condition score 3½ or more) carrying twin foetuses may show inappetance if too much mediocre quality roughage is offered.

The nutrition of pregnant ewe lambs and two-tooths merits special consideration. Generally, these young animals, whether they are ewe lambs in an intensive indoor system or two-tooth ewes in an extensive hill situation, have higher requirements for maintenance of their own body tissues than mature ewes of a similar body weight. Animals in their first pregnancy usually produce fewer lambs than older ewes and a level of feeding between those suggested for twin-bearing and for single-bearing

Figure 5

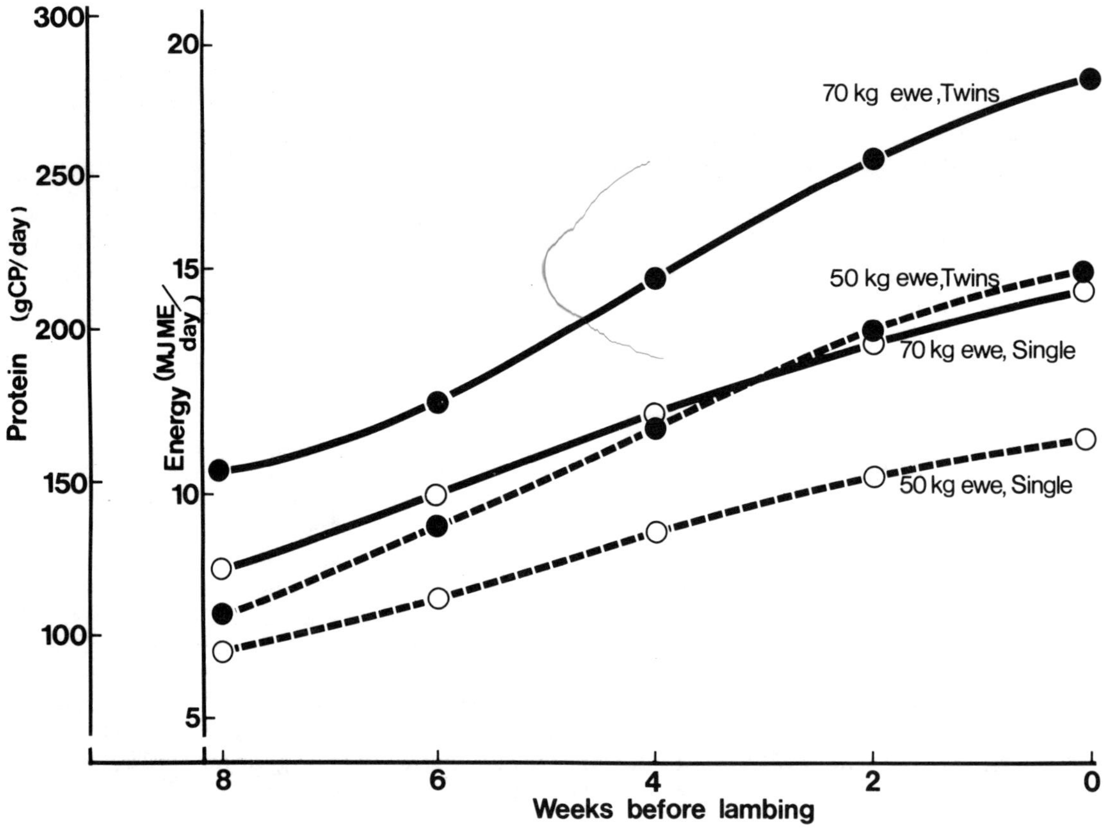
300
250
200
150
100
Protein (gCP/day)
20
15
10
5
Energy (MJ ME/day)
70 kg ewe, Twins
50 kg ewe, Twins
70 kg ewe, Single
50 kg ewe, Single
8
6
4
2
0
Weeks before lambing

Table 1 Examples of amounts of feeding stuffs which will supply the levels of energy and protein suggested in Figure 5 as appropriate during late pregnancy for ewes of two body weights and carrying single and twin foetuses.

Ewe Weight (kg)	Foetuses	Weeks before Lambing									
		8		6		4		2		0	
		Hay (kg)	Conc. (g)	Hay (kg)	Conc. (g)	Hay (kg)	Conc. (g)	Hay (kg)	Conc. (g)	Hay (kg)	Conc. (g)
50	Single	0.90	-	0.90	90	0.90	210	0.90	325	0.90	400
	Twin	0.90	55	0.90	230	0.90	425	0.90	610	0.90	730
70	Single	1.25	-	1.25	55	1.25	215	1.25	360	1.25	460
	Twin	1.25	-	1.25	235	1.25	490	1.25	720	1.25	880

mature ewes would generally be appropriate for young animals carrying single foetuses. Excessive levels of feeding may result in a high proportion of large lambs of above-average birth weight, leading to difficulties at parturition with consequent weakening, or even loss, of both mother and lamb.

Body weight change can also be used to assess the adequacy of nutrition during late pregnancy, although there are difficulties where the ewe herself is losing weight and the foetus is rapidly putting on weight. The lamb comprises approximately 60% of the total weight loss at birth and membranes and fluids the remaining 40%. A weight gain during the final weeks of pregnancy, which is less than the increase in weight of the lamb, the associated membranes and fluids and the increase in weight of the udder, is still compatible with good reproductive performance. As a general rule, ewes carrying single foetuses should increase in body weight by about 10% during the last eight weeks of pregnancy. This order of weight gain would not however be sufficient in ewes with more than one foetus and would lead to poor lamb birth weights. The level of feeding during the final eight weeks should be such as to give an apparent 18% increase in body weight in ewes carrying twins. Ewe lambs and two-tooths carrying single foetuses should gain about 12% over this period. Thus for example a 50 kg during the eight weeks before lambing; a two-tooth of a similar body weight should gain about 6 kg and a 70 kg twin-bearing ewe should increase in weight by about 12 kg.

The adequacy of nutrition in late pregnancy can also be assessed by biochemical parameters which serve as indices of nutritional status. Systems of determining feed inputs based on the analysis of blood samples must necessarily have restricted application in commercial practice but they have certain advantages particularly useful in applied research work and as such they are worthy of mention here. Biochemical parameters provide an immediate measure of nutritional status permitting appropriate action to be taken almost immediately, whereas calculations of weight change give only a retrospective indication of the adequacy of nutrition and it may be too late to counteract the effects of undernourishment.

In experimental work the severity of undernourishment can be measured by analysing blood samples to determine the rate at which the ewe is drawing on body fat. The concentration of non-esterified fatty acids (NEFA) (also referred to as free fatty acids or FFA) and ketone bodies in the blood plasma are related to the rate of fat breakdown and are the most commonly used means of assessing the extent to which the ewe's energy requirements are being met by its energy intake.

Plasma NEFA concentrations are a very sensitive index of moderate degrees of undernourishment but are less useful in situations of prolonged severe undernourishment; plasma ketone concentrations, which show only relatively small increases in moderate undernourishment, are markedly elevated in severe undernourishment and are most responsive at levels where NEFA concentrations become less useful. In experimental work in carefully controlled conditions, both parameters can be used to monitor with precision the whole spectrum of undernourishment. Plasma NEFA concentrations are however very susceptible to certain hormonal changes

Figure 6

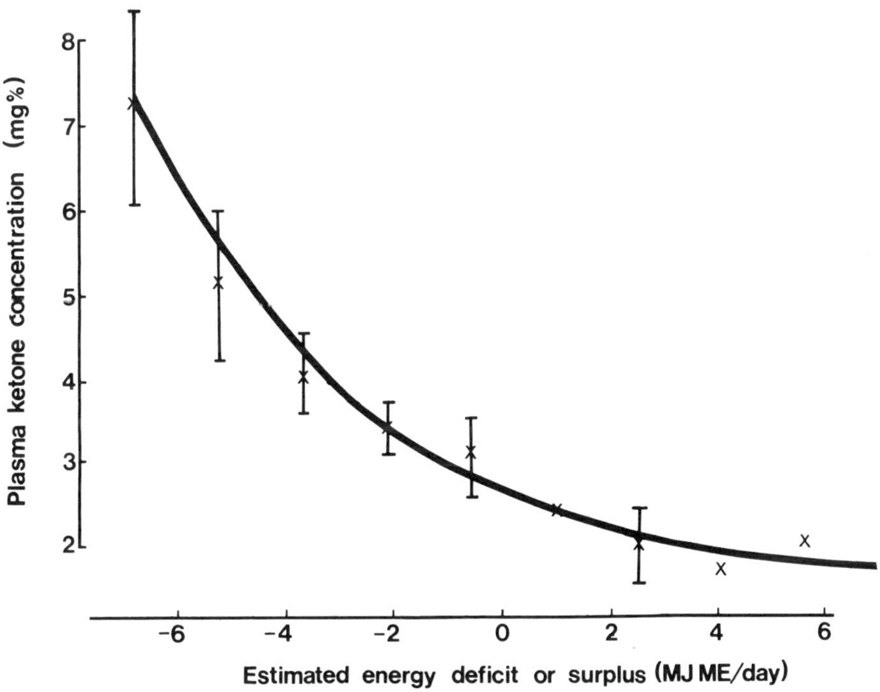

Plasma ketone concentration (mg%)
8
7
6
5
4
3
2
-6
-4
-2
0
2
4
6
Estimated energy deficit or surplus (MJ ME/day)

which follow any disturbance of the animal and are consequently of less use in the field situation with ewes which are unaccustomed to frequent handling or to the procedures associated with blood sampling. Plasma ketone concentrations are less readily affected by such factors and are generally a more reliable index of nutritional state in any work with large numbers of animals under practical conditions.

The technique of using biochemical parameters to control the level of feed inputs to flocks of ewes has now been used for a number of years in the Hill Farming Research Organisation's development programme, in which improved systems of management based on the results of research work are being examined and tested. In each flock in which the technique has been used some 40 ewes, representative of the age structure and previous level of performance of the flock, are blood sampled once a week from about six weeks before the beginning of lambing until one week before lambing.

The levels of feeding of each flock are reviewed in the light of the blood analyses, using a formula based on a relationship derived between plasma ketone concentration and energy status in pregnant ewes (Figure 6) and including an allowance for the energy required for foetal growth over the period of time until the next sampling. By taking account of ewe live weight and expected lambing percentage, the equation allows the technique to be applied over a wide range of conditions. The equation is:

$$x = 10^{-4}\, W^{0.75}\ [4581 \log \frac{(y_1 - 1.57)}{(y_2 - 1.57} + 2.06L + 139],$$

where x = energy increment (MJ ME/day), W = ewe live weight (kg), y_1 = determined plasma ketone concentration (mg/100 ml), y_2 = predetermined plasma ketone concentration (mg/100 ml) and L = expected lambing percentage. Adjustments in levels of feeding are made with the objective of maintaining mean plasma ketone concentrations around 3 mg/100 ml. Although somewhat higher ketone concentrations would be acceptable in individual animals, the distribution of values within a flock is such that a mean of 3 mg/100 ml is considered desirable.

As an alternative to the determination of total blood or plasma ketones, either of the two principal ketone bodies can be used as an index of energy status. Plasma acetoacetate concentration has the advantage over 3-hydroxybutyrate concentration in ruminants in that it occurs only in the undernourished animal, whereas 3-hydroxybutyrate is derived from rumen butyrate as well as from NEFA and is found to some extent in all ruminants. Acetoacetate is however chemically less stable than 3-hydroxybutyrate and reliance on plasma acetoacetate as an index of undernourishment requires sample preparation and handling techniques which are not always possible when working under field conditions. In general, 3-hydroxybutyrate is a more practicable index for most situations. In the Hill Farming Research Organisation we have recently changed from the use of total ketones to 3-hydroxybutyrate and have found that a plasma concentration of 0.78 mmol/l indicates the same moderate degree of undernourishment as is characterized by a plasma ketone concentration of 3 mg/100 ml.

Levels of feed inputs determined as described above for two of the Hill Farming Reearch Organisation's flocks are illustrated in Figure 7. It can be seen that the characteristic pattern of feed increments dictated by the technique was successful in maintaining a relatively constant and moderate degree of undernourishment during late pregnancy. The lambing performances of these flocks were judged to be very satisfactory; mean lamb birth weights were high and lamb mortality below average.

Within any flock there is likely to be considerable variation in the requirements of individual ewes and it will generally be wasteful to give the same level of feeding to all animals. The pattern of feeding whereby the amounts given are increased progressively to match the pattern of foetal growth has already been discussed but improvements in the efficiency of utilisation of feeding stuffs can also be achieved by dividing the flock into groups of ewes which are likely to have similar nutritional requirements. For example, the flock should be divided according to expected date of lambing by the use of coloured crayons on the rams at mating. At about eight weeks before the beginning of lambing, those expected to lamb during the first two weeks should be separated from later lambing ewes and given preferential feeding. It is worthwhile pointing out that, in situations where lambing coincides with the initiation of grass growth, later lambing ewes may obtain a higher proportion of the desired level of intake from grazing and so require lesser amounts of supplementary feed.

Animals in their first pregnancy, whether ewe lambs or two-tooth ewes, should also be separated for preferential feeding as discussed above. As well as having higher requirements in relation to body size and foetal number, these younger animals may be at a disadvantage when competing with older ewes for limited amounts of food. Ewes in poor body condition should also be taken out of the flock and fed more generously during the last six to eight weeks of pregnancy.

Differences in the levels of feeding of single- and twin-bearing ewes have been mentioned. In practice it is seldom possible to separate single- and twin-bearing ewes for differential feeding but, in deciding on a particular level of nutrition, consideration should be given to the expected lambing percentage. In most hill flocks, where few twins are born, the pattern of feeding appropriate to single-bearing ewes should be adopted; in lowground flocks it would generally be necessary to adopt the levels indicated for twin-bearing ewes. Various techniques for determining the number of foetuses a ewe is carrying are being investigated in experimental situations. As yet no technique is available at a cost which would allow it to be widely applied in the field but it is likely that in the future it will be possible to identify single- and multiple-bearing ewes at about eight weeks before lambing. In the meantime, any means of feeding ewes according to their requirements will improve the efficiency of the use of feeding stuffs and, by ensuring that ewes are properly fed, will help to optimise lambing performance.

Figure 7

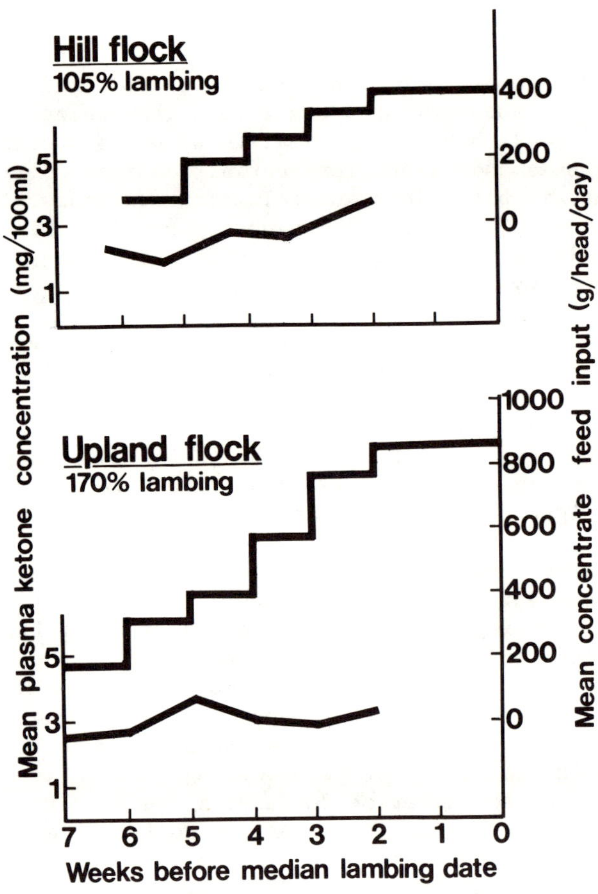

Hill flock
105% lambing
Upland flock
170% lambing
Mean plasma ketone concentration (mg/100ml)
Mean concentrate feed input (g/head/day)
5
3
1
400
200
0
1000
800
600
400
200
0
7
6
5
4
3
2
1
0
Weeks before median lambing date

In conclusion it must be stressed that the nutrition of the pregnant ewe must be considered not in isolation but as part of the whole year-round nutrition of the animal. The immediate effects of nutrition during pregnancy are seen at lambing in factors such as birth weight, lamb vigour and lamb and ewe mortality. Less obvious but nonetheless important effects include the interaction of prepartum nutrition with lactation performance and subsequent recovery of body reserves, both of which are likely to be affected by the body condition of the ewe at lambing and hence by nutrition during pregnancy. These indirect effects are difficult to quantify and even more difficult to express in monetary terms.

At the outset it was emphasised that the principal objective of manipulating the nutrition of the pregnant ewe was an economic one. Economic analyses of the costs of pregnancy nutrition must take account of benefits which accrue outwith the immediate pregnancy period.

APPENDIX 1

BODY CONDITION SCORING

The first stage in body condition scoring is the assessment of the prominence (the degree of sharpness or roundness) of the spinous processes of the lumbar vertebrae - these are the bony points or spikes rising upwards from the backbone. The second step is to assess the prominence of, and degree of cover over, the transverse processes of the vertebrae - these are the horizontal bones coming out from either side of the backbone. Thirdly, judge the extent of the muscular and fatty tissues below the transverse processes by the ease with which the fingers pass under the ends of these bones. Finally, evaluate the fullness of the eye muscle area, and its degree of fat cover, in the angle between the spinous and transverse processes. Animals are then awarded a score according to the following scale:

SCORE 0 Extremely emaciated and on the point of death. It is not possible to detect any muscular or fatty tissue between the skin and the bone.

SCORE 1 The spinous processes are felt to be prominent and sharp. The transverse processes are also sharp, the fingers pass easily under the ends, and it is possible to feel between each process. The eye muscle areas are shallow with no fat cover.

SCORE 2 The spinous processes still feel prominent, but smooth, and individual processes can be felt only as fine corrugations. The transverse processes are smooth and rounded, and it is possible to pass the fingers under the ends with a little pressure. The eye muscle areas are of moderate depth, but have little fat cover.

SCORE 3 The spinous processes are detected only as small elevations; they are smooth and rounded, and individual bones can be felt only with pressure. The transverse processes are smooth and well covered, and firm pressure is required to feel over the ends. The eye muscle areas are full, and have a moderate degree of fat cover.

SCORE 4 The spinous processes can just be detected with pressure as a hard line between the fat-covered eye muscle area. The ends of the transverse processes cannot be felt. The eye muscle areas are full, and have a thick covering of fat.

SCORE 5 The spinous processes cannot be detected even with firm pressure and there is a depression between the layers of fat in the position where the spinous processes would normally be felt. The transverse processes cannot be detected. The eye muscle areas are very full with very thick fat cover. There may be large deposits of fat over the rump and tail

In any flock of sheep it is unlikely that the range of condition will be more than two scores at any one time, or more than three scores over the whole year. In practice, therefore, the range of fatness is narrower than that encompassed by the scoring system, and the scores 0-5 as set out above become inadequate to describe variations within a flock. Faced with this situation it becomes automatic to assess sheep to the nearest half score.

There will always be some assessments falling midway between two points on the scale, even when half scores are used, but for all practical purposes half score divisions are generally considered to be sufficient.

APPENDIX 2

COMPOSITION OF SOME FEEDSTUFFS (ON FRESH BASIS)

	Metabolizable energy (MJ/kg)	Crude protein %
GRAINS		
Oats	10.1	10.5
Barley	11.6	10.0
Maize	12.2	9.0
Wheat	11.9	12.0
HIGH PROTEIN SUPPLEMENTS		
Soya bean meal (extracted)	11.0	45.0
Groundnut meal (decorticated and extracted)	10.3	53.0
Fish meal	9.8	61.0
ROOTS		
Mangolds	1.5	1.1
Swedes	1.4	1.3
Sugar beet pulp (dried)	11.2	9.3
DRIED GRASS		
Very leafy	9.7	18.5
Leafy	9.3	15.0
Early flower	8.7	12.0
HAY		
Cover	7.6	15.5
Ryegrass: leafy	8.5	13.5
early flower	7.5	10.0
seed set	6.5	5.0
STRAW		
Barley	6.3	3.5
GRASS SILAGE		
Leafy	2.1	3.5
Early flower	1.9	3.2
Full flower	1.6	2.9
MOLASSES	10.8	3.5

REFERENCES

Alexander, G. (1974) Birth weight of lamb: influences and consequences. In: Elliott, K.;Knight, J. (Eds) Size at birth. Amsterdam; Associated Scientific Publishers, pp215-245. ISBN 90-219-4030-2.

Boaz, T.G.;Forbes, J.M.;Kilkenny, J.B.;Mudd, C.H.;Robinson, J.J.; Russel, A.J.F.;Treacher, T.T. (1973) Feeding the ewe. Technical Report, Sheep Improvement Service, Meat and Livestock Commission No.2. Bletchley, UK; Meat and Livestock Commission, 52pp.

Rattray, P.V. (1974) Energy requirements for pregnancy in sheep. Proceedings of the New Zealand Society of Animal Production 34,67-77.

Reid, R.L. (1968) The physiopathology of undernourishment in pregnant sheep, with particular reference to pregnancy toxemia. Advances in Veterinary Science 12,163-238.

Robinson, J.J. (1977) The influence of maternal nutrition on ovine foetal growth. Proceedings of the Nutrition Society 36,9-16.

Robinson, J.J.;McDonald, I.;Fraser, C.;Crofts, R.M.J. (1977) Studies on reproduction in prolific ewes. 1. Growth of the products of conception. Journal of Agricultural Science, UK 88,539-552.

Russel, A.J.F.;Doney, J.M.;Reid, R.L. (1967) The use of biochemical parameters in controlling nutritional state in pregnant ewes, and the effect of undernourishment during pregnancy on lamb birth-weight. Journal of Agricultural Science, UK 68,351-358.

Russel, A.J.F.;Doney, J.M.;Reid, R.L. (1967) Energy requirements of the pregnant ewe. Journal of Agricultural Science, UK 68,359-363.

Russel, A.J.F.;Gunn, R.G.;Doney, J.M. (1968) Components of weight loss in pregnant hill ewes during winter. Animal Production 10,43-51.

Russel, A.J.F.;Maxwell, T.J.;Sibbald, A.R.;McDonald, D. (1977) Relationships between energy intake, nutritional state and lamb birth weight in Greyface ewes. Journal of Agricultural Science, UK 89,667-673.

Thomson, W.;Aitken, F.C. (1959) Diet in relation to reproduction and the viability of the young. Part II. Sheep: world survey of reproduction and review of feeding experiments. Commonwealth Bureau of Animal Nutrition Technical Communication No.20. Farnham Royal, Slough, UK; Commonwealth Agricultural Bureaux, 93pp.

THE NUTRITION OF THE LACTATING EWE

DR T.T. TREACHER

The Grassland Research Institute, Hurley, Maidenhead, Berks

In all systems of sheep production the quantity of milk produced by the ewe is of major importance either because of its effect on lamb growth or because the ewe is milked to provide for the manufacture of dairy products. In the first 6 weeks of life the growth of the lamb is largely determined by the amount of milk it obtains from its mother. After 6 weeks, the interactions of the factors determining the lamb's growth become complex. Milk intake has a large, though declining, effect on growth both directly and indirectly by influencing the pattern of increase in solid food intake which, although it commences at 3 weeks of age, only becomes significant after 6 or 7 weeks. Lambs receiving less milk will eat more solid food if it is available. Where ewes are milked it is normal for the lambs to be suckled by their mothers for at least 4 weeks before they are weaned and the ewes are milked. In some cases complete or partial suckling may continue for 3 or 4 months.

There is relatively little information on the nutrition of the milked ewe, probably because, except in Israel and the Roquefort region of France, these ewes are used to exploit natural pastures. Therefore this paper is mainly concerned with the nutrition of the ewe that is suckling its lambs but the principles, with suitable adjustments for level of milk production, will be applicable to the milked ewe.

It is most important to emphasise that the milk production of an animal is determined by a complex interaction of the animal's genotype, its environment, particularly its nutritional environment, and a range of physiological factors, including the animal's age, weight and parity. In this last group of factors must be included the interactions between the ewe and the number, weight and genotypes of the lambs it is suckling, and in the case of the ewe that is milked the particular hand milker or milking machine. The influence on milk yield of most of these factors has been discussed by Louca (1978) and Treacher (1978).

CHARACTERISTICS OF EWES' MILK AND LACTATION CURVES

The composition of ewes' milk is set out in Table 1. The concentration of all constituents is much greater than in cows' milk resulting in the energy content being approximately 70% greater.

The ewe suckling two lambs generally produces 40% more milk than a ewe at the same level of nutrition but suckling a single lamb (Treacher 1978). The number of lambs suckled has a greater effect on milk yield than a wide range of nutrition in both pregnancy (Wallace 1948) and in lactation (Gardner & Hogue 1964).

Table 1 A comparison of the composition (g/100g) of ewes', cows' and goats' milk. Representative figures from Ling (1961)

	Ewe	Range[1]	Cow	Goat
Total solids	18.4	16.3 - 21.8	12.1	13.2
Solids-not-fat	10.9	10.8 - 13.0	8.6	8.7
Fat	7.5	5.1 - 10.0	3.5	4.5
Protein	5.6	4.6 - 6.8	3.25	3.3
Lactose	4.4	4.1 - 4.8	4.6	4.4
Ash	0.87	0.83- 0.97	0.75	0.80
Calcium	0.19	0.19- 0.21	0.12	0.14
Phosphorus	0.15	0.14- 0.17	0.10	0.12
Chlorine	0.14	0.070- 0.126	0.11	0.15
Energy concentration (MJ/kg)	4.4		2.6	3.0

[1]From 9 references cited in Ashton et al (1964)

Differences in the shapes of the lactation curves of ewes suckling single and twin lambs occur from the start of lactation and are illustrated in Figure 1. The yield of ewes suckling twin lambs is generally 50-70% greater in the first week of lactation than that of ewes suckling a single. The yield the increases rapidly to a maximum in the second or third week of lactation. The yield of ewes suckling a single lamb reaches a maximum in the third or fourth week of lactation; this maximum daily yield is not greatly above the initial yield. The rate of decline in yield is greater in ewes suckling twins and by the sixteenth week of lactation there is generally little difference in yield between ewes with single and twin lambs.

These responses occur at reasonably adequate levels of nutrition. Improved nutrition generally results in a very slightly higher maximum daily yield being reached one to two weeks later. Again the rate of decline in yield is slightly higher for ewes with the higher, later maximum daily yield and by the 12-16th week of lactation there is little difference in milk production.

NUTRITION OF THE LACTATING EWE

Experimentation on the lactating ewe has been limited by the difficulty of measuring milk yield when the ewe is suckling. The methods

Figure 1

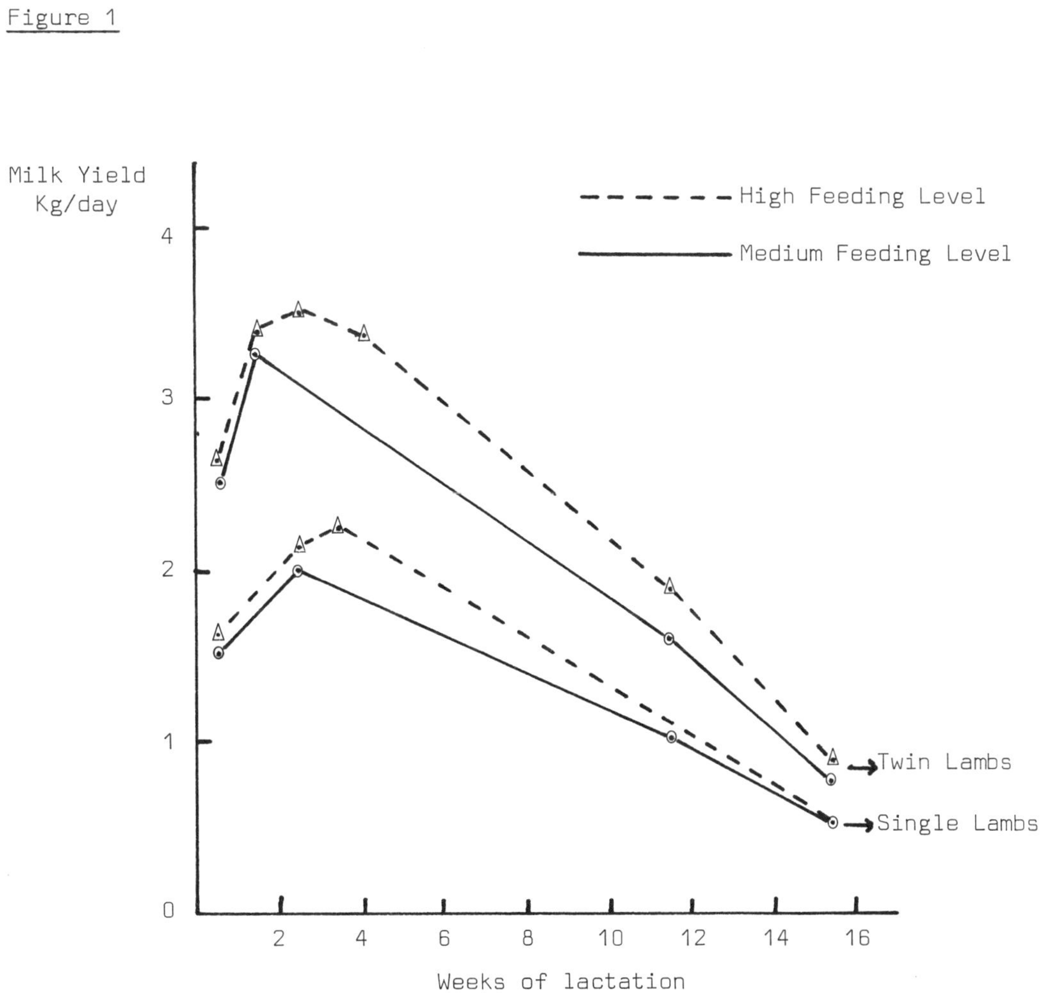
Figure 1
Milk Yield
Kg/day
High Feeding Level
Medium Feeding Level
Twin Lambs
Single Lambs
0
1
2
3
4
2
4
6
8
10
12
14
16
Weeks of lactation

of estimating yield have been reviewed by Boyazoglu (1963). Although the effects of nutrition on lactation in the ewe are well established in principle, the diversity of conditions under which the experimentation has been carried out makes it difficult to describe relationships between feed inputs, milk production and weight change which are of general validity.

FEEDING STANDARDS FOR LACTATING EWES

Energy requirements for ewes during lactation have been published in the UK by MAFF (1975) and Boaz et al (1973) and in the USA by NRC (1975). These requirements have been calculated from estimates of the energy costs of maintenance, lactation and weight change. There are wide variations in the estimates of both the energy requirements for maintenance in the sheep and the efficiency of conversion of metabolizable energy to milk energy. In the MAFF (1975) requirements the efficiency of conversion to milk for dairy cows (62%) was used for the ewe, which agrees closely with the value of 63% calculated by Robinson (1978) from the data of 9 experiments on ewes. The efficiency of utilization of body tissue for the production of milk energy has not been measured in the ewe and a value estimated in the dairy cow (e.g. Moe et al 1971) has to be used.

At present these requirements, set out in Table 2, must be considered as only a very broad guide. They all indicate however how much greater the requirement for metabolizable energy is in lactation than in late pregnancy, when the requirements are given as 13.33 MJ/day in MAFF, 15.0 MJ/day in Boaz et al and 18.4 MJ/day in NRC for a twin bearing ewe of 70 kg.

RESPONSE OF THE EWE TO VARIATION IN ENERGY INTAKE

All animals exhibit a declining response to inputs of nutrients. The response of the lactating ewe to metabolizable energy intake has not been well defined as most experiments have not had treatments spread over an adequate range of nutrition for the response curve to be established. Obviously the size of response measured in a particular experiment will be very much influenced by the relative positions of the energy inputs on the generalized response curve. In some experiments, for example Owen & Ingleton (1963) and Gardner & Hogue (1964), the differences in milk yield were approximately proportional to the levels of intake but in others, e.g. Barnicoat et al (1949) the responses were less than the differences in intake.

Crabtree & Treacher (unpublished data) found a strongly curvilinear response to metabolizable energy intake over the range from 12.5 to 29.3 MJ/day. Yield in the first month of lactation only increased very slightly at intakes above 20 MJ/day, although the rate of live-weight loss was reduced at the higher intake.

Changes in live weight complicate the interpretation of responses to energy intake in lactation. In the first 4-6 weeks of lactation reductions in live weight usually occur and these losses may be large. For example, Gibb & Treacher (1978) found that Scottish Halfbred ewes at pasture lost 320-560 g/day in the first month of lactation and ewes on

Table 2 Requirements of metabolizable energy (MJ/day) for ewes in lactation derived from different feeding systems.

Liveweight kg	Number of lambs suckled	1st month MAFF	1st month Boaz		1st month NRC	2nd month MAFF	2nd month Boaz		2nd month NRC	3rd month MAFF	3rd month Boaz	3rd month NRC
50	1	19.3	17.8	15.7	20.5	18.9	16.2	14.2	Same as in first month	15.7	13.1	15.0
	2	26.6	22.4	20.4	23.6	23.2	19.3	17.2		18.0	16.2	20.5
60	1	20.3	19.1	16.7	22.6	19.9	17.6	15.2		16.7	14.5	16.6
	2	27.6	23.8	21.4	25.5	24.2	20.7	18.2		19.0	17.6	22.6
70	1	21.3	20.5	17.7	24.6	20.9	19.0	16.1		17.7	15.9	18.4
	2	28.6	25.8	22.3	27.5	25.2	22.1	19.2		20.0	19.0	24.6
80	1	22.3	21.9	18.6	25.5	21.9	20.4	17.1		18.7	17.2	19.3
	2	29.6	26.6	23.3	29.5	26.2	23.4	20.2		21.0	20.4	25.5
Assumed milk yield kg/day	1	1.60	1.75	1.75	-	1.55	1.5	1.5		1.14	1.00	-
	2	2.54	2.5	2.5	-	2.10	2.0	2.0		1.43	1.5	-
Rate of weight change	1	0	0	-10%[1]	-25g/day	0	0	-10%[1]		0	0	+45
	2	0	0	-10%[1]	-60/day	0	0	-10%[1]		0	0	+80

[1] in first 6 weeks of lactation

the lowest herbage allowance continued to lose weight throughout the 16 weeks of the experiment. Maintenance or gains in weight in early lactation are usually associated with low milk yields (e.g. Guyer & Dyer 1954).

The energy contributed towards milk production by weight loss is open to question. Live-weight change may not reflect the change in tissue weight and the exact composition and hence energy content of that tissue will not be known.

Changes in intake will result in alteration of gut fill that can partially mask the changes in empty body weight. Hypertrophy of the gut occurs if intake increases in lactation (Fell 1976). It is probable that blood volume decreases in early lactation (Hytten & Leitch 1964) but water retention may increase as fat is mobilized (Cowan et al 1978).

Generally it has been assumed that sheep tissue has an energy content of 20 MJ/kg (MAFF 1975) and that the energy derived from this tissue is utilized with an efficiency of 84% as in the dairy cow (Moe et al 1971). Robinson (1978) prefers the assumption, derived from Gardner & Hogue (1964), that tissue in the ewe has an energy content of 25.5 MJ/kg. Cowan et al (1978) suggested that lactating ewes in negative balance do not lose nitrogen and the entire loss consists of fat with an energy content of approximately 39 MJ/kg. Therefore more information is needed before the contribution of tissue catabolism to milk production can be accurately assessed.

RESPONSE TO VARIATION IN PROTEIN INTAKE

Until recently the protein intake of the lactating ewe was considered to be relatively unimportant as the response to variation in intake of protein was extremely varible. Whiting et al (1952) and Calderon Cortes et al (1977) found that diets with as little as 7% to 10% crude protein had to be fed before the protein content of ewes' milk was reduced. The American NRC (1975) standards suggest that protein concentrations of 10.4 and 11.5% are sufficient for ewes in the first 2 months of lactation suckling single and twin lambs respectively. There is now however increasing evidence that in early lactation, when the ewe's energy requirements are high and the ewe is commonly in negative energy balance, protein intake is important in affecting the partition of nutrients between milk production and weight loss. In the ruminant, as in the monogastric animal, the utilization of protein is affected by energy intake (Broster 1973).

Recent experiments by Robinson et al (1974) and Calderon Cortes et al (1977) have elegantly demonstrated rapid responses to dietary protein concentration in ewes in negative energy balance in early lactation. These experiments confirm the generalized responses of milk production and body weight change to variation in ME and protein intake discussed by Robinson & Forbes (1970) and Robinson (1978).

This generalized relationship from Robinson (1978), shown in Figure 2, illustrates the main interrelationships of energy and protein intake. At a particular level of energy intake and milk production there is a

Figure 2

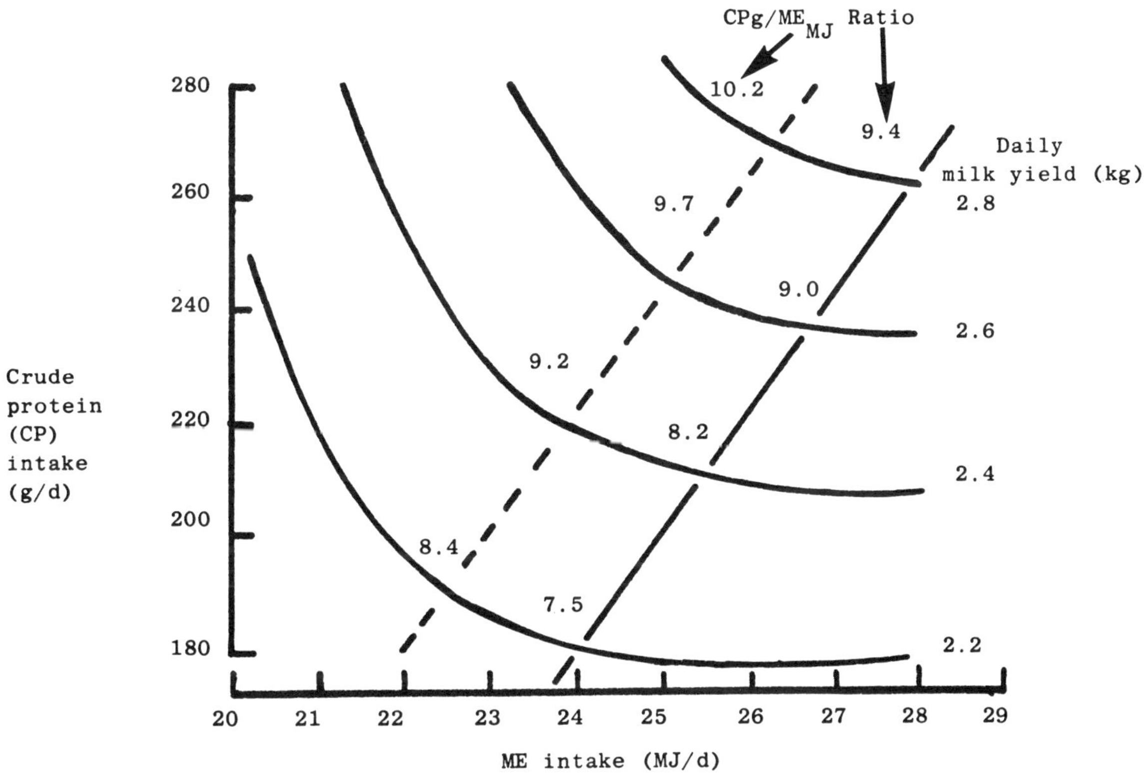

FIG. 2. Response in milk yield to alterations in dietary CP and ME for 70 kg ewes suckling twin lambs. (After Robinson & Forbes 1970)

corresponding minimum protein intake and a reduction below this level reduces milk production. This minimum ratio of CP to ME increases with increasing milk yield. Increasing dietary protein above the minimum requirement without altering the ME intake, i.e. increasing the protein:ME ratio, will increase milk production if the ewe is not producing its potential yield; this will be accompanied by an increase in weight loss.

This model can only give a general outline of the ewes responses to protein and energy intake as the exact amount of protein available to the ewe will be dependent on the amount of microbial protein synthesized in the rumen and the extent to which dietary protein escapes degradation in the rumen; Robinson (1978) gives a detailed discussion of these interrelationships.

EFFECTS OF FEEDING AT DIFFERENT STAGES OF THE PRODUCTIVE CYCLE

The performance of an animal is influenced not only by its current level of nutrition but also by direct or indirect effects of nutrition at previous stages of the productive cycle. In the ewe it is important to consider the effects of nutrition in pregnancy and in early lactation because in pastoral systems levels of intake may be low at these times.

Nutrition in pregnancy

About 95% of the secretory tissue in a ewe's udder is laid down during the last 8 weeks of pregnancy (Denamur 1964) but it is not known if the effects on milk production of undernutrition at this time result from a reduction in the quantity of secretory tissue or from a depression of the overall metabolism and altered endocrine function. Direct effects of feeding in pregnancy on milk production have been demonstrated (Thomson & Thomson 1953, Treacher 1970) when lamb effects were eliminated by cross-fostering of lambs or by machine milking. These effects are reinforced by the direct effect of the lamb's birthweight and vigour on milk production.

Poor nutrition in late pregnancy in addition to reducing weight and vigour of lambs at birth can affect the onset of lactation (McCance & Alexander 1959). There is little milk in the udder at parturition and the onset of lactation is slow, with milk with the high levels of total solids characteristic of colostrum persisting for several days after lambing. The extent to which a reduced milk yield continues over a longer period of lactation is dependent on the severity of under-nutrition in pregnancy and the level of feeding in lactation. If the latter is good, pregnancy nutrition may have little effect on yield (e.g. Owen & Ingleton 1963, Peart 1967). However, Wallace (1948) and Treacher (1970) found effects on yield of severe under-feeding in pregnancy even when feeding in lactation was sufficient for maintenance or gain in liveweight. In this situation a low maximum daily yield is reached early in lactation, often in the first week. There is therefore a tendency for these ewes to lose less weight or even gain weight in lactation (e.g. Guyer & Dyer 1954, Graham 1964).

It is difficult to define exactly the level of nutrition in late pregnancy at which an effect on subsequent milk production can be expected to occur. Unless lamb birthweight is reduced by 25% it is unlikely that milk yield will be reduced. Although it has not been demonstrated experimentally it is possible that the period immediately before parturition is the critical time. Thus a short period of severe under-nutrition, which is not reflected in lamb birthweights, may reduce potential yield.

Nutrition in early lactation

As low availability of herbage at the start of lactation is common in flocks lambing in the spring, attempts have been made to define the length of time for which poor nutrition can occur after lambing without reducing to a major extent yield during the remainder of lactation. Peart (1970) found that yield increased rapidly when ad libitum feeding was introduced after 2 weeks of restricting intake of 53 kg ewes to approximately 20 MJ of ME/day. If the restricted feeding continued to 4 weeks post-lambing the introduction of ad libitum feeding resulted in the maintaining of yield but no actual increase. Coop et al (1972) found that restriction 3 days after lambing had no effect on lamb growth. Restriction for 1 - 1½ weeks had little effect on the growth of single lambs but reduced the growth of twins. This effect was marked if restriction continued to 3 weeks. The growth rate of the lamb was only reduced during the period when the intake of the ewes was restricted. Follow-up studies by Jagusch et al (1972) showed that milk yield increased to follow the normal lactation curve even after a 30 day restriction of intake.

Few experiments have looked at high level feeding followed by lower feeding. The results of Doney & Munro (1962) suggest that even a week of grazing lowland pasture after lambing can stimulate a higher yield, which is reflected in higher growth rates in twin lambs when returned to the hill.

Treacher & Crabtree (1973) demonstrated a residual effect of level of feeding in the period before peak of lactation on the growth of lambs later in lactation. A 73% increase in ME intake of the ewe in the first month of lactation increased lamb growth rate by 24% at that time but the response increased to 34% in the second month of lactation. Part of this increase in growth rate however resulted from higher intake of solid food, as the lambs from the ewes fed the higher intake in the first month were heavier. A subsequent experiment failed to show such a clear residual effect of feeding in the first month of lactation on subsequent yield.

VOLUNTARY INTAKE OF FEED BY LACTATING EWES

Early lactation is the period of highest nutrient requirement in the ewe's production cycle. Although restriction of intake in some instances may not greatly reduce milk production it will result in loss of weight and body reserves. If these losses cannot be regained between weaning and re-mating, subsequent output and particularly reproductive performance may be reduced. It is therefore most important to consider intake patterns and the extent to which they can be influenced by management. Unfortunately this is a very neglected area.

Voluntary intake is considerably higher in lactating ewes than in dry or pregnant ewes and usually rises immediately after lambing, increasing rapidly in the first 2-3 weeks of lactation and then continuing to rise more slowly to reach a maximum 2 or 3 weeks after the maximum daily milk yield is reached (Hadjipieris et al 1966). On diets of low digestibility intake is lower initially and may continue to rise slowly until as late as the twelfth week of lactation (Hadjipieris & Holmes 1966). The actual increae between late pregnancy and the point of maximal intake in lactation can vary between 10 and 63%. It is higher in ewes suckling twin lambs and is generally lower on diets with low digestibility or low protein concentrations.

The limited information on the factors influencing feed intake of the lactating ewe has been discussed by Foot & Tissier (1978) but as the majority of sheep are kept in pastoral systems it is particularly important to consider the intake of herbage by the lactating ewe when grazing. At pasture the patterns of intake are modified by the complex process of food gathering and information is more limited and fragmentary than for conserved feeds. Sheep will select a diet much higher in digestibility and nutrient concentration than the mean sward composition unless the quantity of herbage available is very low. Herbage availability will also affect dry-matter intake but the grazing animal partly compensates by increasing grazing time. The lactating ewe has a higher grazing time than dry or pregnant ewes particularly at high allowances of herbage (Arnold 1963). Comparisons of the intakes by dry, pregnant and lactating ewes on the same pastures (Table 3) suggest that the differences that occur at pasture are of similar magnitude to those on conserved feeds.

The results reported by Foot & Tissier (1978) emphasize the importance of feeding diets with a higher digestibility or nutrient concentration in early lactation to ewes with twin lambs, as dry-matter intake will be higher initially and increase more rapidly, stimulating a higher milk production. With ewes at pasture the importance of the common management practice of making lambing coincide with the expected date of the start of herbage growth can be seen. Clear guidelines do not exist to suggest what feeding strategies should be made if the herbage available appears inadequate. If availability and growth of herbage are both low one must offer concentrates or long forage to raise the total nutrient intake. Even at low herbage availabilities however, intake of herbage will be reduced when concentrates are fed and the response to the concentrate will be less than would be calculated from feeding standards. Young et al (1977) were not able to demonstrate a response of lamb growth rate to concentrate feeding in the first 2 months of lactation until the stocking rate was above 20 ewes plus twin lambs/ha.

Table 3 Daily Digestible Organic Matter Intakes (g/kg live weight) of dry ewes and pregnant or lactating ewes when grazing. (From Treacher & Gibb 1978)

		Physiological condition of ewes			
	Weeks from lambing	Dry	Pregnant	Lactating	Percentage Differences
Arnold & Dudzinski (1967)	Mean of -12, -6 & -3	24.7	24.0		- 3%
	+ 3	25.2		40.6	+61%
	+10	30.2		45.6	+51%
	Mean of -6 & -3	17.6	18.4		+ 4%
	+ 3	16.9		21.5	+27%
	+ 6	18.2		24.1	+32%
	+10	15.2		17.4	+14%
Arnold (1975)	Mean of -6 to -2	13.4	16.3		+22%
	Mean of +2 to +6	16.9		22.0	+30%
Langlands (1977)	Mean of +1 to +14	14.1		26.7	+56%
Gibb & Treacher (1978)	+ 6	20.7		25.8	+24%
	+10	17.5		26.1	+49%
	+16	17.1		22.3	+30%
Cook et al (1961)	+ 8	24.6		32.3	+31%

REFERENCES

Arnold, G.W. (1963) Wool Technology and Sheep Breeding 10,17-19.

Arnold, G.W. (1975) Herbage intake and grazing behaviour in ewes of four breeds at different physiological states. Australian Journal of Agricultural Research 26,1017-1024.

Arnold, G.W.;Dudzinski, M.L. (1967) Studies on the diet of the grazing animal. 2. The effect of physiological status in ewes and pasture availability on herbage intake. Australian Journal of Agricultural Research 18,349-359.

Ashton, W.M.;Owen, J.B.;Ingleton, J.W. (1964) A study of the composition of Clun Forest ewe's milk. Journal of Agricultural Science, UK 63,85-90.

Barnicoat, C.R¨;Logan, A.G.;Grant, A.I. (1949) Milk-secretion studies with New Zealand Romney ewes. II. Milk yields of ewes and factors influencing them. Journal of Agricultural Science, UK 39,47-55.

Boaz, T.G.;Forbes, J.M.;Kilkenny, J.B.;Mudd, C.H.;Robinson, J.J.; Russel, A.J.F.;Treacher, T.T. (1973) Feeding the ewe. Technical Report, Sheep Improvement Service, Meat and Livestock Commission No.2. Bletchley, UK; Meat and Livestock Commission, 52pp.

Boyazoglu, J.G. (1963) Aspects quantitatifs de la production laitière des brebis. 1. Mise au point bibliographique. Annales de Zootechnie 12, 237-296.

Broster, W.H. (1973) Protein-energy interrelationships in growth and lactation of cattle and sheep. Proceedings of the Nutrition Society 32, 115-122.

Calderton Cortes, J.F.;Robinson, J.J.;McHattie, J.;Fraser, C. (1977) The sensitivity of ewe milk to changes in dietary crude protein concentration. Animal Production 24,135.

Cook, C.W.;Mattox, J.E.;Harris, L.E. (1961) Comparative daily consumption and digestibility of summer range forage by wet and dry ewes. Journal of Animal Science 20,866-870.

Coop, I.E.;Clark, V.R.;Claro, D. (1972) Nutrition of the ewe in early lactation. 1. Lamb growth rate. New Zealand Journal of Agricultural Research 15,203-208.

Cowan, R.T.;Robinson, J.J.;Greenhalgh, J.F.D. (1978) Body composition changes in lactating ewes. Animal Production 26,392-393.

Denamur, R. (1964) Les acides nucléiques et les nucléotides libres de la glande mammaire pendant la lactogenèse et la galactopoièse. In: Proceedings of the 2nd International Congress of Endocrinology, London, 1964, Part I, pp434-462.

Doney, J.M.;Munro, J. (1962) The effect of suckling, management and season on sheep milk production as estimated by lamb growth. Animal Production 4,215-220.

Fell, B.F. (1976) The gut in reproduction Annual Report, Rowett Research Institute 32,97-108.

Foot, J.Z.;Tissier, M. (1978) Voluntary intake of feed by lactating ewes. In: Boyazoglu, J.G.;Treacher, T.T. (Eds) Milk production in the ewe. Proceedings of sessions of the Sheep and Goat Commission of the European Association for Animal Production, Brussels, August 1978. European Association for Animal Production Publication No.23, pp66-75. ISBN 0-906562-00-7.

Gardner, R.W.;Hogue, D.E. (1964) Effects of energy intake and number of lambs suckled on milk yield, milk composition and energetic efficiency of lactating ewes. Journal of Animal Science 23,935-942.

Gibb, M.J.;Treacher, T.T. (1978) The effect of herbage allowance on herbage intake and performance of ewes and their twin lambs grazing perennial ryegrass. Journal of Agricultural Science, UK 90,139-147.

Graham, N. McC. (1964) Energy exchanges of pregnant and lactating ewes. Australian Journal of Agricultural Research 15,127-141.

Guyer, P.Q.;Dyer, A.J. (1954) Study of factors affecting sheep production. Missouri Agricultural Experiment Station Research Bulletin No.558,79 pp.

Hadjipieris, G.;Holmes, W. (1966) Studies on feed intake and feed utilization by sheep. 1. The voluntary feed intake of dry, pregnant and lactating ewes. Journal of Agricultural Science, UK 66,217-223.

Hadjipieris, G.;Jones, J.G.W.;Wimble, R.H.;Holmes, W. (1966) Studies on feed intake and feed utilization by sheep. 2. The utilization of feed by ewes. Journal of Agricultural Science, UK 66,341-349.

Hytten, F.E.;Leitch, I. (1964) The physiology of human pregnancy. Oxford; Blackwell Scientific Publications, p 12.

Jagusch, K.T.;Jay, N.P.;Clark, V.R. (1972) Nutrition of the ewe in early lactation. 2. Milk yield. New Zealand Journal of Agricultural Research 15,209-213.

Langlands, J.P. (1977) The intake and production of lactating Merino ewes and their lambs grazed at different stocking rates. Australian Journal of Agricultural Research 28,133-142.

Ling, E.R.;Kon, S.K.;Porter, J.W.G. (1961) The composition of milk and the nutritive value of its components. In: Kon, S.K.;Cowie, A.T. (Eds) Milk: the mammary gland and its secretion. New York and London; Academic Press, Vol.2, 195-263.

Louca, A. (1978) The effects of time of weaning and suckling regime on the lactation performance of sheep and the growth rate of the lambs. In: Boyazoglu, J.G.;Treacher, T.T. (Eds) Milk production in the ewe. Proceedings of sessions of the Sheep and Goat Commission of the European Association for Animal Production, Brussels, August 1978. European Association for Animal Production Publication No.23, pp41-48. ISBN 0-906562-00-7.

McCance, I.;Alexander, G. (1959) The onset of lactation in the Merino ewe and its modification by nutritional factors. Australian Journal of Agricultural Research 10,699-719.

Moe, P.W.;Tyrell, H.F.;Flatt, W.P. (1971) Journal of Dairy Science 54,1.

Owen, J.B.;Ingleton, J.W. (1963) A study of food intake and production in grazing ewes. 2. The interrelationships between food intake and productive output. Journal of Agricultural Science, UK 61,329-340.

Peart, J.N. (1967) The effect of different levels of nutrition during late pregnancy on the subsequent milk production of Blackface ewes and on the growth of their lambs. Journal of Agricultural Science, UK 68, 365-371.

Peart, J.N. (1970) The influence of live weight and body condition on the subsequent milk production of Blackface ewes following a period of undernourishment in early lactation. Journal of Agricultural Science, UK 75,459-469.

Robinson, J.J. (1978) Response of the lactating ewe to variation in energy and protein intake. In: Boyazoglu, J.G.;Treacher, T.T. (Eds) Milk production in the ewe. Proceedings of sessions of the Sheep and Goat Commission of the European Association for Animal Production, Brussels, August 1978. European Association for Animal Production Publication No.23, pp53-65. ISBN 0-906562-00-7.

Robinson, J.J.;Forbes, T.J. (1970) Studies on protein utilization by ewes during lactation. Animal Production 12,601-610.

Robinson, J.J.;Fraser, C.;Gill, J.C.;McHattie, I. (1974) The effect of dietary crude protein concentration and time of weaning on milk production and body-weight change in the ewe. Animal Production 19, 331-339.

Thomson, W.;Thomson, A.M. (1953) Effect of diet on milk yield of the ewe and growth of her lamb. British Journal of Nutrition 7,263-274.

Treacher, T.T. (1970) Effects of nutrition in late pregnancy on subsequent milk production in ewes. Animal Production 12,23-36.

Treacher, T.T. (1978) The effects on milk production of the number of lambs suckled and age, parity and size of ewe. In 'Milk production in the ewe'. Boyazoglu, J.G.;Treacher, T.T. (Eds) Proceedings of sessions of the Sheep and Goat Commission of the European Association for Animal Production, Brussels, August 1978. European Association for Animal Production Publication No.23, pp31-40. ISBN 0-906562-00-7.

Treacher, T.T.;Crabtree, J.R. (1973) The effect of plane of nutrition of the ewe during the first and second months of lactation on the growth rate of twin lambs. Proceedings of the British Society of Animal Production 2,80.

Treacher, T.T.;Gibb, M.J. (1978) Intake of herbage by ewes grazing during lactation. In: Boyazoglu, J.G.;Treacher, T.T. (Eds) Milk production in the ewe. Proceedings of sessions of the Sheep and Goat Commission of the European Association for Animal Production, Brussels, August 1978. European Association for Animal Production Publication No.23, pp76-83. ISBN 0-906562-00-7.

UK, Ministry of Agriculture, Fisheries and Food (1975) Energy allowances and feeding systems for ruminants. Technical Bulletin No.33. London, UK; HMSO, viii+79pp. ISBN 0-11-240894-X.

USA, National Academy of Sciences, National Research Council, Committee on Animal Nutrition (1975) Nutrient requirements of domestic animals. 5. Nutrient requirements of sheep, 5th edn. revised. Washington, D.C., v+72pp. ISBN 0-309-02212-6.

Wallace, L.R. (1948) The growth of lambs before and after birth in relation to the level of nutrition. Journal of Agricultural Science, UK 38, 93-153.

Whiting, F.;Slen, S.B.;Bezeau, L.M. (1952) The quantity and quality of mature ewes' milk as influenced by level of protein in the ration. Scientific Agriculture 32,365-374.

Young, N.E.;Newton, J.E.;Orr, R.J. (1977) The performance and intake of supplemented and unsupplemented ewes at pasture during early lactation. Journal of the British Grassland Society 32,54.

MANAGEMENT AND DISEASE CONTROL IN LAMBING EWES

B. MITCHELL

Animal Diseases Research Association, Moredun Institute,
408 Gilmerton Road, Edinburgh

Productivity and the financial success of sheep management depends on balancing the costs of production with the return from sales. In this crude equation the contribution by the ewe, which is most likely to vary between seasons, is the value of lambs available for sale either as finished carcasses or stores.

Mortality of lambs is predominantly a problem of the perinatal period and it seems logical to direct investment especially towards reducing the death rate associated with the lambing period. The extent of the losses which have to be minimised can be assessed from available survey data indicating that annually in Britain up to 4-5% of lambs are born prematurely or still-born (Watson 1973); 15-20% of all lambs born die before weaning (Haughey 1973, Dennis 1974). Stamp (1967) has attributed 70% of these pre-weaning deaths to the perinatal period.

The aims of management of lambing ewes can be categorised under three main headings:

(i) physical environment and convenience of supervision

(ii) diet and nutrition

(iii) disease control

PHYSICAL ENVIRONMENT AND CONVENIENCE OF SUPERVISION

There are three main lambing situations in Britain:

(i) Intensive flocks for early lamb production (December to February)

(ii) Commercial flocks in lowland areas (March and April)

(iii) Extensive flocks in hill areas (April and May)

The climate throughout the lambing period is usually adverse and shelter provision can be simple or capital intensive, depending on how much one is prepared or able to spend.

A degree of prejudice exists against taking lambing ewes inside in the belief that their exercise is restricted, there is an increase in the occurrence of dystocia, footrot becomes prevalent, outbreaks of pneumonia are exacerbated and, on being returned outside, their desire to forage for food is reduced. Some or even all of these beliefs are not ill-founded but for reasons other than the straight consequence of being housed indoors. On balance the environmental management adopted for the three flock types is very much an individual choice.

Intensive flocks

Housing has become increasingly popular, for some situations throughout the winter period. This practice frees agricultural land for other purposes and permits a growth of fresh grass to be available in spring. In other situations the ewes are accommodated only during the 2 weeks before and after the lambing period.

The buildings vary from capital intensive, specially designed structures to modified, existing buildings such as cattle courts and Dutch barns. Design features which should be incorporated are adequate ventilation from above the sheep to ridge outlets, adequate trough (450-500 mm; 18") and floor space (1.5 sq.m.; 12-15 sq.ft.) per ewe and dry bedding. After lambing, the floor area per animal should be doubled and a separate creep area for feeding the lambs is desirable.

Group sizes of 25-30 ewes are the optimum for a balance between ewe performance and structural costs. Smaller groups undoubtedly settle better into stable hierarchical relationships. Larger groups of pre-parturient ewes can be successful where the cover is used only for overnight dormitory accommodation but wet fleeces, which are often inevitable with this system, demand a sparse density of population or excellent ventilation to prevent the development of unacceptably high humidity with condensation. Ewes with lambs at foot should be committed to either a wholly outdoor or wholly indoor regime. When intermittently outdoors/indoors they accommodate less readily to a stable feeding routine and the lambs appear to acclimatise poorly to the winter weather.

Commercial flocks

Large commercial flocks, often of 300-600 ewes, are handled extensively outdoors. At lambing time they are kept in fields convenient for supervision, preferably sheltered by woodlands, stone dykes or geographical location and well drained to avoid poaching. Temporary shelter in the form of straw bales or windbreaks should be scattered throughout the area to protect the newborn lambs. A central handling area is necessary, with temporary pens to hold ewes for one to five days during fostering of orphan lambs. As part of this 'complex', a mobile hut should be included for the shepherd in which daily essentials such as milk, drugs, elastrator rubber rings, etc can be stored. Also desirable are facilities for boiling water, to encourage the maintenance of hygiene by the shepherd during the obstetrical examination and assistance of ewes, for the sterilization of instruments and for intermittent beverages.

The overnight supervision of these flocks varies. In some, temporary labour is hired and partly used to supervise the ewes overnight. In these situations the lambing ewes are brought at night into small paddocks or illuminated yards for frequent observation. Increasingly, as the cost of labour escalates, the ewes are left unattended during the hours of darkness.

Management of the prelambing group is made much simpler if it can be reduced in size to only those ewes within two weeks of lambing. This is practicable if the fortnightly expected lambing dates are known from sire marking by distinctive colours at mating. Smaller cohorts of lambing ewes can be fed with a diet more appropriate to their stage of gestation and jostling of the heavily pregnant by those early in gestation is avoided.

Within 24-48 hours of lambing, ewes and their lambs usually are moved into one of three fields, one each for single, twin and triplet lambs, to minimise mismothering and stealing by unlambed ewes.

Rationale for supervision

The principal reason for surveillance of ewes during lambing is to minimise lamb and ewe losses due to dystocia. In a survey of Scottish hill flocks, Funn (1968) described the mean prevalence of ewes experiencing difficulty at lambing as being 3.1%. The lamb mortality due to unassisted difficult birth was 2.8% of the lamb crop and where assistance was given to ewes the value was 1.5%. Deaths arise from three main causes:

(i) placental failure because of separation of the cotyledonary areas during prolonged parturition or occlusion of the blood flow in the umbilical cord

(ii) foetal distress due to prolonged impaction in the maternal pelvis or strangulation during incomplete delivery through the vulva

(iii) neonatal asphyxia from occlusion of the nares by foetal membranes and incomplete establishment of respiration due to excess foetal fluids in the nasopharynx and lungs.

Many of these losses can be avoided by timely assistance being given to relieve the dystocia.

The biggest problem is deciding precisely when manual assistance is warranted. This will vary between breeds of sheep and between primiparous and multiparous ewes. Whitelaw & Watchorn (1975) have summarised well the time intervals during normal lambing of a flock of South Country Cheviot sheep; as a general guide if no progress is evident within two hours of the ewe having started lambing, one should examine the presentation of the foetus.

Correction of malpresentations is a skill readily acquired by competent shepherds and veterinary surgeons. Aspects of management worthy of comment are:

(i) avoid moving ewes away from the location where they start lambing until they have a lamb to mother lest the lambing pattern is upset and parturition considerably delayed

(ii) lambs delivered manually should always be presented to the ewe before she is released to avoid the lamb being abandoned

(iii) fostering of lambs on to ewes is achieved most easily in the immediate post-natal period.

Extensive hill flocks

The degree of supervision practicable or desirable in hill flocks has been an issue widely debated for many decades. The location of ewes during pregnancy and at parturition is dictated predominantly by the system of nutrition which has been adopted. If little or no supplementary feeding is being offered, the ewes must necessarily forage for their maintenance diet and lamb in the environment where they find themselves. More commonly the ewes are brought off the hill areas to more sheltered inbye fields when supplementary food is available. Lambing is essentially unsupervised except for daytime attention to ewes which are seen to be in distress. In some flocks the degree of attention is comparable to that for commercial lowground ewes but this is only practicable where adequate inbye ground is available on the hill unit.

DIET AND NUTRITION

Two conditions worthy of mention are prolapse of the vagina and pregnancy toxaemia in late pregnancy. Prolapse of the vaginal wall, often including the cervix and the bladder, is exacerbated by increased intra-abdominal pressure sometimes associated with the excessive feeding of turnips or highly fermentable foods. Pregnancy toxaemia at one time was claimed to be a consequence of inadequate exercise but is now associated with low levels of nutrition, particularly in ewes with twin and triplet pregnancies.

DISEASE CONTROL

The concepts of disease control have been approached from three main aspects:

(i) Intra-uterine infections; abortion, weak lambs, control

(ii) Hygiene of lambing areas

(iii) Vaccination programmes and immune status.

Abortions in sheep occur predominantly towards the end of pregnancy through infections such as chlamydia, toxoplasma and border disease; although a proportion abort dead foetuses, many ewes produce both dead and live lambs at term or prematurely. Frequently lambs born alive are weak and infected by the aetiological organisms. In the context of management the primary concern is identifying the causal organism. Action subsequently is directed at minimising extension of disease to susceptible pregnant ewes in the flock. All aborted ewes should be segregated from the lambing flock, mainly for reasons of hygiene but also to prevent barren ewes jostling heavily pregnant ewes and new born lambs. All placentae and dead lambs should be removed to avoid reservoirs of infection. In explosive outbreaks it is not possible to predict the ewes which are already incubating disease but in large flocks and those having a long lambing season, the ewes not immediately due to lamb should be shifted to a clean environment. In the face of an abortion "storm" due to campylobacter fetus, immediate vaccination of the pregnant ewes has been shown to reduce significantly the severity of the outbreak (Gilmour et al 1975).

A decision which must be taken is what to do with the ewes which have aborted. The practice of exchanging nurse ewes between flocks should be discouraged unless one is happy that their lambs died for purely physical reasons and the ewes will not introduce infection to a susceptible flock. Within a flock, lambs can be fostered on to ewes which have milk but their status as excretors of infection should be carefully considered. Salmonella, brucella and campylobacter infections can each be carried over to a further season and ewes harbouring these organisms should be segregated until culled. Border disease is similarly thought to be transmissible during subsequent seasons and should be considered in the same way.

Ewes which have recovered from chlamydia and toxoplasma infections are immune for life and may be retained safely for further breeding. Chlamydia infections are transmitted to susceptible females who will abort at their next pregnancy. Thus enzootic abortion ewes should never be mixed with maiden hoggs or ewes which have not previously been exposed to chlamydia. Susceptible females should be vaccinated before mating with ovine enzootic abortion vaccine (Wellcome).

Toxoplasma infection similarly is transmitted to susceptible females and if pregnant they may well abort. If non-pregnant they become immune following infection and are protected before their next pregnancy. Thus, electively, aborted ewes should be mixed with maiden hoggs deliberately to infect and immunise them while non-pregnant. If mixing is postponed till mating time there may be a problem during the subsequent pregnancy.

Hygiene of lambing areas

A serious consequence of folding ewes in a "lambing area" is the build-up of infection in the soil, bedding and associated woodwork. This is a serious hazard during the lambing period and can carry disease over from a previous season. Common infections are:

(i) staphylococci, streptococci, corynebacteria and *Erysipelothrix rhusiopathiae*, associated in the ewe with metritis and mastitis and in the lamb with umbilical and joint infections

(ii) E. coli, rotovirus and Clostridium welchii (type B), associated with "watery mouth". neonatal diarrhoea and lamb dysentery

(iii) other clostridial organisms and fusiforms, which are sources of tetanus, enterotoxaemia, footrot, etc in both ewes and lambs.

To avoid such infections, lambing areas preferably should be temporary in construction and location. During extended lambing sesons, successive groups of ewes ideally should have freshly erected lambing areas in a place not used in a previous season or, if indoors, in a building where the floors have been thoroughly cleaned and the woodwork creosoted since previous exposure to infection.

Many commercial flockmasters choose to lamb their ewes outdoors despite adverse climatic conditions rather than annually endure serious problems of ewe and lamb mortality associated with contaminated premises.

Vaccination programmes and immune status

An important concept in the management of lambing ewes is their immune status and the transfer of antibodies to their lambs. Natural passive immunity in sheep is achieved exclusively by the transfer in colostrum of antibodies secreted during the first few days of lactation and absorbed efficiently by the gut of the lambs, up to 24 hours of age but preferably during the first 6-12 hours. The ability of lambs to survive and acclimatise to the microbial flora, with which the adult sheep flock is in ecological balance, is determined in large measure by the adequacy of their immune globulin levels. Lambs having low gamma globulin levels are particularly susceptible to neonatal infections.

Specific infections against which it is important that lambs have specific antibodies are clostridia in all situations and louping-ill in tick infested areas. Other infections such as rotavirus and E. coli will be discussed in detail by subsequent speakers. Management of the lambing ewe should be directed at ensuring by a vaccination programme that she has high antibody levels in her colostrum.

Equally important is the need to ensure that new born lambs ingest adequate quantities of colostrum. For this purpose colostrum should be stored, frozen if necessary, ready to be given to orphan lambs or to those whose mothers have inadequate secretion to satisfy the needs of twins and triplets.

Clostridial vaccination in Britain can be approached by providing for isolated diseases such as pulpy-kidney/enterotoxaemia and tetanus. More commonly however a vaccine having eight components is used. Antigens included are Clostridium welchii types B, C and D; Clostridium oedematiens types B and D; Clostridium chauvoei; Clostridium septicum and Clostridium tetani.

The Wellcome Foundation Ltd manufactures two vaccines; 'Covexin 8' (Wellcome) is an aqueous preparation for subcutaneous use, and 'Omnivax' (Wellcome) contains similar antigens but is prepared with an oily adjuvant and is administered intraperitoneally. Booster doses for the latter are achieved by 'Omnivax CVA' (companion vaccine aqueous - Wellcome) administered subcutaneously. The timetable of administration is simplified in Table 1.

There are many areas of management of lambing ewes where decisions have to be pragmatic rather than ideal. Vagaries of the weather, availability of foodstuffs and above all unexpected outbreaks of disease each can dictate the need for a dramatic change of course. Much basic knowledge is lacking in our understanding of the epidemiology of uterine infections and how best to manage the available resources of breeding stock and vaccination programmes.

Table 1: Clostridial vaccination - Volume, route of administration and timetable of vaccination of sheep with 'Covexin 8' (Wellcome) and with 'Omnivax' and 'Omnivax CVA' (Wellcome).

<u>'Covexin 8' (Wellcome)</u>

First dose[1]	5 ml subcutaneously
Second dose 6 weeks later	2 ml subcutaneously
Booster dose	2 ml subcutaneously

The booster dose is advocated at 6 month intervals or annually if for low risk situations. The booster dose (or second dose in situations of primary vaccination) should be given 2 weeks before lambing to achieve maximum colostrum antibody levels.

<u>'Omnivax' (Wellcome)</u>	oily adjuvant
First (and only) dose[1]	2 ml intraperitoneally

Protection 2 years.

Booster dose	2 ml 'Omnivax CVA' s/c

<u>'Omnivax CVA'</u> (companion vaccine aqueous - Wellcome). Used as a booster dose at 2 year intervals after primary vaccination by 'Omnivax'. The booster dose may be given 12 to 2 weeks before lambing to achieve maximum colostrum antibody levels.

[1] For lambs which have been passively protected, the "first dose" should not be given till they are 3-4 months old.

REFERENCES

Dennis, S.M. (1974) Perinatal lamb mortality in Western Australia. I. General procedures and results. 2. Non-infectious conditions. Australian Veterinary Journal 50,443-449;450-453.

Gilmour, N.J.L.;Thompson, D.A.;Fraser, J. (1975) Vaccination against vibrio (Campylobacter) fetus infection in sheep in late pregnancy. Veterinary Record 96,129-131.

Gunn, R.G. (1968) A note on difficult birth in Scottish hill flocks. Animal Production 10,213-215.

Haughey, K.G. (1973) Birth injury to the central nervous system as a component of perinatal lamb mortality. New South Wales Veterinary Proceedings 9,29-30.

Houston, D.C.;Maddox, J.G. (1974) Causes of mortality among young Scottish Blackface lambs. Veterinary Record 95,575.

Stamp, J.T. (1967) Perinatal loss in lambs with particular reference to diagnosis. Veterinary Record 81,530-534.

Watson, W.A. (1973) The prevention and control of infectious ovine abortion. British Veterinary Journal 129,309-314.

Whitelaw, A.;Watchorn, P. (1975) An investigation into dystocia in the South Country Cheviot flock. Veterinary Record 97,489-492.

SCRAPIE DISEASE OF SHEEP

DR J.T. STAMP

Animal Diseases Research Association, Moredun Institute, 408 Gilmerton Road, Edinburgh

Virologists considering viral diseases have been mainly concerned with acute febrile disease in which the causative virus enters the body where it multiplies rapidly so that symptoms appear after incubation periods of some two days to several weeks. Host defences are thereafter mobilised and unless death of the host has occurred the infecting agent is eliminated and convalescence begins.

Scrapie of sheep is however a very different kind of infectious disease and since much is now known of its unusual features such knowledge is being increasingly used in attempts to elucidate other chronic degenerative diseases of animals and of man such as mink encephalopathy, kuru, Creutzfeldt-Jacob disease and senile dementia.

Scrapie is a naturally-occurring disease of sheep and rarely of goats in many parts of the world including the UK. Scrapie in sheep is of considerable economic importance to many sheep farmers and it has been recognised for at least two hundred years. It is a subacute or chronic disorder of the central nervous system although its name is derived from the pronounced rubbing and scratching of the skin which is an obvious feature of the clinical picture in most affected sheep. The natural disease develops progressively over months or years following a long incubation period of years. Death is inevitable once obvious clinical signs have developed. Scrapie is obviously a disease of adult animals with a peak incidence at three to four years; the earliest cases are seen at fourteen months old and continue to occur at all later ages, even beyond ten years old.

No specific lesions have been demonstrated in any organ other than the central nervous system. The histological lesions in the natural disease in sheep are mainly in the grey matter of subcortical brain centres, are bilaterally symmetrical and consist of a progressive vacuolating degeneration of neurones, status spongiosus and hypertrophy of glial cells. The extent and degree of the lesions vary considerably as will be discussed later but, qualitatively, there is little variation.

TRANSMISSION

Natural

It has long been recognised that a familial pattern exists in natural scrapie of sheep and it is known that when affected ewes are mated with affected rams the resulting offspring nearly always become affected. Reciprocal cross matings show that the status of the mother is much more important than that of the sire. Vertical transmission therefore is the most important means of dissemination but lateral transmission also occurs.

In neither vertical nor lateral transmission is the exact mechanism known but the disease is not dependent upon milk-borne transmission nor indeed may it be milk-borne at all.

Experimental

The cause of scrapie has not been determined but an experimental disese can be produced in sheep, goats, mice, hamsters, rats, gerbils and mink by one or other or all of the following routes - intracerebral, intraperitoneal and oral - using many different tissues from naturally- or experimentally-affected animals. The disease can be passaged in series. The lesions in the experimental disease are in general terms similar to those seen in naturally-affected sheep.

AGENT

No method has been found which will separate the transmissable agent from host tissue so that it has not been concentrated or purified in vitro. Greatest activity is always present in the CNS in clinically-affected animals. The agent is, in biological terms, very resistant to heat and many chemicals including formaldehyde, ultra-violet and ionising radiation. DNase, RNase, lipase, phospholipase, protease, etc have little effect on the titre of preparations.

The target size to irradiation suggests that the active particle has a molecular weight of about 1.5×10^5 and it is therefore considered by some workers to be much too small to contain nucleic acid capable of coding for the usual viral protein of animal viruses. However, plant viruses have been described in which protein is not an essential component in their replication, in which case the small size of the active particle of scrapie does not preclude a nucleic acid component.

Electron microscopy has not identified any biological agent.

Extensive tissue and cell culture work has given no evidence of cytopathic change but the agent does replicate although at no greater rate than the cell division in the culture. There is no release of agent from the cultured cells.

For many and different reasons it is not possible to draw any definite conclusions from the results of work done on the biochemical changes in scrapie-affected tissues.

Many ingenious theories, mainly concerned with explaining agent replication without the necessity for nucleic acid, have been postulated regarding the nature of the scrapie agent. Supporting experimental evidence is not sufficient to confirm any one of them. The evidence of agent variation in fact refutes several of these theories.

HOST RESISTANCE

Specific immune

A feature of both natural and experimental scrapie disease is the complete absence of any demonstrable specific immune processes and the scrapie agent has not been shown to be antigenic nor does it induce synthesis of interferon. Affected animals are capable of immunological response and of interferon production when stimulated with other agents.

Genetic

Because of the very variable occurrence and incidence of the disease at the farm, breed and experimental level, it has been frequently postulated that some sheep must possess a genetic resistance to scrapie but evidence of this in the natural disease has not been adequately demonstrated largely because maternal transmission complicates analysis of this type of data. However, under experimental conditions, sheep have been shown to be either susceptible or resistant to specific strains of scrapie agent and that this variability is under genetic control.

The experiment in which this was demonstrated consisted of mating scrapie-free Cheviot sheep on an isolated farm. After each year's lambing the parents were transferred to the laboratory where their susceptibility or resistance to the SSBP-1 scrapie agent given subcutaneously was determined. On the farm rams were chosen each year and used on the basis of their parents' susceptibility, one line of sheep being selected for an increased incidence and the other for a decreased incidence of experimental scrapie when challenged with SSBP-1 agent.

The results indicate a considerable and progressive divergence in susceptibility, the susceptibility in this case behaving as a genetically-dominant gene. It must not be assumed however that the susceptibility and resistance of these two lines of sheep would remain the same if they were challenged by scrapie material other than SSBP-1. In sheep there is as yet no clear evidence one way or the other on this point although preliminary findings suggest that the variation would be different. There is also good evidence that the sheep resistant to subcutaneous inoculation with SSBP-1 are not all resistant to intracerebral inoculation with the same material.

The failure of the scrapie-infected host to develop any immunological reaction along with only poorly developed in vitro culture techniques has given no scope for orthodox microbiological methods in epidemiology and pathogenesis studies. However, some progress has been made using incubation period, biological assay of activity and brain lesion profiles as variables capable of measurement.

Although considerably more progress has been made using scrapie in mice as the model it is more appropriate that the findings in sheep be considered first since the natural disease occurs in this species.

INCIDENCE

It is very clear from observation of the natural disease on commercial farms that in a scrapie-contaminated environment some sheep do not develop overt clinical signs during their lifetime.

On some farms only one or two cases occur each year in a flock of several hundred sheep but on other farms as many as half the sheep die of scrapie before the end of their expected economic life which is on average about six to seven years.

INCUBATION PERIOD

Since the size and nature of the natural challenge, the route of infection and the genetic make-up of the host are largely uncontrolled and unknown in field situations, it is quite impossible to give valid reasons in terms of incubation period etc for the considerable variation in age at occurrence or for the peak incidence being at three to four years.

In the experimental disease in sheep, even where the origin of the inoculum and the dose and route of inoculation can be defined, it has not been possible to utilise variation in incubation period for assay other purposes, even though these are well established methods in mouse-scrapie. Indeed it is even impossible, with commercially available sheep, to know what incidence of disease is likely to occur in any one batch of inoculated sheep even when they are allowed to survive to a very old age. It has been suggested that in many sheep the incubation period may be longer than the natural life-span of the sheep and that this is the reason for their apparent resistance. Assay, using mice, is however still too unreliable to test this hypothesis by the demonstration of scrapie agent in the tissues of such ageing sheep.

LESION PROFILE

So far as lesion profile is concerned it is known that the degree, extent and distribution of the degenerative lesions in the CNS of sheep can be different.

Variation has been found in comparisons of natural scrapie in different breeds, between natural and experimental scrapie in a single breed, in experimental scrapie produced by inoculum from different sources and in experimental scrapie produced by different stages of passage in random-bred mice.

It is not possible however to determine whether the variations in lesion pattern are due to differing agent strains or to differing genotypes of sheep, neither of which could be defined at the time when the observations were made.

Probably the only meaningful finding in studies of distribution of lesions in sheep is that in Cheviot sheep, experimentally inoculated subcutaneously with the SSBP-1 inoculum; degeneration and vacuolation of the neurons and ground substance of the brain are, in many cases, almost absent even when the clinical disease is advanced. This would suggest that the visible microscopic neuronal damage is not the direct cause of the clinical symtoms and the eventual death of the animal. However, with the same genotype of sheep which develop minimal lesions with SSBP-1 it has been shown that extensive lesions occur if a different agent is injected subcutaneously.

This finding poses the problem of whether some strains of scrapie agent may produce no spongy degeneration yet kill the host. The present position clearly makes histological diagnosis difficult and argues against use of the term 'spongy encephalopathies' as an appropriate name for the group of diseases which includes scrapie.

MOUSE TRANSMISSION

In contrast to the sheep, use of the mouse as an experimental animal has advanced knowledge considerably. It is now generally assumed that the mouse is readily infected by sheep scrapie and that the incidence is about 100%. This is not always so however for it is still not possible to passage scrapie from some cases of natural scrapie in sheep to mice or in other instances the incidence may be very low and the incubation period very long. Two types of explanation are possible for this; one is that some type of species-adaptation occurs, implying that some component of host tissue is a necessary part of the infective unit, or, secondly, that sheep are infected with various strains of scrapie agent of which only some are transmissible to mice and that certain sheep are infected with several strains of agent. These alternatives are not mutually exclusive but there is more evidence for the second possibility than the first.

When the agent is passaged once in the mouse, and the correct inbred strain of mouse used, the incidence is 100% except at diminishing dilutions of the agent (which vary according to the route of inoculation and mouse genotype) and such models have been successfully used to achieve relatively accurate and reproducible quantitative results based upon the length of incubation period as the unit of measurement.

The incubation period can be shown to vary with the titre of inoculum, the genotype of the mouse, the route of inoculation and the source and type of the inoculum. In general, the incubation period is lengthened when the recipient and donor species are different, when a smaller dose of agent is given or when a peripheral route of infection is used rather than the intracerebral one.

There is no evidence of vertical transmission of scrapie in the mouse and evidence of true contagion, even to cage mates, is unsatisfactory. No evidence of specific immune processes has been found in mice.

INCUBATION PERIOD AND <u>SINC</u> GENE

As a result of very careful genetic studies it hs been determined that in mice a single gene known as <u>sinc</u> controls the incubation period of a wide range of scrapie agents. Two alleles are known and have been designated s7 and p7. A good example of the behaviour of this gene can be seen with a strain of scrapie known as ME7 agent originally derived from a Suffolk sheep. When this agent is inoculated into s7s7 mice a short incubation period follows but when inoculated into p7p7 mice (VM inbreds) a long incubation follows. The reverse is the case when the scrapie agent used is 22A originally derived from a Cheviot sheep. Nevertheless, it is certain that the one gene is operating in all cases.

The type of allelic interaction can range from absence of dominance through partial dominance of either allele to full dominance and even overdominance when different strains of scrapie are used. The gene exerts a control over the incubation period whether the inoculum is given by the intraperitoneal or by the intracerebral route. The site or sites at which the scrapie agents replicate are not known but they are probably limited in number. This suggestion is supported by the experimental findings which show that one strain of scrapie agent when established in the host tissues lengthens the incubation period of another strain introduced later.

AGENT VARIATION

Incubation period

The gene sinc has been used to show that there is extensive genetic variation between agents derived from various sources both natural and experimental.

For each defined agent the ratio of incubation period in inbred mice homozygous for the p7 allele (e.g. VM mice) to that for mice homozygous for the s7 allele (e.g. C57BL mice) is relatively constant over a wide range of doses of each agent.

Since the length of incubation period in a single inbred strain of mice, although relatively constant for any one agent, is still dose-dependent it is essential to know the dose response curve and titre of each agent before any deductions can be made from observed incubation periods.

Again it must be emphasised that these methods can at present only be used if the agents being examined have already been mouse-passaged. They are still of only very limited value for the examinatioin of agents taken directly from sheep tissues.

Lesion profile

Strain variation can also be demonstrated by comparing the distribution and degree of grey matter lesions in inbred strains of mice following intracerebral inoculation with mouse-passaged scrapie from different sources. A useful and original technique based on scoring the intensity of vacuolation in nine standard areas of the brain has been developed for this purpose and the findings are referred to as 'brain lesion profiles'. For any specified genotype of mouse and route of injection, the shape and overall height of the lesion profile is specific for each strain of scrapie agent which has so far been tested and this holds true over a wide range of doses of agent. Two types of inbred mice, C57BL and VM, are now in standard use for agent strain identification from lesion profiles; it is clear however from the results in these two strains and their F_1 cross that there is genetic control of profile shape but this is complex and is not known to involve the *sinc* gene.

Bilateral symmetry of lesions

Although it is generally accepted that the CNS lesions of scrapie are bilaterally symmetrical it has recently been reported that certain defined agents cause grey matter lesions which deviate consistently from a bilaterally symmetrical distribution. Results so far only relate to intracerebral injections and whether the asymmetry will occur with peripheral injections is not yet known. There is known to be some broad relationship between the site of injection in the brain and the subsequent asymmetry. This finding suggests that the genesis of the grey matter lesions can have both immediate and long-term components.

PATHOGENESIS

Agent replication

Scrapie has frequently been called 'a slow virus infection' or referred to as a disease in which clinical signs only appear after a long 'latent' period. However, the obtaining of factual evidence as to what actually happens to the agent when it enters host tissues has only been possible in the experimental disease in the mouse. The findings show that use of the term 'latent' in connection with scrapie is incorrect because the preclinical phase is one in which pathogenesis is going ahead in a continuous controlled sequence and there is no inactive 'latent' phase awaiting some environmental 'trigger'.

Following inoculation by any route multiplication is first detected in lymphoreticular tissue, particularly the spleen, and a maximal spleen concentration can be reached as early as four to six weeks after injection. High titre is reached earlier in mice homozygous to the s7 allele than in mice homozygous to the p7 allele when the agent used is ME7. The maximum concentration of agent in the brain is reached much later than in the spleen: again, in the case of ME7, this occurs earlier in mice homozygous for the s7 allele than in those homozygous for the p7 allele.

The difference in time at which maximum concentration is reached does not appear to depend on the difference in speed replication but rather by a delay in the start of replication and it would seem that the *sinc* gene controls this time of start of replication.

Route of inoculation

There has been debate as to whether scrapie agent multiplies in the brain or whether the agent first replicates in the lymphoreticular system, including the spleen, and is then transported to the brain even after an intracerebral inoculation.

It is known that even when the inoculation is intracerebral there is a relatively rapid increase of agent in the spleen where it replicates to maximum titre long before similar titres are reached in the brain. However, the fact that the incubation period following extraneural inoculation is always longer than that following inoculation of the same dose of the same agent into the brain suggests very strongly that replication in the brain must follow from some small remnant in the brain of the original inoculum. If this is so then there must be some reason for the very long interval between injection and the first appearance of rapid multiplication in the CNS, an interval which does not occur before replication in the spleen takes place. Further confirmation that replication in the spleen, though rapid, is irrelevant to pathogenesis following intracerebral injection comes from the finding that splenectomy or congenital absence of the spleen increases the length of the incubation period by some 20% following intraperitoneal inoculation, but in no way influences the incubation period following intracerebral injection.

Leson profile

Additional evidence that the replication following intracerebral inoculation is not dependent upon prior multiplication in extraneural tissue is that the brain lesion profile is quite different following an intracerebral inoculation than that following an intraperitoneal one, even if sterile brain trauma is implicated at the same time as the intraperitoneal inoculation is made.

SOME FEATURES OF SCRAPIE IN SHEEP SIGNIFICANT IN FIELD WORK

As has already been stated death due to scrapie occurs most frequently between 2 and 5 years of age. Scrapie in older sheep is not often seen because husbandry practices severely limit the number of older sheep kept on a farm. There is considerable evidence that the majority of sheep developing scrapie between 2 to 5 years of age have been infected since early life and the agent has replicated within the sheep during the whole of this period without causing any apparent symptoms. Thus length of incubation period and the age of the sheep at death are virtually the same. Symptoms of scrapie are often considered to be specific for diagnosis and this is frequently the case but in some sheep clinical signs are by no means typical and can be very misleading. For example pruritus and debilitation may be absent, affected sheep merely showing incoordination of movement. Other affected sheep may die quickly after collapse and paralysis.

Although scrapie has occurred in most areas of the world for long periods of time the disease has apparently never become endemic in New Zealand or Australia despite the earlier frequent importation of sheep from countries where scrapie is endemic. In other instances however apparent freedom from scrapie is often more apparent than real since the disease is so widely feared that suspect sheep are slaughtered and no diagnosis obtained. Such secrecy makes recognition and hence control of the disease difficult as there is no specific diagnostic test for the living animal to confirm or refute freedom from scrapie.

It has long been recognised that a familiar pattern often exists in flocks affected with natural scrapie. It is very unfortunate that the working out of the epidemiology of scrapie has been bedevilled by the general acceptance that because related individuals develop scrapie the disease must be entirely carried by a defective gene and hence the occurrence of cases fit a pattern of Mendelian gene segregation. There are, of course, reasons other than defective genes why relatives can be affected by the same disease and so far as scrapie is concerned it is now well established that maternal transmission of an independent scrapie agent accounts for much of the familiar pattern seen in the field. Ewes which eventually die of scrapie frequently have several crops of lambs before they show clinical signs of the disease and the evidence indicates that the earlier born lambs are equally liable to develop scrapie as those born during the clinical period of the disease. At what time the lambs become infected is not well established; whether the embryo is infected or whether it is during or immediately after the birth process is not clear. That the scrapie agent is present in the placenta of affected sheep is however well established so that maternal transmission almost certainly occurs early in life or in the uterus. It is not known whether maternal transmission occurs in goats.

Until comparatively recently it was not generally accepted that scrapie is an infectious disease and indeed even now the majority of flockmasters and some investigators do not consider it to be so. Nevertheless there is much indirect evidence that scrapie must be infectious and irrefutable direct evidence has recently been obtained. However for a disease to be infectious the causal agent must leave the infected host and enter a new clean host in some way. The scrapie agent has up to the present not been detected in any body secretion except in the placenta nor has much work been directed towards detecting the agent in pasture. Earlier Icelandic work on scrape (Rida) eradication programmes does however strongly suggest that infection either persists on pasture for very long periods of time or in an intermediate host other than sheep or goats. The fact that rams are frequently thought to be infective in the field while non-pregnant ewes have been shown to be contagious experimentally suggests that the placenta is not always the source of infection. More is known regarding routes of entry of infection; oral dosing and conjunctival contamination are both efficient routes as also is wound contamination.

Although maternal transmission and contagion are the means of infection in the field it is very clear that not all sheep develop scrapie even when the opportunity of infection is present. As already indicated a "resistance" and susceptibility factor genetically controlled has been demonstrated in sheep experimentally infected but under field conditions there are uncontrolled variants which make a similar analysis impossible. Such variants include the scrapie agent strains present in a flock over a number of years, the degree of maternal transmission which can vary considerably, the weight of infection present at any one time, the age of the sheep when exposed to lateral transmission, and whether there are strains of scrapie agent present in the sheep which in themselves may not cause clinical signs but nevertheless "vaccinate" the sheep against more lethal strains.

Much has been written about the different strains of scrapie agent in mice but only recently have studies been reported on the isolation of agents from natural scrapie in sheep, information that is very necessary when considering the epidemiology of the natural disease. Present findings indicate that ME7 agent is widely disseminated in many breeds while infection of sheep with more than one strain of agent is fairly common, the mixture of strains often resulting from contagion. Lastly some strains of scrapie in sheep cannot be transmitted to mice.

In countries where scrapie is endemic control of scrapie on a farm (but not eradication) depends upon rigorous culling of affected maternal lines. In the USA where scrapie has probably not become endemic extensive slaughter of flocks and the tracing and slaughter of source flocks has probably done much to prevent the disease becoming endemic but the cost has been considerable. Currently, New Zealand has a problem of imported scrapie and it will be interesting to see if anything can be salvaged or whether complete elimination of all imported sheep and their contacts and offspring will be necessary.

Following upon the successful genetic selection techniques in sheep at both the Moredun and Compton Institutes it is tempting to think that similar techniques could be applied in the field for the control of scrapie but it is unlikely that sheep selected as resistant against one or more strains of scrapie will be of any value when faced with other and perhaps new strains of scrapie. There is also an unknown risk that resistant sheep could be slowly incubating the disease over their span of life rather than resisting infection and so disseminating infection to the environment.

Scrapie is, of course, not the only slow virus infection for in recent years a number of other diseases of both man and animals have been shown to have similar features to those of scrapie and are also transmissible by inoculation to primates, hamsters, etc. Kuru of New Guinea, Creutzfeld Jakob disease, familial Alzheimer's disease and probably other senile and presenile dementias of man along with transmissible milk encephalopathy are presently being investigated. The demonstration of amyloid and senile plaques produced at will in the brains of certain mouse strains inoculated with certain scrapie agents could be a model for an even wider field of dementias in man. Concern about the similarity of scrapie to these disorders of man has recently caused the USDA to ban the use for human food not only meat of sheep affected with scrapie but also meat of related or contact sheep or goats. Although there is no evidence at all that sheep meat causes these diseases in man, the fear of such a possibility could have serious consequences for the sheep industries of countries where scrapie is endemic.

REFERENCES

Brotherston, J.G.;Renwick, C.C.;Stamp, J.T. Zlotnik, I.;Pattison, I.H. (1968) Spread of scrapie by contact to goats and sheep. Journal of Comparative Pathology 78,9-17.

Chandler, R.L. (1961) Encephalopathy in mice produced by inoculation with scrapie brain material. Lancet i,1378-1379.

Cuille, J.;Chelle, P.L. (1936) La maladie dite tremblante du mouton est-elle inoculable? Comptes Rendus Hebdomadaires des Séances de l'Académie des Sciences 203,1552-1554.

Dickinson, A.G. (1975) Host-pathogen interactions in scrapie. Genetics 79, Suppl. 387-395.

Dickinson, A.G.;Stamp, J.T.;Renwick, C.C.;Rennie, J.C. (1968) Some factors controlling the incidence of scrapie in Cheviot sheep injected with a Cheviot-passaged scrapie agent. Journal of Comparative Pathology 78, 313-321.

Dickinson, A.G,;Stamp, J.T.;Renwick, C.C. (1974) Maternal and lateral transmission of scrapie in sheep. Journal of Comparative Pathology 84, No.1, 19-25.

Dickinson, A.G.;Young, G.B.;Renwick, C.C. (1966) Scrapie: Experiments involving maternal transmission in sheep. In 'Report of Scrapie Seminar, 1964' [Report No. ARS 91-53] US Department of Agriculture pp244-247.

Dickinson, A.G.;Young, G.B.;Stamp, J.T.;Renwick, C.C. (1964) A note on the distribution of scrapie in sheep of different ages. Animal Production 6,375-377.

Dickinson, A.G.;Young, G.B.;Stamp, J.T.;Renwick, C.C. (1965) An analysis of natural scrapie in Suffolk sheep. Heredity 20,485-503.

Gordon, W.S. (1946) Advances in veterinary research. Louping-ill, tick-borne fever and scrapie. Veterinary Record 58,516-518.

Gordon, W.S. (1959) Discussion on studies in scrapie. Veterinary Record 69,1324-1328.

Gordon, W.S.;Brownlee, A.;Wilson, D.R. (1939) Studies in louping-ill, tick-borne fever and scrapie. Proceedings of III International Congress on Microbiology, New York pp362-363.

Greig, J.R. (1940) Scrapie. Observations on the transmission of the disease by mediate contact. Veterinary Journal 96,203-206.

Greig, J.R. (1950) Scrapie in sheep. Journal of Comparative Pathology 60,263-266.

Hourrigan, J.L.;Klingsporn, A.L.;McDaniel, H.A.;Riemenschneider, M.N. (1969) Natural scrapie in a goat. Journal of the American Veterinary Medical Association 154,538-539.

Mackay, J.M.K. (1968) Detection of the scrapie agent in tissues of normal mice. Nature UK 219,182-183.

Mackay, J.M.K.;Smith, W.;Stamp, J.T. (1960) Experimental scrapie: some recent work. Veterinary Record 72,1002-1006.

Nussbaum, R.E.;Henderson, W.M.;Pattison, I.H.;Elcock, N.V.;Davies, D.C. (1975) The establishment of sheep flocks of predictable susceptibility to experimental scrapie. Research in Veterinary Science 18, No.1,49-58.

Palmer, A.C. (1959) Scrapie, a review of the literature. Veterinary Review and Annotations 5,1-15.

Palsson, P.A.;Sigurdsson, B. (1959) Rida. En langsom, progredierende, infektiös nervesygdom hos far. Proceedings of VIII Nordisk Veterinary Congress, Helsinki pp179-191.

Pattison, I.H. (1957) Myopathy in sheep. Lancet i,104-105.

Pattison, I.H.;Hoare, M.N.;Jebbett, J.N.;Watson, W.A. (1974) Further observations on the production of scrapie in sheep by oral dosing with foetal membranes from scrapie-affected sheep. British Veterinary Journal 130,65-67.

Pattison, I.H.;Millson, G.C. (1961) Experimental transmission of scrapie to goats and sheep by oral route. Journal of Comparative Pathology 71,171-176.

Stamp, J.T.;Brotherston, J.G.;Slotnik, I.;Mackay, J.M.K.;Smith, W. (1959) Further studies on scrapie. Journal of Comparative Pathology 69,268-280.

Wilson, D.R.;Anderson, R.D";Smith, W. (1950) Studies in scrapie. Journal of Comparative Pathology 60,267-282.

Zlotnik, I.;Rennie, J.C. (1962) The pathology of the brain of mice inoculated with tissues from scrapie sheep. Journal of Comparative Pathology 72,360-365.

NON-PARASITIC RESPIRATORY DISEASE OF SHEEP

DR W.B. MARTIN

Animal Diseases Research Association, Moredun Institute, 408 Gilmerton Road Edinburgh

Data on the prevalence of respiratory disease in sheep throughout the world are scanty. Some severe outbreaks have been recorded but the incidence of respiratory disease is generally unknown. In Britain pneumonia is recognised as being an important cause of disease and death though no accurate assessment of losses from non-parasitic respiratory disease is available.

In an attempt to throw light on the problem of sheep pneumonia, the changes present in the lungs of sheep have been classified (Stamp & Nisbet 1963). As the lung can only react to assault in a limited number of ways, this type of classification has only been partially successful, and sheep pneumonias are usually described as acute pasteurella pneumonias or the more chronic proliferative type (so-called "atypical"). Knowledge of the agents involved in respiratory disease of sheep has improved considerably over the past decade and we now know more of the role that viruses, mycoplasmas and bacteria play in this complicated disease problem. In this paper I shall summarise the situation, making use of the research work undertaken by my colleagues at Moredun Institute.

There is no doubt that, in Britain, Pasteurella haemolytica is at present the most important organism in respiratory disease of sheep. It can be found as a commensal on the nasal mucosa and tonsils of many sheep but under some circumstances P. haemolytica can become a lethal organism. We have no evidence over many years that P. multocida is an important pathogen in sheep in the UK.

Most sheep flocks in Britain contain sheep which are carriers of one or more serotypes of P. haemolytica, though the number of sheep carrying these organisms varies throughout the year, as does the distribution of serotypes (Biberstein et al 1970). All strains of P. haemolytica can be divided into two biotypes, i.e. those which ferment arabinose (A-types) and others which ferment trehalose (T-types). A-biotype strains, of which there are nine serotypes, are present in the tonsils of normal sheep (Gilmour et al 1974).

Biotype-A strains are usually recovered from cases of pneumonia and septicaemia in lambs under three months of age, whereas biotype-T strains are generally isolated from septicaemias of more mature sheep (Biberstein & Thompson 1966). Of the known serotypes, A2 and the T strains are most frequently encountered in disease situations in Britain.

The experimental reproduction of pasteurella pneumonia has been recognised as being very difficult and over many years workers have failed in their attempts to reproduce pneumonia in a repeatable way. Using Caesarian-derived, colostrum-deprived lambs, free from specific pathogens

(SPF), it has been possible to produce pneumonia with an aerosol of P. haemolytica of biotype A but only, at best, in less than 50% of lambs (Gilmour et al 1975). However, when parainfluenza 3 virus is given before administering the P. haemolytica, a much higher percentage of lambs develop clinical and even fatal pneumonia (Sharp et al 1978). Before dealing with this combined infection I should like to outline briefly the virus infections of sheep, particularly those found in Britain.

Several viruses have been recovered from sheep, which may be considered to be capable of infecting the respiratory tract, though not all have been shown to do so. These viruses include parainfluenza type 3, ovine adenoviruses types 1-5, a bovine adenovirus, reoviruses types 1-3, enteroviruses, a herpesvirus and a retrovirus. There is no evidence at present that maedi-visna occurs in Britain.*

Despite the claim by Belak & Palfi (1974) from Hungary that interstitial pneumonia was produced experimentally in lambs infected with reovirus 1, it is likely that reoviruses are generally unimportant as causes of disease in sheep. Reovirus 2 has been isolated from the nasal passages of one lamb at the same time as contact lambs showed rising antibody titres to reovirus, but none of the lambs showed any signs of disease (Snodgrass et al 1976). At the present state of our knowledge there is also no evidence that enteroviruses cause important respiratory disease. Antibodies in sheep to respiratory syncytial virus have been recorded but no virus has been isolated.

Antibodies to adenoviruses have been found in a varying percentage of sheep depending on the author, country and test used. In Scotland Sharp (1977) found antibodies to adenovirus type 1 (21%), type 2 (40%), type 3 (68%) and type 4 (71%) in sheep. Little is known of the pathogenicity of these viruses. The bovine adenovirus was isolated from sheep with signs of a "catarrhal pneumonia" and of which 10% died (Belak & Palfi 1974). In experiments at Moredun Institute ovine adenovirus type 4 caused both mild pulmonary eodema and focal hepatic necrosis (Rushton & Sharp 1976). Under certain circumstances adenoviruses may possibly predispose lambs to more serious secondary infections, such as occurred when type 4 virus was isolated originally from sheep dying of penumonia.

The most prevalent virus infection of the respiratory tract of sheep throughout the world would appear to be that caused by parainfluenza 3 (PI3). Antibodies to this virus are present in the majority of sheep, probably in all countries of the world. PI3 can be recovered from sheep with respiratory disease and experimentally, though serious pneumonia has sometimes been produced, the virus has generally been shown to cause a mild respiratory infection and in some lambs also small areas of consolidation in the lungs. The penumonic lesions on histological examination show hyperplasia of bronchiolar epithelium, infiltration of inter-alveolar septa and cellular exudate in the bronchiolar lumen. A

---

* Since giving this paper the presence in Britain of maedi-visna virus has been reported from a flock known to contain sheep imported from Europe.

transient but pathognomonic feature is the presence of acidophilic intracytoplasmic inclusions in the bronchiolar epithelial cells, but these are only present during the first week of infection.

The transient, mild respiratory infection produced by PI3 can be converted into a severe pneumonia when *P. haemolytica* bacteria are inhaled a few days later. The technique we have used to produce this form of pasteurella pneumonia is to inject SPF lambs intra-tracheally with PI3 virus and, seven days later, expose them to an aerosol of *P. haemolytica* of biotype A. Most of the exposed lambs develop a pneumonia which is identical to that evident in the field situation (Sharp et al 1978).

This method of infection is being used to study the pathogenesis of pneumonic pasteurellosis and allows the prophylactic value of vaccines to be assessed accurately.

A third category of micro-organisms associated with respiratory tract infections are the mycoplasma. Two mycoplasmas, *M. ovipneumoniae* and *M. arginini* are commonly associated with the respiratory tract of sheep in Britain but five other identified species, viz *M. conjunctivae, M. mycoides* subspecies *mycoides, M. agalactia, Acholeplasma laidlawii* and ureaplasmas, have also been recovered from the respiratory tract of sheep throughout the world. The role of mycoplasmas in sheep pneumonias is still not clear and more information is required.

M. arginini as a single infection in SPF lambs has not been shown to be pathogenic. It may be that together with more virulent respiratory micro-organisms it may alter or enhance respiratory infections, but little experimental evidence exists for the role of *M. arginini* at present.

M. ovipneumoniae will cause lung damage in lambs (St. George et al 1971, Foggie et al 1976). Histology of areas of lung infected with *M. ovipneumoniae* has shown the presence of inflammatory cells, mainly macrophages, in alveoli and peribronchiolar cuffing. However, these changes are difficult to reproduce repeatedly. For this reason experiments at Moredun have moved towards examining the possible synergistic effects of *M. ovipneumoniae* and *P. haemolytica* in particular. These experiments have not been completed but there is an indication that these organisms, acting together, produce more serious clinical and pathological changes than either alone.

A study of the mycoplasma infections of lambs in an intensive indoor lambing unit was made by Jones (1978) who found that *M. ovipneumoniae* was present in the nasal passages of only 3% of lambs up to 5 days of age, but the number of lambs infected increased with age so that by the time the lambs were between 100-114 days old 92% were infected. *M. arginini* was isolated from a small but fluctuating percentage of lambs during the 4 months of observation. Mild, clinically-apparent respiratory disease could be found in some lambs when they were 6-14 days old and the numbers affected increased so that, at 25 and 55 days of age, 10-25% of lambs showed signs and by about 3.5-4 months old over 50% of the lambs were affected. In the ewes in the flock only slight nasal discharge and coughing was ever noted.

Serum antibody titres in the lambs to M. ovipneumoniae declined from birth to 45-55 days after which they rose to plateau at about 140 days.

Thirty-five lambs were examined *post-mortem* but no correlation could be obtained between the presence of *M. arginini* and histological changes in lungs. Nor was it clear that the presence of *M. ovipneumoniae* in the lungs correlated with the lung changes of lymphoid hyperplasia and interstitial thickening. Though numbers were small the histological changes observed appeared to be more common in lambs killed late in the survey.

One chronic pneumonia which is present in sheep in Britain is pulmonary adenomatosis (jaagsiekte). This is a contagious adenocarcinoma in which the alveolar cells are replaced with cuboidal or columnar cells. The outcome is fatal for those sheep which develop clinical signs. Generally sheep are about three years old or more when they show signs and the disease affects a variety of breeds.

From the tumours a herpesvirus can be isolated (Mackay 1969) but this virus alone is incapable of producing the disease. More recently a particle containing reverse-transcriptase (a retrovirus) has been isolated from the tumour. Present experimental evidence suggests that both viruses act together to cause the tumour formation and disease (Martin et al 1976).

REFERENCES

Belák, S.;Pálfi, V. (1974) Pneumoenteritis of lambs caused by adenovirus. *Acta Veterinaria Academiae Scientarium Hungaricae* 24,327-328.

Biberstein, E.L.;Shreeve, B.J.;Thompson, D.A. (1970) Variation in carrier rates of P. haemolytica in sheep flocks. I. Normal flocks. *Journal of Comparative Pathology* 80, 499-507.

Biberstein, E.L.;Thompson, D.A. (1966) Epidemiological studies on P. haemolytica in sheep. *Journal of Comparative Pathology* 76,83-94.

Foggie, A.;Jones, G.E.;Buxton, D. (1976) The experimental infection of SPF lambs with Mycoplasma ovipneumoniae. *Research in Veterinary Science* 21,(1), 28-35.

Gilmour, N.J.L.;Thompson, D.A.;Fraser, J. (1974) The recovery of P. haemolytica from the tonsils of adult sheep. *Research in Veterinary Science* 17,(3), 413-414.

Gilmour, N.J.L.;Thompson, D.A.;Smith, W.D.;Angus, K.W. (1975) Experimental infection of lambs with an aerosol of P. sp haemolytica. *Research in Veterinary Science* 18,340-341.

Jones, G.E. (1978) PhD Thesis, Edinburgh.

Mackay, J.M.K. (1969) Tissue culture studies of sheep pulmonary adenomatosis (jaagsiekte). *Journal of Comparative Pathology* 79,147-154.

Martin, W.B.;Scott, F.M.M.;Sharp, J.M.;Angus, K.W.;Norval, M. (1976) Experimental production of sheep pulmonary adenomatosis (jaagsiekte). Nature UK 164,183-184.

Rushton, B.;Sharp, J.M. (1976) Pathology of ovine adenovirus type 4 infection in SPF lambs: pulmonary and hepatic lesions. Journal of Pathology 121, 163-167.

St George, T.D.;Sullivan, N.D.;Love, J.A.;Horsfall, N. (1971) Experimental transmission of pneumonia in sheep with a Mycoplasma isolated from pneumonic sheep lung. Australian Veterinary Journal 47,282-283.

Sharp, J.M. (1977) PhD Thesis, Edinburgh.

Sharp, J.M.;Gilmour, N.J.L.;Thompson, D.A.;Rushton, B. (1978) Experimental infection of specific pathogen-free lambs with parainfluenza virus type 3 and Pasteurella haemolytica. Journal of Comparative Pathology 88,237-243.

Snodgrass, D.R.;Burrells, C.;Wells, P.W. (1976) Isolation of reovirus type 2 from respiratory tract of sheep. Archives of Virology 52,(1/2) 143-149.

Stamp, J.T.;Nisbet, D.I. (1963) Pneumonia of sheep. Journal of Comparative Pathology 73,319-328.

UK, Ministry of Agriculture, Fisheries and Food (1976) Veterinary investigation diagnosis analysis report.

THE ARTIFICIAL REARING OF LAMBS

DR. P.D. PENNING

Grassland Research Institute, Hurley, Maidenhead, Berks

INTRODUCTION

The new born lamb requires a liquid diet and intake of solid food does not begin until lambs are about three weeks old (Penning et al 1973). If the lamb is not suckled by the ewe it must be provided with an alternative source of milk or milk substitute, which it must be trained to drink or suck.

The composition of the milk substitute used and the quantity given, the weaning age and the composition of the solid feeds offered all affect the performance of the lamb and its transition from a non-functional to a functional ruminant. An understanding of the complex factors controlling these relationships is necessary in order to devise successful systems for the artificial rearing of lambs.

NUTRIENT REQUIREMENTS OF THE LIQUID-FED LAMB

Colostrum

Colostrum provides the lamb with passive immunity to disease and is also a concentrated nutrient source. Campbell (1974) showed that 8 out of 10 lambs given ewes' milk and then dosed with a pathogenic strain of E.coli died whereas for lambs that had received colostrum only 2 out of 10 died. Table 1 also illustrates this point; some lambs that do not receive colostrum may survive but with lower growth rates than those given colostrum.

If the lamb obtains an inadequate amount of colostrum from its mother it is essential to feed it ewes' colostrum or a substitute. Ewes' colostrum can be deep frozen and stored until required for feeding. From Table 1 it can be seen that as little as 8g of colostrum per kg of birth weight will give some protection but it is suggested that 24g per kg would give better results. Nedkvitne & Haugland (1970) found 0.4 to 0.8 litres of cows' colostrum fed within 18 hours of birth was an adequate substitute although it might not provide immunity to sheep-specific diseases. Gamble et al (1938) have also suggested that blood serum from sheep could be an adequate substitute for colostrum.

Table 1 The effect of intake of ovine colostrum on survival and live-weight gain of artificially reared lambs

	Colostrum fed (g/kg birth weight) 0	8	16	24	Mean	± SE	Significance of difference
Number of lambs per treatment	6	6	6	6	-	-	-
Number of lambs surviving to 32.5 kg live weight	3	5	6	6	-	-	-
Daily live weight gain (g) 0-28 days	181[1]	282	313	323	284	27.3	**
Number of days to reach slaughter weight (32.5 kg)	155[1]	147[1]	135	123	138	11.3	NS

[1] survivors only

Southdown x Dorset Horn Lambs
(Penning, unpublished data)

Generally the characteristics of colostrum have disappeared from ewes' milk by two to four days after lambing (Dattilo 1969, Halliday 1970) and the gut of the lamb becomes impermeable to protein, including immunoglobulin, during the second day of life (Mason et al 1930, McCarthy & McDougall 1953). For these reasons, if a lamb is to be artificially reared there is no point in leaving it to suckle for longer than 48 hours. In fact there are disadvantages, as Bryant & Own (1971) have shown that the longer a lamb sucks its mother the longer it takes to learn to feed itself when reared artifically.

Vitamins and minerals

Although there is little information on the vitamin and mineral requirements of the young lamb, it appears that the requirements given by the A.R.C. (UK 1965) form a reasonable basis for calculating vitamin and mineral inclusions in milk substitutes. However Walker (1972) found that the values given by A.R.C. (UK 1965) for calcium requirements tended to be high but his results agreed with those of A.R.C. for phosphorus.

Vitamins, minerals and trace elements have been added to milk substitutes to give concentrations similar to those in ewes' milk (Walker

1975, Gibney & Walker 1977, Penning et al 1977b) and the supplements listed by these workers provide useful guidelines when formulating diets.

Protein, Fat and Carbohydrate

The three main constituents of ewes' milk are protein, fat and lactose which comprise approximately 30, 38 and 25% of the dry matter respectively. Lamb milk substitutes are however usually based on dried skimmed cows' milk with added animal and vegetable fats. Dried skimmed cows' milk contains about 39% protein and 53% lactose whereas ewes' milk on the same basis has 50% protein and 42% lactose. It is difficult to discuss any of these individual components in isolation as level of intake, energy concentration and type of fat or protein in the diet may affect the diets' overall utilisation as well as the utilisation of the individual constituent.

Milk Protein

In general, the proportion of protein in the diet controls the rate of live-weight gain at any particular level of energy intake (Walker & Faichney 1964a, Walker & Cook 1967, Black 1974). Thus, protein requirements are usually expressed as the percentage of the total energy in the diet derived from protein (PE).

An example of the effects of increasing PE in the diet is given in Table 2. It shows that increasing PE increased the apparent digestibility coefficients of all the constituents of the diet, its metabolizability and hence the growth rate of the lambs.

Table 2 The effect of protein concentration on the apparent digestibility coefficients (%) of nutrients in a milk substitute, metabolizability (%) and lamb growth rate (g/day)

Digestibility coefficients	Energy from Protein (%)			SE of means	Significance of differences
	20	25	30		
Dry Matter	92.1	93.8	94.5	0.66	NS
N	92.7	94.0	94.9	0.41	**
Fat	88.9	92.7	93.7	1.19	*
Lactose	99.7	99.7	98.8	0.03	NS
Ash	77.3	81.9	82.4	1.27	*
Energy	91.1	93.1	94.3	0.82	*
Metabolizability	90.1	90.2	92.9	0.88	*
Growth Rate	263	266	280	9.3	NS

For Finn x Dorset Horn lambs given a daily allowance 7.95 (MJ) gross energy over the first 3 weeks of life (Penning 1975).

Quality of milk protein

Patureau-Mirand & Thériez (1976) have suggested requirements for some of the essential amino acids. They concluded that the requirements varied with level of feeding and age of the lamb and suggested that the DL methionine and lysine content of protein derived from cows' milk may be inadequate for fast-growing lambs.

Heat damage to whey proteins, during the manufacture of milk substitutes, has an important effect on lamb performance. The use of non-casein nitrogen (NCN) as a measure of heat damage has been discussed by Penning et al (1977b); low values of NCN indicate heat damage. A good spray-dried skim milk powder, based on cows' milk, has an NCN value of about 20%. The effects of varying the NCN content of a milk substitute from 3 to 17% are shown in Table 3. Increasing protein quality (NCN value) generally increases the digestibility coefficients for the dietary constituents, metabolizability and lamb growth rate.

Table 3 The effects of heat damage, measured as non-casein nitrogen expressed as a percentage of total Nitrogen, on the apparent digestibility coefficients (%) of the constituents of a milk substitute, metabolizability (%) and lamb growth rate (g/day)

Apparent digestibility of:-	Non-casein nitrogen % 3	10	17	Mean	S.E.	Significance of difference
Dry Matter	94.9	96.8	96.8	96.2	0.45	**
N	94.1	95.2	96.0	95.1	0.31	**
Fat	92.2	95.6	95.2	94.3	1.37	NS
Lactose	99.9	99.9	99.9	99.9	0.01	NS
Ash	80.1	87.6	91.5	86.4	1.12	***
Energy	94.7	96.5	96.2	95.8	0.55	NS
Metabolizability	91.9	93.8	93.3	93.0	0.59	*
Growth Rate	273	290	306	290	3.8	***

For Finn x Dorset Horn lambs given a daily allowance of 7.95(MJ) gross energy over the first 3 weeks of life (Penning 1975)

Non-milk protein sources

Rape seed, yeasts, dried fish hydrolysate and soybean flour have been used to replace partially milk protein in milk substitutes for lambs (Gorrill et al 1976, Thériez et al 1975, Gorrill 1974, Drori & Jordan 1974). This partial replacement of milk protein has usually resulted in decreased lamb growth rate and N retention. Problems of settling out and abomasal bloat have also been found with rape seed and soya protein, and problems of reduced intake with fish hydrolysate.

Walker (1975) and Walker & Kirk (1975) have reared lambs on diets in which all the protein has been supplied by eggs or isolated soya protein. The nitrogen balance and performance of the lambs was generally lower than for animals given diets containing milk protein.

Soliman & Ørskov (1976) reared lambs on diets where milk proteins were replaced by undried fish protein hydrolysate (FPH), prepared from fresh cod fillets. In two consecutive 15-day periods, lambs given diets containing FPH had live-weight gains 29% and 8% lower than those given milk proteins.

It would appear that it is possible to replace milk protein with protein from other sources but these proteins usually result in reduced lamb performance. As yet, milk substitutes which contain non-milk proteins have not been tested on large numbers of lambs and are not commercially available. There appears at present to be little financial incentive to replace milk protein because of the EEC 'skim-milk mountain' and the associated subsidies offered to encourage the use of skim milk powder in animal feeds.

Fat

Walker & Stokes (1970) examined the digestibilities of a number of fats incorporated at 30% of the dry matter of milk substitutes. The values ranged from 99 to 97% with coconut, groundnut and cottonseed oils to 90% with lard and 54% with rapeseed oil. This value for rape oil may have been associated with high erucic acid content. Gorrill & Walker (1974) found that rapeseed oil with a low erucic acid content (0.4%) was better digested (94%) than rapeseed oil containing 16 or 39% erucic acid (74 and 82%) when incorporated in lamb diets. Thériez et al (1973) found digestibility values of 96, 92 and 91% for coconut, palm oil and tallow incorporated in milk substitutes at 25% of the dry matter. Penning et al (1977a) reported digestibilities of 98.5 and 84.9% for butter fat or a mixture of beef tallow (66%) and coconut oil (33%) incorporated into milk substitutes at 30% of the dry matter. Penning et al (1977a) and Thériez et al (1973) found that intakes tended to be higher for diets containing the more digestible fats and Bouchard & Brisson (1970) found a higher intake for a diet containing tallow than for one with corn oil. These workers also found that lower fat digestibilities were associated with decreased digestibilities of other dietary components.

Little difference for lamb performance has been found for diets with fat contents in the range of 20 to 30% of the dry matter (Pinot & Teissier 1965, Gorrill et al 1975) but Owen & Davies (1970) found that lamb growth rate decreased when fat content of the diet was increased to 40%.

It is very difficult to discuss the digestibility of fat in isolation and Table 4 illustrates this point well. The digestibility coefficients of five fats are shown. These are affected by the source of protein and the protein concentration in the diet. It is interesting to note that butter fat is least affected by these changes but the digestibility coefficients for groundnut oil vary from 76 to 91% with changes in protein energy concentration in the diet.

Digestibility of fat by calves has been shown to increase when a high proportion of the fat globules is less than 4 um in diameter (Warner 1970). Although there is no experimental evidence, small fat globule size has been assumed to be of importance in the manufacture of milk substitutes for lambs.

Carbohydrates

Walker (1959) showed that in the new born lamb lactose activity is high and remains high throughout the first five weeks of life. The activity of amylase and maltase is low and only increases slightly over the first five weeks of life. Thus the young lamb is only able to digest lactose and its two constituent hexoses, glucose and galactose, and very limited quantities of starch. It cannot digest sucrose in the first five weeks of life.

Soliman et al (1977) successfully reared lambs from 7-days old on diets where lactose had been replaced by partially pre-digested starch. The practical implications of this work however have still to be explored as the lambs were fed every 3 hours at a relatively low level of intake.

Penning (1975) has suggested that the isoenergetic replacement of fat by lactose will have little effect on lamb performance provided the proportion of the energy supplied by protein is similar between diets. The energy values for fat, protein and lactose are approximately 39, 24 and 16 kJ/b DM respectively. It can be seen that replacing fat by lactose decreases the energy concentration of the diet and to achieve a similar energy intake absolute intake of the diet must increase. Penning (1975) has shown however that for lambs offered milk substitutes ad libitum decreasing energy concentration of the diet decreases energy intake.

Excessive intake of lactose causes diarrhoea in the lamb (Walker & Faichney 1964b). These workers reported a linear relationship between faecal dry matter and hexose equivalent intake. From their equation it is possible to calculate that a daily intake of lactose above 14.6 g/kg live weight will cause diarrhoea (faeces with less than 15% DM). Penning (1975) has shown however that the values given by Walker & Faichney (1964b) appear to be low. Lambs given 170 g or more of lactose per day tended to exhibit diarrhoea. This agrees with the findings of Walker (1959) who calculated that a one-week old lamb could hydrolyse 168 g of lactose per day. Therefore when replacing lactose by fat the absolute intake of lactose is a more important factor to be considered than the actual lactose fat ratio.

FEEDING METHODS AND MANAGEMENT

Milk feeding equipment

Many types of feeder have been used successfully and some have been

Table 4 Coefficients of apparent fat digestibility by lambs (%)
(Values in the same row (a-c) or in the same column (A,B) without a common subscript differ significantly, $P < 0.05$)

Energy from Protein %	Protein Source	Fat Source: Butter	Olive	Safflower	Groundnut	Coconut	SE
		Effect of source of Protein					
-	Skimmed milk	97.5aA	92.8abA	94.1aA	86.7bA	86.5bA	± 0.016
-	Soybean Protein	92.9aB	82.4bB	80.5bB	78.3bB	77.8bB	± 0.016
		Effect of protein concentration					
10.0	-	95.5aA	86.3ba	88.9bA	76.0cA	77.1cA	± 0.015
25.0	-	94.8aA	89.4abA	85.2bA	91.1abB	89.0ab/b	± 0.011
		Effect of source of fat					
-	-	95.2a	87.6b	87.5b	82.5c	82.2c	± 0.011

(Gibney & Walker 1977)

discussed and illustrated by Spedding (1970), M.L.C. (1976) and Treacher & Penning (1971). The type of feeder selected will depend on the number of lambs and the method of rearing.

Although lambs can be trained to drink from buckets or bowls (Gamble et al 1938, Owen & Davies 1965, Gorrill et al 1971), feeders which allow the lamb to suck milk substitute through teats appear to be currently used.

Level of dry matter in reconstituted milk substitute

Milk substitutes for lambs are generally reconstituted to give dry-matter contents of 16 to 20% (similar to that found in ewes' milk) but the effects of a wider range of dry-matter contents have been studied. Large (1965) measured intakes of milk substitutes reconstituted at dry-matter levels of 10, 15, 20 and 25% and fed warm to appetite four times daily. Molénat & Thériez (1974) measured intakes of milk substitutes reconstituted at 10, 16, 22 and 28% dry matter offered to lambs ad libitum. An example of the effects of altering DM concentration is given in Table 5.

Table 5 The effects of dry matter concentration of milk substitute on the performance of lambs offered cold milk substitute ad libitum over 28 days

	Dry matter concentration of milk substitute (%)					Significance of difference
	10	16	22	28	SE$\pm$	
Daily intake of liquid milk substitute (kg)	3.33	2.03	1.85	1.40	0.215	**
Daily intake of milk substitute dry matter (kg)	0.333	0.324	0.406	0.392	0.0334	NS
Total intake of drinking water (kg)	0.713	0.428	1.328	3.773	0.1572	***
Daily live-weight gain (kg)	0.307	0.334	0.391	0.387	0.0235	NS
Daily intake of concentrates (kg)	0.009	0.012	0.014	0.010	0.003	NS

Dorset Down x Finn x Dorset Horn lambs
(Corcuera & Penning unpublished)

Lambs were offered diets ad libitum containing 10, 16, 22 and 28% DM. Increasing DM concentration decreased the intake of liquid milk and caused an increase in the intake of drinking water. In this experiment varying DM concentration had no effect on the digestibility of the diet or on overall performance of the lambs from birth to slaughter.

Milk temperature and feeding pattern

Penning et al (1977a) have shown that the temperature at which milk substitute is offered has no effect on its digestibility but, when lambs were offered milk substitute in four daily feeds to appetite at either 34°C or 5°C, they tended to have a greater intake of warm than cold milk substitute. Lambs offered milk substitute ad libitum at ambient temperature consumed 20% more than those fed four times daily with warm mik substitute.

Walker et al (1967) and Penning et al (1977a) showed that feeding frequency had no effect on digestibility or nitrogen retentinon and live-weight gain was not affected when the same quantity of feed was given.

Gibb & Penning (1972) compared the effects of environmental temperatures of 7, 14 and 21°C on the performance of lambs offered cold milk substitute ad libitum and at a restricted level of intake. Increasing temperature reduced voluntary intake but live-weight gain remained constant. For lambs at the restricted level of intake live-weight gain increased with increasing temperature.

Weaning age and allowance of milk substitute

Lambs start to eat measurable quantities of solid food at three weeks of age (Owen 1969, Penning et al 1973). These results suggest that three weeks is the minimum age for weaning to concentrate diets. The concentrate diets mentioned here refer to a barley-soya-protein pellet containing 16% CP. Brown (1964) suggested lambs should be six weeks old before weaning on to pasture alone. This is confirmed by Langlands & Donald (1975) who found that 50% of lambs weaned onto pasture at 3 weeks of age died. The intake of the surviving lambs had only risen to 70 g DM/day at 30 days of age and unweaned lambs consumed 38 g of herbage DM/day at 30 days of age.

Owen (1969) placed great emphasis on reducing the quantity of milk substitute fed to lambs because of its high price. Davies & Owen (1967), Owen et al (1969), Owen & Davies (1970), Davies (1970) and Morgan & Owen (1972a & b, 1973) have examined the effects of restricting total milk substitute intake to 5 to 9 kg DM on the growth of lambs. All these experiments showed little benefit in feeding more than 5 kg of milk substitute. However Owen et al (1969), Morgan & Owen (1972a) and Penning & Treacher (1975) showed that lambs given a small quantity of milk substitute tended to have a lower growth rate and poorer food conversion efficiency after weaning. The design of most of these experiments made it difficult to separate the effects of total milk substitute intake, daily allowance of milk substitute and weaning age. However Penning & Treacher

(1975) and Hunter et al (1972) have shown that weaning age can affect the performance of lambs.

Table 6 shows the results of an experiment in which some lambs received the same total quantity of milk substitute distributed in different ways (i.e. Total Milk intake = Daily Allowance of milk x Number of days given).

Table 6 Mean daily live-weight gains (before and after weaning) from 1 to 48 days of age (g per lamb)

	Weaning age days					
	21		29		40	
Daily Milk Substitute DM intake (g per lamb)	Before	After	Before	After	Before	After
180	203	307	221	329	247	405
248	254	292	276	362	287	392
343	326	212	321	338	326	206

SE of treatment means ±18.4

(Penning et al 1973)

In general, lambs that receive a high allowance of milk substitute and are weaned abruptly will suffer a greater check in growth at weaning than those receiving a lower daily allowance, and lambs weaned at an early age will suffer a more severe check than those weaned later.

CONCLUSIONS

Formulation of milk subtitute diets

Protein

Protein should supply 25 to 30% of the dietary energy. Where milk protein is used, a minimum NCN value should be 16% of total N. Milk protein can be partially replaced by protein from other sources but lamb performance and N retention will generally be reduced.

Fat

A wide range of fats can be used and should constitute 20 to 30% of the dietary DM. Rapeseed oil containing high levels of erucic acid should not be fed.

Carbohydrates

Lactose or glucose are the only carbohydrates at present recommended. In order to avoid diarrhoea daily lactose intake should not exceed 160 g/lamb.

Feeding and Management

(i) The temperature at which the milk substitute is fed does not appear to be important. For restricted feeding warm milk substitute is recommended.

(ii) Milk substitute should be reconstituted to 16 to 20% DM.

(iii) Milk substitute can be offered ad libitum or if a restricted level of intake is required an allowance of 250g DM per day will give good results.

(iv) A minimum weaning age of four weeks is suggested.

(v) Lambs should not receive fresh herbage as the sole feed until they are at least six weeks old.

REFERENCES

Black, J.L. (1974) Manipulation of body composition through nutrition. Proceedings of the Australian Society of Animal Production 10,211-218.

Bouchard, R.;Brisson, G.J. (1970) Tallow versus corn oil in a milk substitute for artificially reared lambs and digestibility of diet constituents. Journal of Animal Science 31,804-809.

Brown, T.H. (1964) The early weaning of lambs. Journal of Agricultural Science,UK 63,191-204.

Bryant, M.J.;Owen, E. (1971) Milk substitute intake of artificially reared lambs during the first days of life as affected by time of removal from the dam and frequency of feeding. Animal Production 13,337-341.

Campbell, S.G. (1974) Experimental colostrum deprivation in lambs. British Veterinary Journal 130,538-543.

Dattilo, M. (1969) L'allevamento degli agnelli con l'imiego di sostituti del latte e di una allevatrice maccanica. 1. Caratteristiche dei sostituti del latte. Alimentazione Animale 13,71-79.

Davies, D.A.R. (1970) Performance of lambs in artificial rearing systems. Animal Production 12,367.

Davies, D.A.R.;Owen, J.B. (1967) The intensive rearing of lambs. 1. Some factors affecting performance in the liquid feeding period. Animal Production 9,501-508.

Drori, D.;Jordon, R.M. (1974) Effects of soyflour in milk replacer followed by a liquid protein supplement on the growth of young lambs. Journal of Animal Science 39,97-101.

Gamble, J.A.;Earle, I.P.;Howe, P.E. (1938) Rearing foals and lambs by the use of colostrum, blood serum, and substitute milks. US Department of Agriculture Technical Bulletin No.661, 34pp.

Gibb, M.J.;Penning, P.D. (1972) The effects of environmental temperature° and feeding pattern on the performance of artificially-reared lambs fed cold milk substitute. Animal Productin 15,177-182.

Gibney, M.J.;Walker, D.M. (1977) Milk replacers for preruminant lambs: protein and fat interactions. Australian Journal of Agricultural Research 28,703-712.

Gorrill, A.D.L.;Cameron, C.D.T.;Nicholson, J.W.G. (1971) Growth, digestibility and nitrogen retention by lambs fed liquid diets containing milk and soybean proteins. Canadian Journal of Animal Science 51, 663-667.

Gorrill, A.D.L.;Jones, J.D.;Nicholson, J.W.G. (1974) The nutritional value and trypsin inhibitor content of processed soybeans for lamb milk replacers. Canadian Journal of Animal Science 54,337-346.

Gorrill, A.D.L.;MacIntyre, T.M.;Nicholson, J.W.G. (1975) Effects of fat levels in milk replacers and protein levels in creep feed on the performance of artificially reared lambs. Canadian Journal of Animal Science 55,377-381.

Gorrill, A.D.L.;Seoane, J.R.;Jones, J.D.;Nicholson, J.W.G. (1976) Nutrient digestion and nitrogen retention by lambs fed milk replacers containing solvent-extracted or full-fat products from different rapeseed cultivars. Canadian Journal of Animal Science 56,401-408.

Gorrill, A.D.L.;Walker, D.M. (1974) Rapeseed oils low or high in erucic acid in milk replacers for lambs: their effects on growth, digestion, nitrogen balance and internal organs. Canadian Journal of Animal Science 54, 411-418.

Halliday, R. (1970) Protein concentrations in colostra from Finnish Landrace x Scottish Blackface ewes during the first week of lactation and in sera from the ewes and their lambs on the third day of lactation. Journal of Agricultural Science, UK 74,103-106.

Hunter, G.L.;Vosloo, L.P.;Grobbelaar, J.;Niekerk, C.H. van (1972) Pre- amd post-weaning performance of artificially reared lambs. South African Journal of Animal Science 2,75-78.

Langlands, J.P.;Donald, G.E. (1975) The intakes and growth rates of grazing Border Leicester x Merino lambs weaned at 21, 49 and 77 days. Animal Production 21,175-181.

Large, R.V. (1965) The effect of concentration of milk substitute on performance of artificially reared lambs. Animal Production 7,325-332.

Mason, J.H.;Dalling, T.;Gordon, W.S. (1930) Transmission of maternal immunity. Journal of Pathology and Bacteriology 33,783-797.

McCarthy, E.F.;McDougall, E.I. (1953) Absorption of immune globulin by the young lamb after ingestion of colostrum. Biochemical Journal 55,177-182.

Molénat, G.;Thériez, M. (1974) L'allaitement artificiel des agneaux. 5. Comparaison de différentes concentrations du lait de remplacement. Annales de Zootechnie 23,491-502.

Morgan, J.A.;Owen, J.B. (1972a) The nutrition of artificially reared lambs. 1. The effect of different feeding methods applied at three stages of growth. Animal Production 15,285-292.

Morgan, J.A.;Owen, J.B. (1972b) The nutrition of artificially reared lambs. 2. The effect of feed restriction at three stages of growth on growth and carcass composition. Animal Production 15,293-300.

Morgan, J.A.;Owen, J.B. (1973) The nutrition of artificially reared lambs. 3. The effect of sex on the performance and carcass composition of lambs° subjected to different nutritional treatments. Animal Production 16, 49-57.

Nedkvitne, J.J.;Haugland, E. (1970) Kunstig oppal av lam. Nordisk Jordbruksforskning 52,70-71.

Owen, J.B. (1969) The artificial rearing of lambs. Veterinary Record 85, 372-374.

Owen, J.B.;Davies, D.A.R. (1965) Artificial rearing of lambs. Agriculture, London 72,54-57.

Owen, J.B.;Davies, D.A.R. (1970) Milk replacers in the artificial rearing of lambs. Journal of the Science of Food and Agriculture 21,340-341.

Owen, J.B.;Davies, D.A.R.;Ridgman, W.J. (1969) The effects of varying the quantity and distribution of liquid feed in lambs reared artificially. Animal Production 11,1-9.

Patureau-Mirand, P.;Thériez, M. (1976) Amino acid requirements of preruminant lambs. In: 2èmes Journées de la recherche ovine et caprine. Paris; I.N.R.A. & I.T.O.V.I.C., pp241-253.

Penning, I.M. (1975) The nutrition of the liquid-fed lamb. Thesis, Reading University, UK.

Penning, P.D.;Cottrell, K.M.;Treacher, T.T. (1973) The effects of quantity and distribution of milk substitute on the performance of artificially reared lambs to forty-eight days of age. Animal Production 17,179-186.

Penning, P.D.;Penning, I.M.;Treacher, T.T. (1977) The effect of temperature and method of feeding on the digestibility of two milk substitutes and on the performance of lambs. Journal of Agricultural Science, UK 88, 579-589.

Penning, P.D.;Penning, I.M.;Treacher, T.T. (1978) The effects of heat treatment and protein quantity on digestibility and utilization of milk substitutes by lambs. Journal of Agricultural Science, UK 90,221-228.

Penning, P.D.;Treacher, T.T. (1975) The effects of quantity and distribution of milk substitute on the performance and carcass measurements of artificially reared lambs. Animal Production 20,111-121.

Pinot, R.;Teissier, J.H. (with Bauchau, P.) (1965) L'allaitement artificiel des agneaux. 1. Comparaison entre différents laits de remplacement et le lait de brebis. Annales de Zootechnie 14,261-278.

Soliman, H.S.;Ørskov, E.R. (1976) Utilization of fish-protein hydrolysate for artificial rearing of lambs. Proceedings of the Nutrition Society 35,91A-92A.

Soliman, H.S. ;Ørskov, E.R.;Smart, R.I. (1977) Milk replacers based on non-milk constituents for lambs. Proceedings of the Nutrition Society 36,52A.

Spedding, C.R.W. (1970) Sheep production and grazing management, 2nd edn. London, UK;Baillière Tindall and Cassell, xi+435pp. ISBN 0-7020-0326-3.

Thériez, M.;Molénat, G.;Daniel, M.;Aurousseau, B. (1973) L'allaitement artificiel des agneaux. 3. Influence de la nature des matières grasses incorporées dans l'aliment d'allaitement. Annales de Zootechnie 22, 185-197.

Thériez, M.;Protais, M.;Molénat, G. (1974) L'allaitement artificiel des agneaux. 4. Comparaison de différentes sources de matières azotées en replacement de la poudre de lait. Annales de Zootechnie 23,325-341.

Treacher, T.T.;Penning, P.D. (1971) The development of an automatic system for rearing lambs artificially. Outlook on Agriculture 6,227-231.

UK, Agricultural Research Council (1965) The nutrient requirements of farm livestock. No.2. Ruminants. Technical reviews and summaries. London, UK; Agricultural Research Council, xi+264 pp.

UK, Meat and Livestock Commission, Sheep Improvement Service (1976) Artificial rearing of lambs. Technical Report. Bletchley, Milton Keynes, UK; Meat and Livestock Commission, 35 pp.

Walker, D.M. (1959) The develpment of the digestive system of the young animal. 3. Carbohydrates enzyme development in the young lamb. Journal of Agricultural Science, UK 53,374-380.

Walker, D.M. (1972) Calcium and phosphorus retention by the milk-fed lamb, with estimates of the endogenous losses. Journal of Agricultural Science, UK 79,171-179.

Walker, D.M. (1975) Utilization of whole egg and its components in milk replacers for preruminant lambs. Australian Journal of Agricultural Research 26,599-614.

Walker, D.M.;Cook, L.J. (1967) Nitrogen balance studies with the milk-fed lamb. 4. Effect of different nitrogen and sulphur intakes on live-weight gain and wool growth and on nitrogen and sulphur balances. British Journal of Nutrition 21, 237-256.

Walker, D.M.; Cook, L.J.;Jagusch, K.T. (1967) Nitrogen balance studies with the milk-fed lamb. 5. Effect of frequency of feeding. British Journal of Nutrition 21,275-287.

Walker, D.M.;Faichney, G.J. (1964a) Nitrogen balance studies with the milk-fed lamb. 3. Effect of different nitrogen intakes on growth and nitrogen balance. British Journal of Nutrition 18,295-306.

Walker, D.M.;Faichney, G.J. (1964b) Nutritional diarrhoea in the milk-fed lamb and its relation to the intake of sugar. British Journal of Nutrition 18,209-215.

Walker, D.M.;Kirk, R.D. (1975) The utilization by preruminant lambs of milk replacers containing isolated soya bean protein. Australian Journal of Agricultural Research 26,1025-1035.

Walker, D.M.;Stokes, G.B. (1970) The nutritive value of fat in the diet of the milk-fed lamb. 1. The apparent and corrected digestibilities of different dietary fats and of their constituent fatty acids. British Journal of Nutrition 24,425-433.

Warner, R.G. (1970) High energy diets for the young ruminant. In: Swan, H.;Lewis, D. (Eds) Proceedings of the University of Nottingham Fourth Nutrition Conference for Feed Manufacturers. London, UK; J. & A. Churchill, pp2-19. ISBN 0-7000-1469-1.

DISEASES OF LAMBS

DR. J.A. WATT

Veterinary Investigation Centre, Bush Estate, Penicuik, Midlothian

In most sheep rearing areas in the world and most certainly in the UK the lamb crop is the most important source of revenue. Any factor which reduces the return from this sector is therefore of considerable economic importance and in recognition of this a considerable proportion of research resources have been directed to the control of diseases of lambs. This research has in turn highlighted the fact that simple 'disease' as we usually strictly interpret the word, while important and meriting the attention directed to it, is of less significance than losses due to factors, which should be more properly classified as managerial. Even for diseases which are properly classified as infective, managerial factors play a fundamental role in their causation and not infrequently prevention and control is most readily achieved by managerial means.

The results of numerous surveys highlight the fact that the greatest losses of lambs occur in the neonatal period (Stamp 1967, Howe 1976) but an analysis of 'cause of death' shows that only some 15-20% of such losses can be properly ascribed to disease (McFarlane 1961, Stamp 1967, Whitelaw 1976). In other words, at least 80% are the result of managerial or environmental factors (Moule 1954). These figures do not of course allow for outbreaks of a specific disease which may cause heavy losses in the individual flock but such incidents, generally isolated and contained, do not greatly influence the overall picture.

There is wide variation in the levels of loss given in this 'neonatal' group, the available figures being in most cases expressed as averages of a number of flocks. Even when accompanied by deviation figures, it is difficult if not impossible to determine the number of flocks in each 'loss level', i.e. high, medium, low, which is an important basic fact in assessing the size of the problem within groups having the same basic environment and management system. One important deduction can be made and that is the level of loss at which it is practical to aim; HFRO have reduced losses in hill sheep flocks from 17.5 to 6.5% on one farm and from 16 to 9% on another. Similarly an MLC report shows losses of 6 to 20% in upland flocks, the lower figure being recorded on farms where management and supervision were of a high order. Obviously this figure can be regarded as a realistic target.

A further illustration of the importance of the neonatal period is shown in an MLC survey of lowground flocks where, of the 8% of lambs which died before weaning, 5.2% died in the first seven days of life. These figures are from well managed and co-operative farms and the overall losses are appreciably higher at 11 to 15%.

Although only a comparatively small proportion of the losses can be strictly accounted for by disease, I have dwelt on these figures at some length as it is of utmost importance to identify the narrow confines of the field to be discussed.

After the neonatal period, which one could fix arbitrarily as 14 days post natum, losses from actual death are low in the overall picture, again with the proviso that isolated incidents can result in heavy losses on the individual farm. Various surveys indicate that in the UK losses to weaning are on average some 2 to 5% and these can in general be ascribed to disease in the broad sense of the word. Again however it must be emphasised that averages of this type, while indicating the overall significance, do not take account of individual farm loss and the wide deviation giving this average. Ignoring therefore the immediate neonatal period with weak lambs, absence of milk, low birth weights, exposure, etc, let us look at the more specific diseases which cause loss.

The most common and earliest diseases seen in the post natal period are the enteric diseases, ranging from the readily recognised and specific lamb dysentery to the more complicated colibacillosis, watery mouth, viral infections, etc (Watt 1965).

Lamb dysentery is one of the well documented group of diseases caused by the anaerobic Clostridia. Susceptibility is short lived, extending to some 14 days. Its existence however gives me the opportunity to introduce the subject of immunisation against most members of the group. There is no doubt that considerable sporadic losses are due to clostridia, causing enterotoxaemia, Cl. oedematiens infections, blackquarter, etc, and one can confidently assert that flock immunisation with a multiple vaccine is an economic and strongly advised procedure in all flocks. Given such immunisation, the gamma globulin passed to the lamb in the colostrum will protect it effectively.

Colostrum is generally accepted as essential in the protection of the lamb against a number of diseases but there are certain criteria which must be met. The quantity must be sufficient, the quality must be high and the gamma globulin must contain the specific antibodies to combat potential pathogens in the lamb's environment. Recent work has shown that these factors cannot be accepted as being invariably present. There is a considerable variation even within a single group of ewes.

Colibacillosis and other enteric diseases are favoured by intensification and with the growth of this practice it is essential to ensure that, during the natal period, management systems are adopted to control these diseases (Watt 1965). Apart from inadequate colostrum the factors of most significance are close contact of neonates with older lambs and the build up of infection by using the same enclosure over too long a period with resultant heavy challenge and exposure of the neonate to specific pathogens. All these factors can be controlled by adoption of proper managerial principles. Recently interest has been directed to the possible role of viruses in enteric infection in lambs and this is the subject of a separate paper.

Other infections gain entry by the navel, e.g. streptococci and staphylococci which localise in the joints and internal organs causing local sepsis - hence "joint ill" and "navel ill". Fusiformis necrophorus is a not infrequent invader at this time though its effects may not be apparent for up to 10 days. The source of infection is believed to be infected feet in the ewes; good management practice would ensure that no such ewes were admitted to the lambing pen. General prophylactic measures include attention to hygiene, clean bedding - not sawdust, and the dressing of the navel with iodine or antibiotic. Congenital swayback is a condition which manifests itself at or soon after birth, the clinical signs ranging from a slight ataxia or apparent muscular weakness to an inability to rise. More severely affected lambs submitted to the laboratory without a good history may lead to a diagnosis of starvation, correct in fact but singularly unhelpful in practice. Fortunately the classical brain cavitation is commonly present in this type of this deficiency. The disease most commonly erupts as a problem when marginal land is improved by liming and fertilising, which radically reduces the available copper to the ewe (Reith[1970]). The intricate relationship between early growth of pasture, involving climate and fertilisation, the amount and source of supplementary feeding and the prevalence of swayback is now well recognised. Prophylaxis is by injection or dosing of the ewes some 4 to 6 weeks before lambing. Where the disease is diagnosed early in lambing dosing of ewes still to lamb has proved beneficial.

The spread of contagious pustular dermatitis (orf, contagious ecthyma) is favoured by intensification. The condition affects the lips of the lambs but occasionally spreads into the mouth and gums. The ewes are not usually badly affected but the lesions, which develop on the teats and udders, cause them to refuse to suckle the lambs and not infrequently results in mastitis. The lambs try to steal milk from healthy ewes at the boxes and this is one of the main factors in the spread of the condition. Careful and frequent examination of the ewes and lambs and removal of infected animals is helpful in controlling the condition but presents practical difficulties in operation. Affected fields should be strictly isolated and if the lambing field seems to be involved it should be abandoned. Vaccination may be helpful but there is little objective evidence on which to base a definite opinion.

Tick infested hill land in Scotland presents its own problems in the form of tick pyaemia and louping ill. Pyaemia is the result of the effect of tick-borne fever on the immunological mechanism of the lamb. As well as the recognised pyaemia, TBF can result in outbreaks of pneumonia in the lambs. Unfortunately no protection is passed to the lambs in colostrum and, while the pyrexia of TBF is readily controlled, chemotherapy has no effect on the immuno-suppressive aspect of the disease. Control is however possible by the use of long acting penicillins given at the first signs of disease (Watt & Foster 1968).

Louping ill can on occasion cause very heavy losses in young lambs. Fortunately an effective vaccine is available and properly protected ewes will ensure colostral protection for their offspring.

With the exception of these particular diseases, lambs have now passed the immediate post natal danger period and though the peak of loss has passed "production" conditions now assume major importance. These are the parasitic infections which, though not generally causing a high mortality, result in a considerable loss in production (Reid & Armour 1978). Exceptions to this are delayed swayback and enterotoxaemia. The former frequently appears without any warning; neonatal cases are seen though the frequency with which 'delayed' cases, i.e. cases first noticed at 3 to 6 weeks of age, follow the neonatal incidents is an additional reason for dosing when a neonatal incident is diagnosed. The prevalence of this condition is governed by the same factors described for congenital swayback.

Enterotoxaemia or pulpy kidney disease may occur at any time, even in very young lambs, but in this country reaches its highest incidence in June when the combination of rich grass growth, appreciable intake by the lamb and full milk on the ewe occurs. Under modern conditions little can be suggested on the management side to control this; indeed losses occur on the natural herbage of the hill and control is dependent on a properly designed flock vaccination programme. In the occasional emergencies antiserum gives protection. As protection from the dam can last 9 to 10 weeks, it is important to remember that within the group the quality and quantity of the colostrum intake will have varied between lambs and thus the duration of protection will also vary.

Coccidiosis is a condition which presents certain difficulties in diagnosis. It undoubtedly occurs as a sometimes fatal disease in 4 to 8 week old lambs and would appear to be a significant contributory cause of diarrhoea in older lambs. The diagnostic difficulty is that large numbers of coccidial oocysts can be demonstrated in the faeces of normal lambs on grazing in June and July. In housed lambs the disease can be controlled by the use of coccidiostats. Grazing lambs in July in Scotland are not infrequently affected by a persistent diarrhoea which has all the hallmarks of PGE and in which initially high worm egg counts can be demonstrated. Anthelmintics fail to clear the condition and as coccidia continue to be excreted the clinical signs may be ascribed to coccidiosis. Most of these cases are however the result of nematodirus infestation followed by the return of the lambs to infected pasture after dosing. Quite small numbers of larvae under these conditions can perpetuate the enteric condition. It is essential where clinical disease is being treated that affected lambs be removed from infected pastures on dosing.

Nematodirus infection itself is now a very different proposition from that presented when no effective remedy existed. Its appearance in southern Scotland was spectacular both in its effects and in the accuracy of the timing, the last week in May and the first week in June, which never varied by more than a few days. While new drugs give excellent control it is of utmost importance that dosings be carefully timed and aimed not to help exclusively the lambs being dosed but to protect the pasture against infection for the sake of next year's lambs.

While nematodirus presents the most spectacular manifestation of parasitism, the most important cause of loss of production next to neonatal loss is that generally caused by worms. Dosing alone, even with modern highly efficient anthelmintics, will not give the desired high level of control unless combined with well planned grazing management. This entails planning for two years ahead at all times; a 'skeleton' system evolved at the East of Scotland College of Agriculture is given in Figure 1.

Where such a system has not been evolved or modified, a basic control principle is that when lambs are moved to 'clean' grazing they should be dosed on moving. By this simple precaution the grazing is kept clean for long periods and it obviates the need for subsequent dosing to control the development of clinical disease with its associated loss of production.

At the end of the summer season when lambs are being finished for sale with supplementary feeding and grazing on stubble, rape, etc, pulpy kidney disease tends to reappear in unprotected flocks, accompanied by acidosis and acute pasteurellosis.

The last two conditions appear to be closely related and are usually associated with the grazing of barley stubble. Both are characterised by death after a very short period of illness and frequently lambs are not observed to have been ill at all. All three conditions are readily controlled by moving the group to poor pasture. The likelihood of outbreaks occurring can be greatly reduced by introducing the lambs gradually to the richer grazing and providing a run-off where adequate roughage is available.

Figure 1 Control by Pasture Management

The following examples of grazing and conservation routines can provide solutions in specific cases. In each case the hay (or silage) aftermath is used by lambs at weaning.

1. Permanent grass - cattle and sheep

Year 1 Year 2 Year 3

either

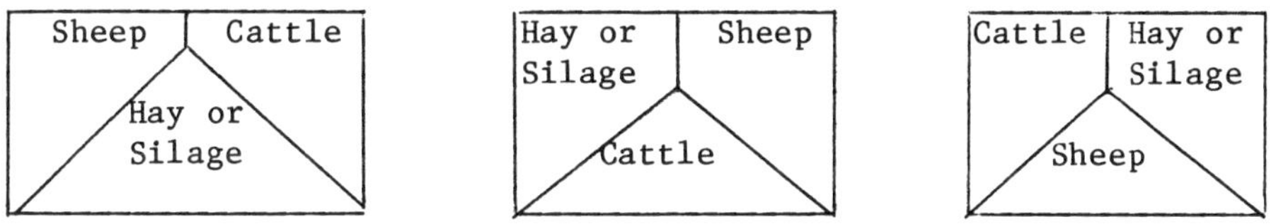

or

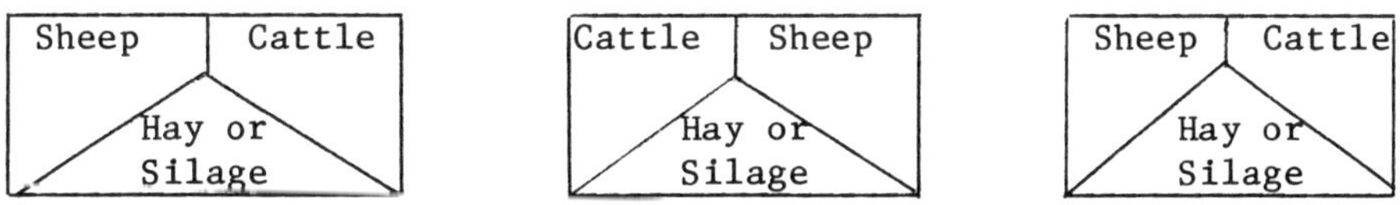

2. Temporary grass - cattle and sheep or all sheep

	Undersown stubble	1st year grass	2nd year grass	3rd year grass
More sheep* than cattle	dosed ewes	sheep	cattle	½ sheep ½ hay then lambs
More cattle* than sheep	weaned lambs and/or dosed ewes	cattle	sheep	½ cattle ½ hay then lambs
All sheep	dosed ewes	sheep	½ hay then lambs ½ cropped	all cropped

* in terms of grazing requirements.

REFERENCES

Howe, K.S. (1976) The cost of mortality in sheep production in the UK 1971-74. Agricultural Economics Unit (University of Exeter) [Report No.198.]

McFarlane, D. (1961) Perinatal lamb losses. Australian Veterinary Journal 37,105-109.

Moule, G.R. (1954) Observations on mortality amongst lambs in Queensland. Australian Veterinary Journal 30,153-171.

Reid, J.F.S.;Armour, J. (1978) An economic appraisal of helminth parasites in sheep. Veterinary Record 102,4-7.

Reith, J.W.S. (1970) Soil factors influencing the trace element of herbage. In Mills, C.F. (Ed) 'Trace element metabolism in animals. Proceedings of the WAAP/IBP International Symposium, Aberdeen 1969'. pp410-412.

Stamp, J.T. (1967) Perinatal loss in lambs with particular reference to diagnosis. Veterinary Record 81,530-534.

Watt, J.A. (1965) Abortion and neonatal losses in sheep. Veterinary Record 77,637-641.

Watt, J.A.;Foster, W.N.M. (1968) Benzathine penicillin as a prophylactic in tick pyaemia. Veterinary Record 83,507-508.

Whitelaw, A. (1976) A survey of perinatal losses associated with intensive hill sheep farming. Veterinary Annual 16,60-65.

DIARRHOEA IN NEONATAL LAMBS

DR. D.R. SNODGRASS

Animal Diseases Research Association, Moredun Institute, 408 Gilmerton Road, Edinburgh

Diarrhoea is the principal infectious condition affecting young lambs. It is a feature of intensive lambing systems where small paddocks or indoor lambing areas are subject to a build-up of disease as the lambing season progresses. Lamb dysentery, caused by Clostridium perfringens type B, is a cause of severe diarrhoea and death in very young lambs. Its prevalence has declined markedly due to efficient vaccination programmes but there is a continuing common occurrence of diarrhoea in young lambs.

It is this neglected area of disease that we are currently investigating, defining our area of interest as those lambs under 3 weeks old. Information available at the start was meagre. 'Colibacillosis' has been a common label for neonatal diarrhoea in animals. The evidence for the involvement of Escherichia coli in lamb diarrhoea is merely that E. coli is commonly isolated from intestinal contents and faeces of diarrhoeic lambs. A very few strains have been shown to possess enteroxtoxins, and to be pathogenic for lambs (Smith & Halls 1967, Shaw 1971). E. coli can be isolated with equal frequency from normal lambs; there have been no general attempts to characterise field isolates for either enterotoxicity or enteropathogenicity and there is at the moment no basis on which to make an assessment of the significance of E. coli in lambs.

The work on which we have results so far is confined to the virology aspects. Viruses have been isolated from the enteric tract of sheep for many years by conventional virological techniques i.e. tissue culture for enteroviruses, reoviruses and adenoviruses. These viruses are probably of little overall significance.

The recent use of direct electron miscroscopic examination of faeces has revolutionised our knowledge of enteric viruses and in lambs the following viruses have recently been described; rotavirus, astrovirus, coronavirus (McNulty et al 1976, Snodgrass et al 1976, Snodgrass & Gray 1977).

The astrovirus and coronavirus are of unknown distribution and significance but there is a considerable body of information on rotaviruses. They have been reported from diarrhoeic lambs only in Scotland and Northern Ireland but, by analogy with rotaviruses from other species, are likely to be worldwide in distribution. In our surveys in Scotland, neutralising antibody occurs in most adult sheep. Rotaviruses were detected in faeces from 22 of 87 diarrhoeic lambs (25%) under 3 weeks old, from 13 of 28 farms (46%) in a survey carried out in spring 1976 (Snodgrass et al 1977). With the more sensitive diagnostic techiques now at our disposal, it is likely this figure could be increased.

Clearly rotaviruses occur commonly in neonatal diarrhoea but what is their significance? In all animal species from which they have been recovered, rotaviruses induce diarrhoea and pathological lesions. In gnotobiotic or colostrum deprived lambs, we consistently produce a clinical picture of diarrhoea, apathy and anorexia, with a mortality of 5-10%, suggesting that rotaviruses are a significant cause of diarrhoea in neonatal lambs. Diagnostic techiques for rotavirus infection centred originally on electron microscopy but newer techniques have since been developed.

At Moredun Institute we have concentrated on two main approaches to rotavirus infections; studying immunological mechanisms, and pathogenesis. In our experience, most adult sheep have circulating rotavirus antibody, which is passed to the suckling lambs through colostrum. Thus although most neonatal lambs have rotavirus antibody in their serum, many are still susceptible to rotavirus infection. This has naturally caused the protective role of colostrum to be questioned. Experimentally, we have confirmed that colostrum is not always protective and, when it does protect, acts through its continued presence in the gut. This can be achieved either by giving very large doses of colostrum, which may leave a residue in the gut for a few days, or by continued feeding of first-day colostrum as part of the diet (Snodgrass & Wells 1976).

If a ewe could be stimulated to produce rotavirus antibodies in her colostrum and milk for a prolonged period after parturition, it would ensure the continued presence of antibody in the gut of the suckling lamb, thus achieving protection against rotavirus infections. Unfortunately, there is a normal rapid decline in immunoglobulin and antibody in milk after lambing. At Moredun a group of 5 ewes were vaccinated before tupping with an inactivated lamb rotavirus preparation and a second group received a placebo vaccine. There was a significant increase in serum antibody titre to rotavirus which was maintained throughout pregnancy. After lambing, the vaccinated ewes had very high colostral and milk antibody titres throughout the 10-day sampling period, which were sufficient to protect any lamb against rotavirus infection during this period. This suggests that vaccination of the ewe with an inactivated rotavirus vaccine may be possible.

The other aspect of rotaviral infection on which we have concentrated has been pathogenesis. Lamb rotavirus infects primarily the columnar absorptive epithelium lining the small intestinal villi. A very rapid virus multiplication ensues, followed by epithelial destruction and shedding, which leads to villus atrophy. The small intestine is thus functionally deficient in both digestion and absorption, producing a diarrhoea which continues until the lesions have completely healed.

The importance of rotaviruses as aetiological agents is established but further work is necessary to ascertain the significance of other viral and bacterial agents. If rotaviruses do prove to be the major single cause of lamb diarrhoea, as seems likely, there are long term prospects of controlling the disease through vaccination of the ewe.

REFERENCES

McNulty, M.S.;Allan, G.M.;Pearson, G.R.;McFerran, J.B.;Curran, W.L.; McCracken, R.M. (1976) Reovirus-like agent (rotavirus) from lambs. Infection and Immunity 14,1332-1338.

Shaw, W.B. (1971) Escherichia coli in newborn lambs. British Veterinary Journal 127,214-219.

Smith, H.W.;Halls, S. (1967) I. Observations by the ligated intestinal segment and oral inoculation methods on Escherichia coli infections in pigs, calves, lambs and rabbits. II. Studies on Escherichia coli enterotoxim. Journal of Pathology and Bacteriology 93,499-529.

Snodgrass, D.R.;Gray, E.W. (1977) Detection and transmission of 30 nm virus particles (astroviruses) in faeces of lambs with diarrhoea. Archives of Virology 55,287-291.

Snodgrass, D.R.;Herring, J.A.;Linklater, K.A.;Dyson, D.A. (1977) A survey of rotaviruses in sheep in Scotland. Veterinary Record 100,341.

Snodgrass, D.R.;Smith, W.;Gray, E.W.;Herring, J.A. (1976) A rotavirus in lambs with diarrhoea. Research in Veterinary Science 20,113-114.

Snodgrass, D.R.;Wells, P.W. (1976) Rotavirus infection in lambs: studies on passive protection. Archives of Virology 52,201-205.

UTILIZATION OF GRASSLAND FOR SHEEP PRODUCTION

DR J. HODGSON

Hill Farming Research Organisation, Bush Estate, Penicuik, Midlothian

INTRODUCTION

With this topic it is useful to consider sheep production as a series of three inter-linked processes:

(i) Herbage production

(ii) Herbage consumption

(iii) The conversion of ingested herbage into animal products.

It follows that:

(iv) Efficiency of harvesting $= \frac{\text{Weight of herbage consumed}}{\text{Weight of herbage produced}}$

(v) Efficiency of conversion $= \frac{\text{Weight of animal product}}{\text{Weight of herbage consumed}}$

(vi) The efficiency of utilization of grassland for sheep production is then the product of (iv) and (v).

Our objective is to consider some of the factors influencing the efficiency of these processes and the way they interact to affect the overall efficiency of animal production from grassland. Examples will be taken mainly, but not exclusively, from lowland temperate grassland.

HERBAGE PRODUCTION

In cutting trials the amount of herbage harvested usually increases as the interval between cuts increases. This evidence formed the basis for recommendations on systems of rotational grazing management. Under grazing conditions, the indications are that production is in fact quite insensitive to variations in grazing management (McFeely et al 1975) and also to a surprisingly wide range of stocking rates (Langlands & Bennett 1973). This would not necessarily be true for crops like lucerne, where the growing points are susceptible to grazing, or for native vegetation of low productive potential.

It is not clear why grazed swards do not respond in the way which would be predicted from the results of cutting trials. It is important however to distinguish between herbage growth - the amount of new plant material produced - and the amount ultimately harvested either by cutting or grazing. The rate of accumulation of herbage in a sward (herbage production) is the balance between the growth of new material and the senescence and decay of existing material; growth and senescence are seldom measured in the field. Variations in grazing management can affect growth and senescence in different ways (Table 1), in addition to any effect they may have upon the efficiency of the harvesting process itself.

The nutritive value of what is grown is not much affected by grazing management unless there are marked changes in the botanical composition of the sward, but the nutritive value of the sward as a whole can be greatly influenced by the accumulation of residues from previous grazings.

The effects of fertiliser use on herbage production and nutritive value are well documented (e.g. Behaeghe & Carlier 1973), and will not be dealt with here.

Table 1 The effect of cutting management on the amounts of herbage grown and lost through senescence and decay, and the amount of green material harvested (g DM/m^2), between 13 July and 5 October. From Morris (1970).

	Treatment					
	Clipped frequently			Cut to 12 cm every 2 wks	Cut to 3 cm every 4 wks	S.E. treatment mean
	High LAI	Medium LAI	Low LAI			
Gross aerial Growth	709	703	656	604	531	± 50
Death of green leaf	457	385	331	320	198	± 34
Harvested green material	644	728	728	707	770	± 20

Note: Values for death of green leaf are probably over-estimates; values for material harvested include initial harvest, and are therefore not strictly comparable with the estimates for growth and death (see Morris 1970). Values within lines are comparable.

HERBAGE CONSUMPTION

The voluntary herbage intake of grazing animals can be strongly influenced by the weight of herbage per unit area, its distribution and its digestibility, all of which may be widely manipulated by grazing management. Variation in these three parameters can explain much but not all the differences in intake between plant species or varieties (see Hodgson 1977) and between seasons of the year (Marsh 1975a).

In cattle, intake appears to be linearly related to the digestibility of the diet over the full range of values measured in any particular experiment (Figure 1). This suggests that the herbage intake of grazing animals is not often restricted by chemostatic limits (Balch & Campling 1969). There is no strictly comparable evidence for grazing sheep but there is no reason to think that their responses will differ.

Intake tends to increase at a declining rate as either the weight (Figure 2) or the height (Figure 3) of the sward increases, and may even start to decline again beyond some optimum value of either variable. The size of individual mouthfuls of herbage tends to fall once the density of material in the grazed horizon of a sward falls below 25 kg DM/ha/cm (Figure 4), and the short-term rate of herbage intake falls steadily with increasing proximity to ground level (Figure 5), thus demonstrating the behavioural links between sward structure and intake. Responses for cattle and sheep are similar.

It is not easy to disentangle the separate effects on intake of variations in herbage weight, structure and digestibility because they vary together in relation to the maturity of the sward. So far as we can see at present the effects are additive. Therefore in order to maximise nutrient intake, a sward with a high density of young leaf at the surface is required and animals should not be forced to graze too close to ground level.

EFFICIENCY OF GRAZING

On any given sward, herbage intake will be progressively reduced as the herbage allowance - the weight of herbage per animal or preferably per unit live weight - decreases. This relationship is clearly demonstrated for rotational grazing managements (Figure 6) but under set-stocking it tends to be confused by parallel effects upon the weight and digestibility of the herbage. The explanation for this effect probably also lies in grazing behaviour since the herbage allowance will determine the rate at which animals graze down into limiting horizons within the sward.

The <u>proportion</u> of the herbage which is actually consumed increases progressively as the allowance is reduced (Figure 7). Note that this is not the efficiency of grazing as defined in the introduction, where intake was related to herbage production over time rather than to the weight of standing herbage at a point in time. It is clearly difficult to achieve a high herbage intake per animal <u>and</u> efficient defoliation of the sward, though there is some room for manoeuvre in the middle sections of the curves where changes are still relatively slow.

The digestibility of the herbage eaten also tends to decline progressively as the herbage allowance becomes limiting, forcing animals to graze further into the sward (Figure 8). Thus the effects of declining intake and declining digestibility upon nutrient intake are additive.

Fig.1. Relationship between the digestibility of diet selected (%OMD) and herbage intake (g OM/kg LW) of lactating cows (---) and and growing calves (——). From Hodgson 1977. (see original paper for explanation of reference numbers.)

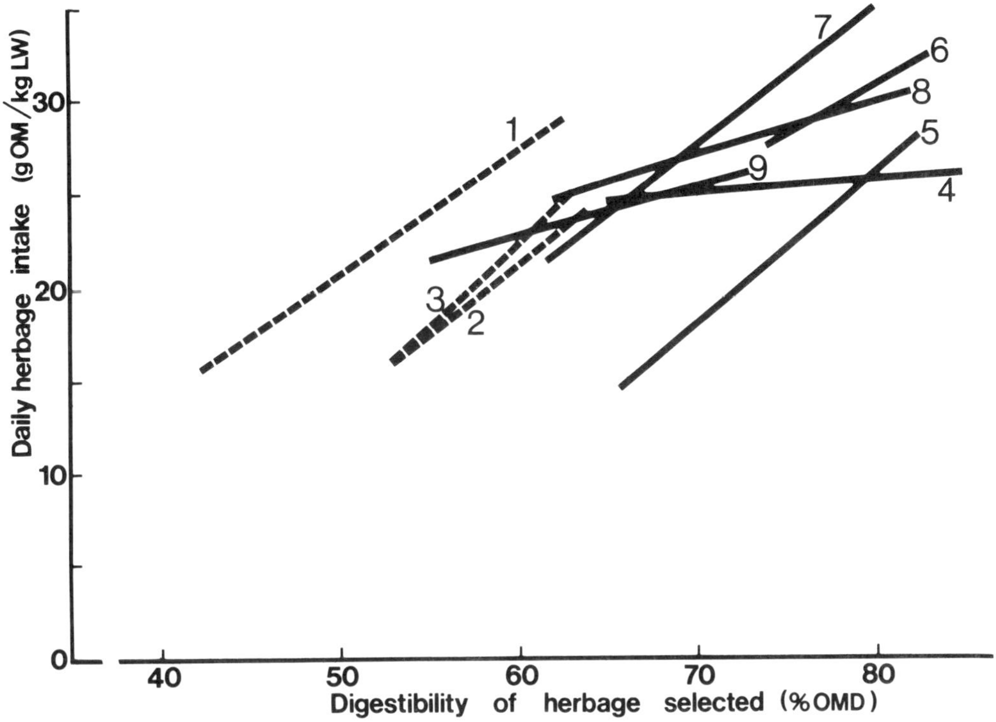

CONVERSION OF INGESTED HERBAGE INTO ANIMAL PRODUCT

The results shown in Figure 9, which illustrate the relationship between herbage intake and lamb growth rate, are derived from current feeding standards (Ministry of Agriculture, Fisheries and Food 1975). Growth rate increases at a roughly constant rate as herbage intake increases. This means that the relationship between herbage allowance and animal performance for example will be similar in shape to the relationship between allowance and intake shown in Figure 6 (see Gibb & Treacher 1976). However, the relationship between intake and food conversion efficiency $\left(= \frac{\text{Weight gain}}{\text{Feed intake}}\right)$ is sharply curvilinear because of changes in the proportion of nutrients used for maintenance and gain.

GRAZING MANAGEMENT AND ANIMAL PRODUCTION

We can now examine briefly some aspects of the influence of the above processes on the productivity of grazing systems.

Grazing Management

On lowland temperate grasslands the choice of grazing system appears to have little influence on animal performance except perhaps at very high stocking rates (Marsh 1975b). The reasons for this are now becoming clearer. There is no good reason to expect that herbage production will be substantially greater under rotational grazing than under set-stocking. Herbage intakes and hence levels of animal performance are likely to be similar under the two systems of management at equal stocking rates, though the factors controlling intake in the two situations may differ.

These arguments would not necessarily apply to natural grasslands of low productive potential, though even in this case there is often no good evidence for differences in production between grazing managements run at the same stocking rate.

Stocking Rate

The relationship between stocking rate (animals per hectare) and the growth of individual animals over a grazing season is usually negative and linear (Figure 10), though the shape of the response curve is influenced by the duration of the period of grazing and by the growth potential of the animals concerned (Hodgson 1975). Most of the available evidence relates to growing lambs or cattle and there is little information for dairy cows, ewes and lambs, or cows and calves, where the ability of the lactating female to utilise body reserves as a buffer against periods of nutritional inadequacy may influence the shape of the response.

Wool growth is much less sensitive to variations in nutrition than either growth or lactation and changes relatively slowly as stocking rate increases (Hamilton & Bath 1970).

If we accept that animal performance is linearly related to stocking rate over a wide range, the implication is that it is unrealistic to seek

Fig.2. Relationship between amount of herbage on offer (kg DM/ha) and herbage intake of grazing ewes. From Arnold & Dudzinski 1967.

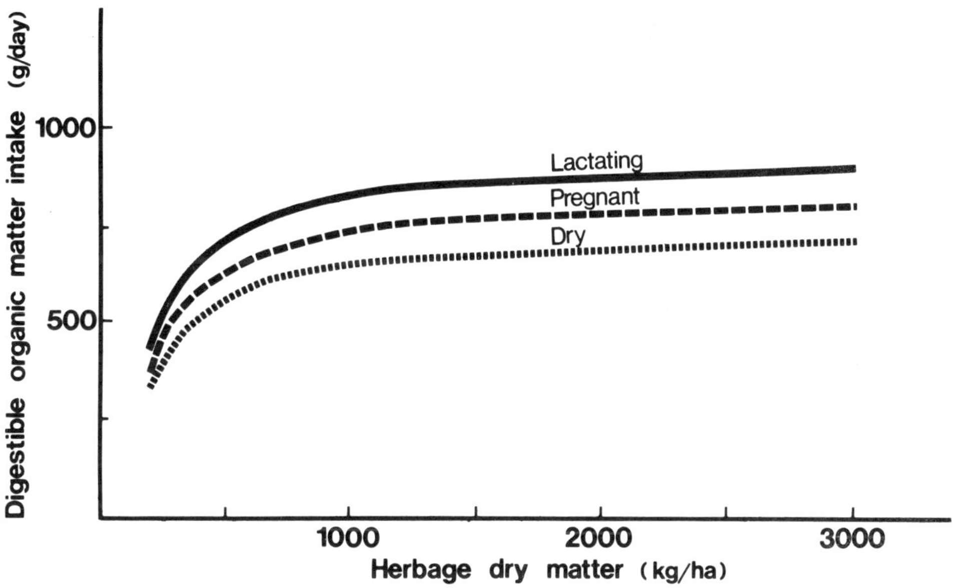

the stocking rate which maximises production per animal. Also, production per unit area is relatively insensitive to variations in stocking rate on both sides of the rate which maximises output (Figure 10; c.f. Mott 1960). At this point the performance of individual animals is only about half the possible maximum, and the stocking rate maximising gross margin or profit per hectare is almost invariably lower than that at which output per hectare is maximal (e.g. Tribe & Lloyd 1962).

Herbage Production and Animal Production

From the conclusions above, it follows that an increase in herbage production, following the use of for instance more nitrogen fertiliser, will not be accompanied by an increase in the output of animal products unless the animals are already operating on sensitive sectors of the intake response curves (Figures 2-7), and/or the stocking rate is increased at the same time (Marsh 1975b). These two conditions are not often realised in practice; this accounts in large measure for the often observed, low responses in animal output to nitrogenous fertiliser (Holmes 1968).

Seasonal variations in the rate of herbage production are extreme in most climates and usually far exceed variations in the nutrient requirements of a population of grazing animals. The discrepancy is smaller in the case of a flock of breeding ewes than in most other systems of animal production from grass (Spedding 1977). Even if we discount the winter period when rates of herbage production are usually very low, there are still critical periods of feed shortage around the time of lambing and often in mid-summer, which can seriously limit animal performance.

The period of feed shortage in spring can be overcome to some extent by the use of grass varieties chosen for early growth, such as Italian ryegrass, or by delaying lambing. Both of these strategies however tend to exacerbate the effects of the depression in herbage production in mid-season, when the lamb's dependence on grass is increasing rapidly. Because of their different pattern of growth the legumes can help to meet this deficiency, either grown with grass, e.g. White Clover, or as a special crop, e.g. Lucerne. The inclusion of legumes in the diet also results in a substantial increase in lamb growth rates (Thomson 1977).

Although it is conventional in temperate climates to cut herbage surplus to immediate requirements in the spring for conservation, this may not be justified in financial terms in drier climates (Hutchinson 1971).

CONCLUSION

It is still difficult to determine the most sensible balance between the conflicting demands for high intake per animal, and hence high efficiency of food conversion, on the one hand and high efficiency of grazing on the other. A proper understanding of the individual conversion processes outlined above and the way they interact to influence the efficiency of animal production from grassland will provide the best basis for the development of improved systems of grazing management in the future.

Fig. 3. Relationship between length of tiller and rate of pasture intake. From Allden & Whittaker 1970.

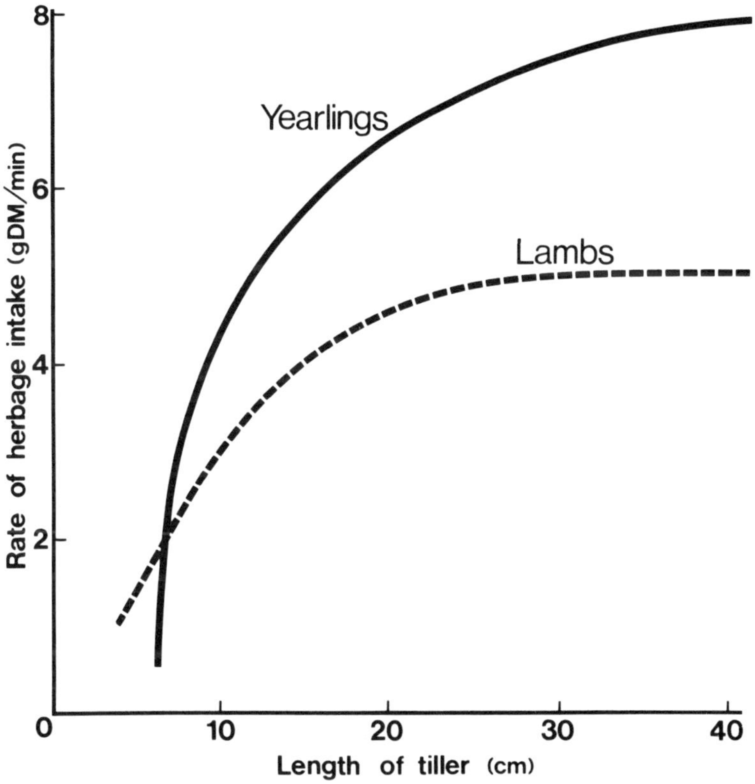

Fig. 4. Relationship between density of herbage in the surface horizons of the sward and size of individual mouthfuls of herbage taken by grazing cows. Derived from Stobbs 1975.

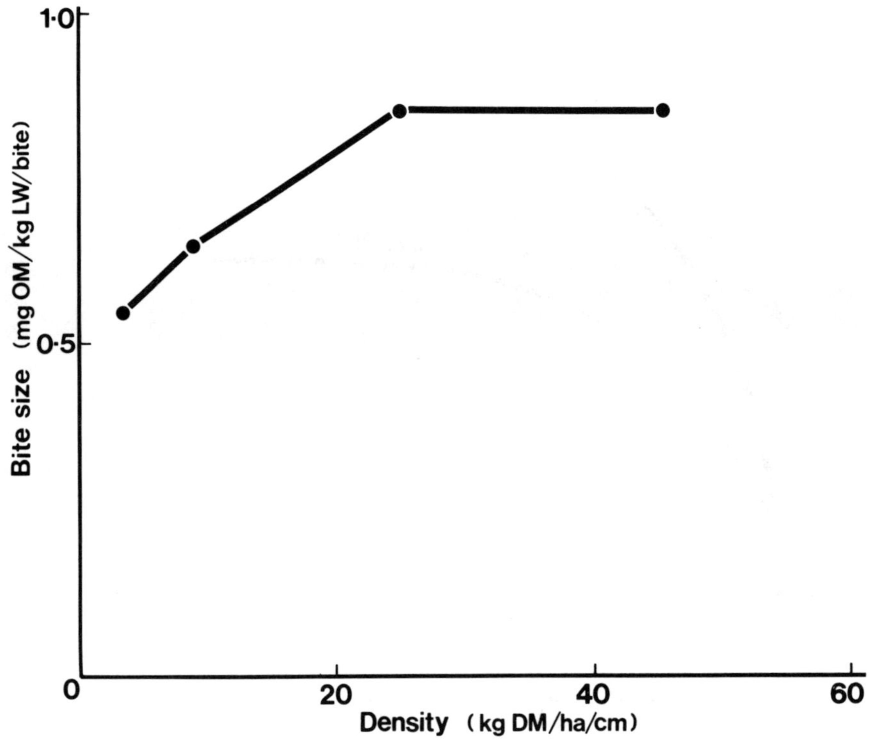

Fig.5. Relationship between height of grazed horizon above groundlevel and rate of herbage intake by grazing calves. From Hodgson 1977.

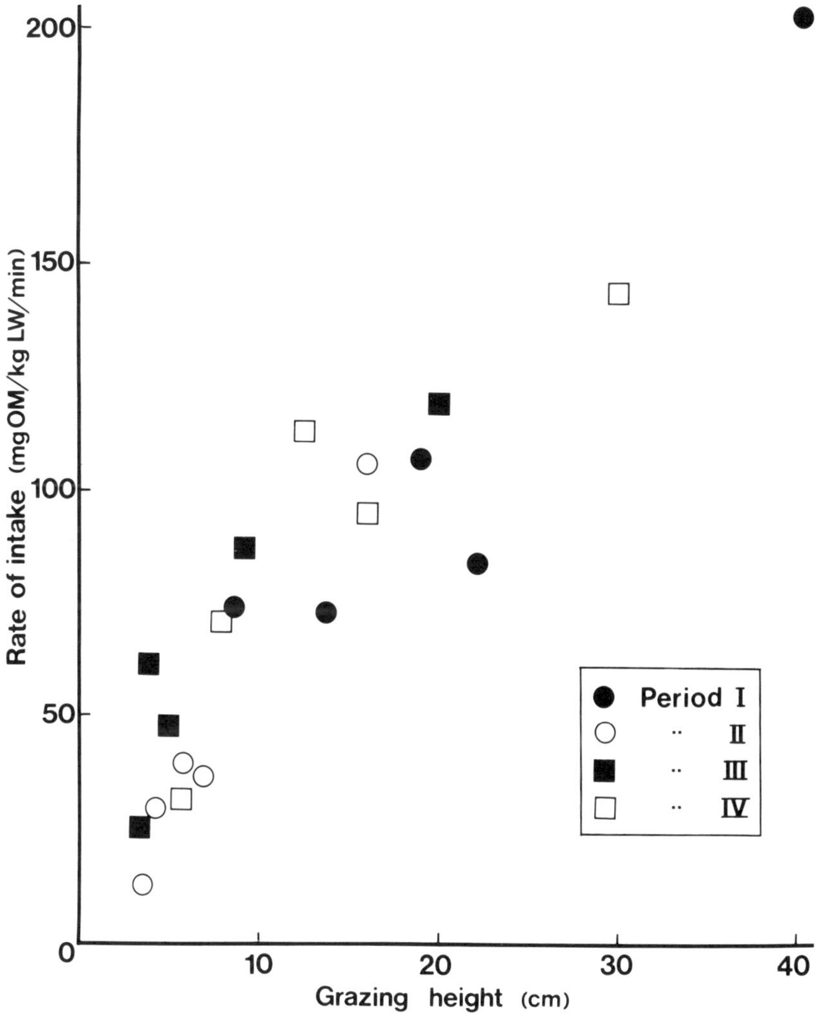

Fig. 6. Relationship between herbage allowance and herbage intake of grazing lambs. From Gibb & Treacher 1976.

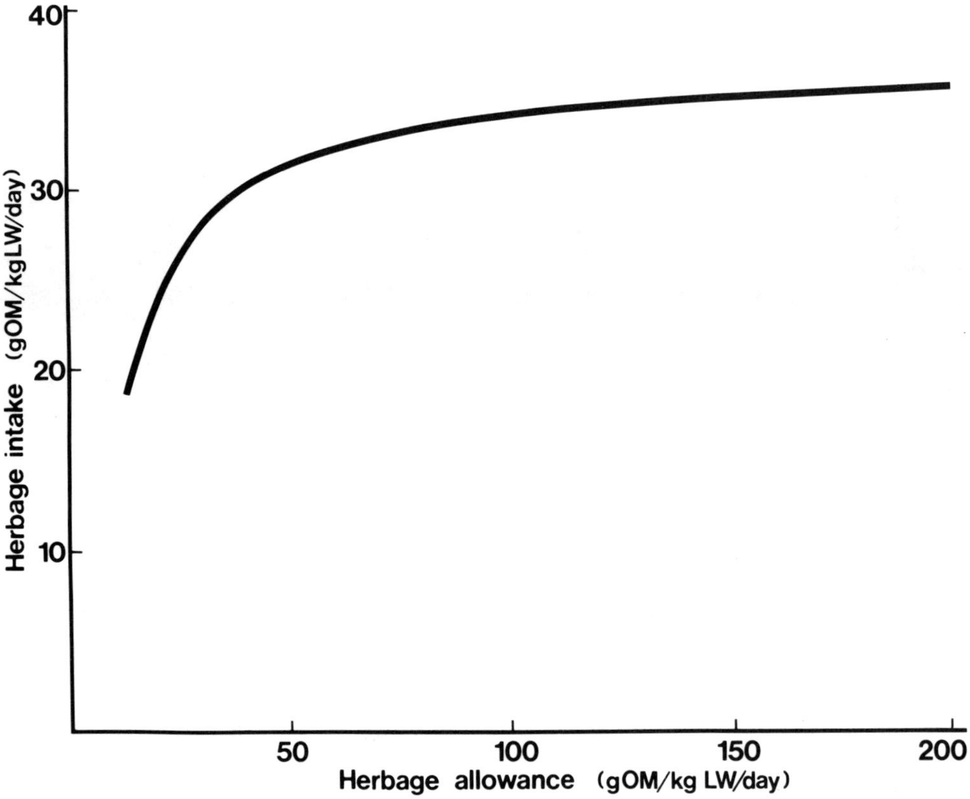

Fig.7. Relationship between herbage allowance and (a) the amount of herbage eaten per animal (% of maximum) and (b) the proportion of the daily allowance consumed (%).
Derived from Gibb & Treacher 1976, after Leaver 1976.

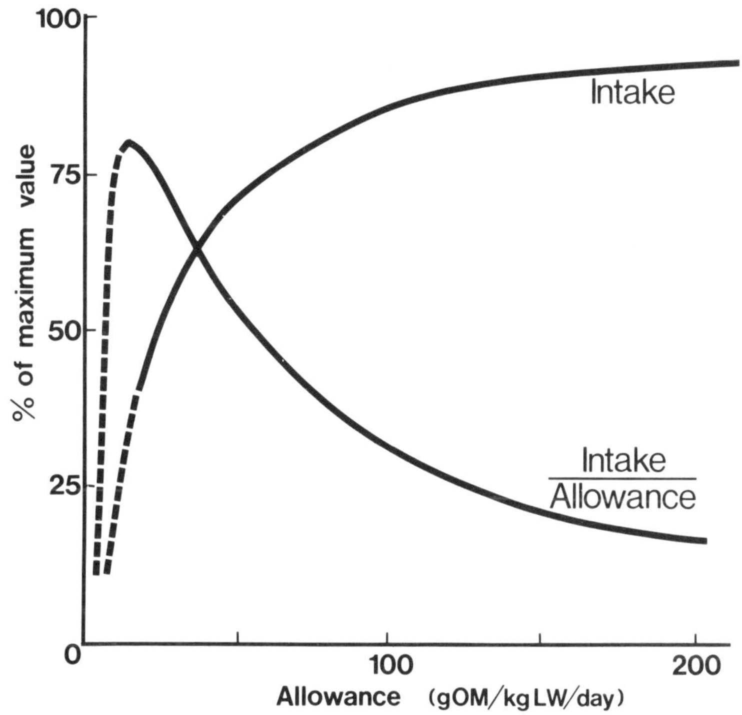

Fig.8. The digestibility of herbage selected by calves from different horizons within strip-grazed swards (% OMD). Hodgson & Jamieson, unpublished.

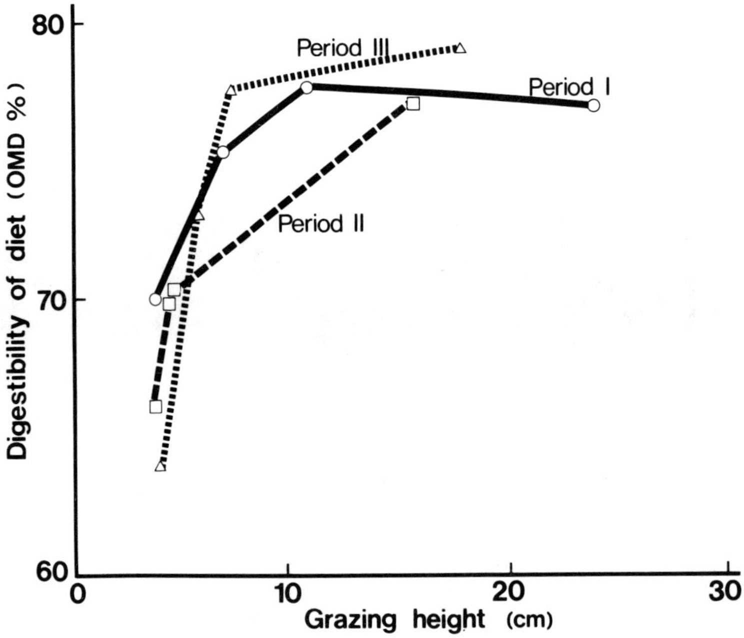

Fig.9. Herbage intake, growth rate and the efficiency of food conversion in lambs. Derived from MAFF 1975. This example is for a 20kg lamb eating herbage of 74% OMD; for derivation of results see reference quoted.

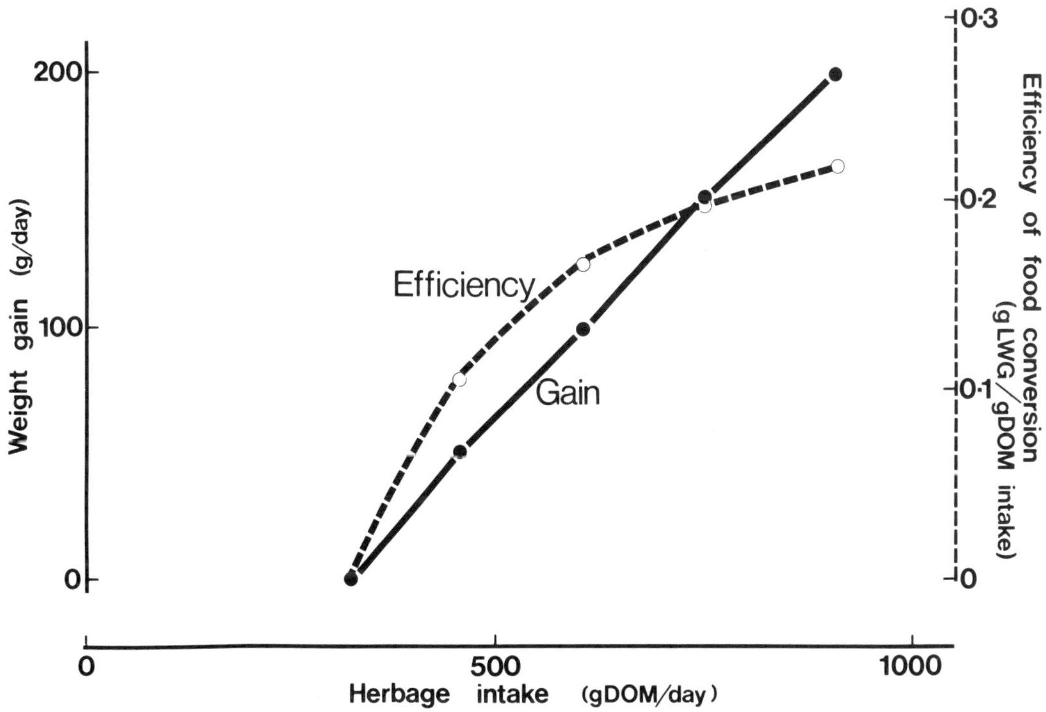

Fig. 10. Relationship between stocking rate (animals/ha) and (a) performance of individual animals or (b) output per ha. From Jones & Sandland 1974.

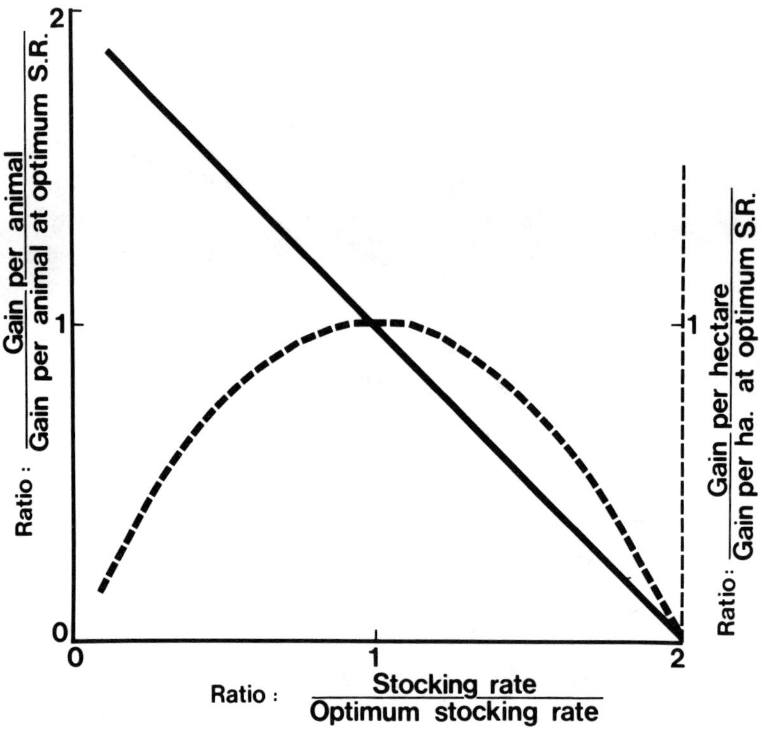

REFERENCES

Allden, W.G.;Whittaker, I.A.M. (1970) The determinants of herbage intake by grazing sheep: the interrelationship of factors influencing herbage intake and availability. Australian Journal of Agricultural Research 21,755-766.

Arnold, G.W.;Dudzinski, M.L. (1967) Studies on the diet of the grazing animal. II. The effect of physiological status in ewes and pasture availability on herbage intake. Australian Journal of Agricultural Research 18,349-359.

Balch, C.C.;Campling, R.C. (1969) Voluntary intake of food. In: Lenkeit, W. Breirem, K. and Grasemann, E. (Eds) 'Handbuch der Tiernahrung'. Hamburg: Paul Parey, Pt 1,554-579.

Behaeghe, T.J.;Carlier, L.A. (1973) Influence of nitrogen levels on quality and yield of herbage under mowing and grazing conditions. Plant Husbandry 28,52-66.

Gibb, M.J.;Treacher, T.T. (1976) The effect of herbage allowance on herbage intake and performance of lambs grazing perennial ryegrass and red clover swards. Journal of Agricultural Science, Cambridge 86,355-365.

Hamilton, D.;Bath, J.G. (1970) Performance of sheep and cattle grazed separately and together. Australian Journal of Experimental Agricultural Animal Husbandry 10,19-26.

Hodgson, J. (1975) The influence of grazing pressure and stocking rate on herbage intake and animal performance. In: 'Pasture utilization by the grazing animal'. British Grassland Society Occasional Symposium No.8, pp93-103.

Hodgson, J. (1977) Factors limiting herbage intake by the grazing animal. Proceedings, International Meeting on Animal Products from Temperate Grassland Dublin, 1977, 70-75

Holmes, W. (1968) The use of nitrogen in the management of pasture for cattle. Herbage Abstracts 38,265-277.

Hutchinson, K.J. (1971) Productivity and energy flow in grazing/fodder conservation systems. Herbage Abstracts 41,1-10.

Jones, R.J.;Sandland, R.L. (1974) The relation between animal gain and stocking rate in grazing trials: Derivation of a model from experimental results. Journal of Agricultural Science, Cambridge 83,335-342.

Langlands, J.P.;Bennett, I.L. (1973) Stocking intensity and pastoral production. I. Changes in the soil and vegetation of a sown pasture grazed by sheep at different stocking rates. Journal of Agricultural Science, Cambridge 81,193-204.

Leaver, D.J. (1976) Utilisation of grassland by dairy cows. In: Swan, H. and Broster, W.H. (Eds) 'Principles of cattle production'. Proceedings, Nottingham Easter School, 1975 307-327.

Marsh, R. (1975a) A comparison between spring and autumn pasture for beef cattle at equal grazing pressures. Journal of the British Grassland Society 30,165-170.

Marsh, R. (1975b) Systems of grazing management for beef cattle. In: 'Pasture utilisation by the grazing animal'. British Grassland Society Occasional Symosium No.8, 119-128.

McFeely, P.C.;Browne, D.;Carty, O. (1975) Effect of grazing interval and stocking rate on milk production and pasture yield. Irish Journal of Agricultural Research 14,309-319.

Morris, R.M. (1970) The use of cutting treatments designed to simulate defoliation by sheep. Journal of the British Grassland Society 25,198-206.

Mott, G.O. (1960) Grazing pressure and the measurement of pasture production. Proceedings, 8th International Grassland Congress 606-611.

Spedding, C.R.W. (1971) Grassland ecology. Oxford: Oxford University Press, p152.

Stobbs, T.H. (1975) The effect of plant structure on the intake of tropical pasture. III. Influence of fertilizer nitrogen on the size of bite harvested by Jersey cows grazing Setaria anceps cv. Kazungula swards. Australian Journal of Agricultural Research 26,997-1007.

Thomson, D.J. (1977) The role of legumes in improving the quality of forage diets. Proceedings, International Meeting on Animal Production from Temperate Grassland Dublin, 1977.

Tribe, D.E.;Lloyd, A.G. (1962) Effect of stocking rate on the efficiency of fat lamb production. Journal of the Australian Institute of Agricultural Science 1962, 28,274-278.

UK, Ministry of Agriculture, Fisheries and Food, Department of Agriculture and Fisheries for Scotland and Department of Agriculture for Northern Ireland (1975). Energy allowances and feeding systems for ruminants. Technical Bulletin 33.

CONSERVED FORAGES IN SHEEP PRODUCTION

DR R.J. WILKINS

The Grassland Research Institute, Hurley, Maidenhead, Berks

Sheep production in most countries is heavily dependent on the use of grass; in Britain some 90% of the nutrients supplied to sheep come from fresh and conserved grass. A further feature of many sheep industries is the attempt made to synchronize as far as possible the times of highest demand for feed with periods of rapid grass growth. This reflects the generally lower cost of grazed rather than conserved grass and the generally higher feed value of grazed grass. An example of a pattern of grass growth and nutrient demand is given in Figure 1 and applies to ewes with twin lambs stocked at 12 ewes/ha in England (Spedding 1967). The ability to synchronize feed requirements with grass growth will clearly vary with climatic conditions and the need for conserved forage, or a feed other than grazed grass, increases as the period of active grass growth declines.

This paper considers (i) the major factors that determine the efficiency of forage conservation methods and their suitability for use in sheep enterprises, and (ii) the ability of conserved feeds to satisfy the requirements of sheep in particular physiological conditions.

FORAGE CONSERVATION METHODS

The conservation of forage as either hay or silage is discussed here. High-temperature drying is an extremely efficient method of conservation but is also costly. The use of dehydrated forage as a replacement for concentrates is discussed by Tayler & Wilkins (1976).

Hay

Hay is the conserved forage which is used most widely in sheep feeding. The nutritive value is affected both by the quality of the crop at cutting and the weather during the field curing period. The results given in Table 1 are from experiments carried out with grass in the south of England (Shepperson 1960) and are average values from nine experiments.

Table 1 The effect of haymaking on crude protein content and in vivo dry matter digestibility

	Crude protein (%)	In vivo dry matter digestibility (%)
Grass at cutting	10.7	72.5
Barn-dried hay	10.7	69.1
Field-cured hay	9.0	64.4

From Shepperson (1960)

Figure 1

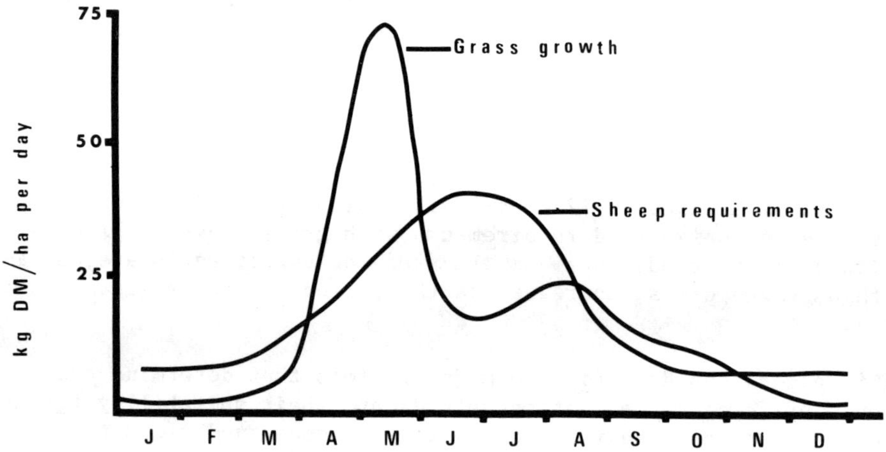
75
50
25
kg DM/ha per day
Grass growth
Sheep requirements
J
F
M
A
M
J
J
A
S
O
N
D

Field-cured hay was considerably lower than the grass at cutting in both digestibility and crude-protein content; the fall in quality was greatest when the cut crop was in the field for the longest period. Barn-dried hay, removed from the field at 40% moisture content, was similar in crude-protein content although on average 3.4% units lower in digestibility than the crop at cutting. Some breakdown of the true protein occurs through plant enzyme activity during hay making but this is unlikely to result in marked reduction in protein value. Although there is a lack of direct comparisons, the voluntary intake of hay appears to be of the same order as that of fresh grass of similar digestibility.

Some 95% of the water in grass at cutting must be removed before it can be stored aerobically as hay, without moulding and deterioration occurring.

Hay-making is clearly most attractive in regions in which spells of hot dry weather are common during the period in which there is available a large quantity of forage of the required nutritive value, generally in continental rather than maritime areas.

The rate at which water is lost from grass after cutting has been related to the vapour pressure deficit (Spatz et al 1970) and to potential evapo-transpiration (Roy & Wilkins 1969). In the latter study the evaporation from the cut grass on the day of cutting was 50% of the potential as indicated by evaporation from a free water surface, but this rate fell to 8% of the potential at around 50% moisture content (fresh basis) and then still further. It is this increase in resistance to water loss that makes hay-making such a risky conservation method in humid areas and leads to periods in the field of up to two weeks and, in some cases, complete loss of the cut crop.

Research in hay-making in humid areas has centred around attempts to reduce the period of field-curing. This can be done by removing the material from the field at higher than normal moisture contents or by increasing field drying rates. The results of Spatz et al (1970) indicated that only 50% of the accumulated vapour pressure deficit required for complete field curing was needed if the grass was removed from the field at 40% moisture content. Clearly removal from the field at this moisture content would make a large contribution to reducing weather risk.

Barn drying methods are well developed for chopped hay and hay in conventional bales but are at present little used in Britain. It is now established that large rectangular bales of about 5 m^3 can be dried by ventilation (Nellist 1976).

Work on the use of chemical preservatives to enable material to be stored at high moisture content holds great promise. Research at Rothamsted Experimental Station has established in small scale conditions that storage of moist hay is satisfactory, if it contains 1.2 g propionic acid/100 g water in the hay (Lacey pers. comm.). Unfortunately in practice there are problems in obtaining both evenness in application of the chemical and in prevention of loss of the chemical through volatilisation. Methods of application have been improved (Klinner & Holden 1977, National

Institute of Agricultural Engineering 1977) and losses are much less with ammonium propionate than with propionic acid. In a number of experiments, Klinner & Holden (1977) found a mean application loss of 37% for ammonium propionate compared with 61% for propionic acid. The capital investment in equipment and the labour requirements in preservation of moist hay with chemicals are much less than in barn drying and it is likely that the use of hay preservatives will have considerable impact in hay making in the next few years.

There has been less progress in attempts to accelerate drying in the field. The use of mechanical treatments after cutting has been thoroughly discussed by Wieneke (1972) and Klinner & Shepperson (1975). Unfortunately treatments such as flail mowing which maximise the rate of drying may give excessive fragmentation of the crop with the result that loss of dry matter is increased. This has led to an interest in conditioning methods which bruise the crop without causing fragmentation. A mower-conditioner, developed at the National Institute of Agricultural Engineering, is of particular interest. The crop is conditioned with flails directly after cutting and is then deposited into a swath through which air penetrates easily (Klinner 1975). Mower-conditioners with a similar working principle have been produced commercially. This approach is most successful with stemmy crops, the structural strength of which helps prevent the swath from settling and interrupting air flow. With more leafy crops which do not form such a well-structured swath, it may be better to spread the crop to intercept the maximum solar radiation. Chemicals such as formic acid which damage the cuticle of the plant may result in marked increases in the drying rate of single leaves in controlled conditions (Thaine & Harris 1973) but, when used in the field, they have not reduced the time required to reach a safe moisture content for storage (Tetlow et al 1975).

Different species vary in their suitability for hay-making. Tall fescue dries more quickly than perennial ryegrass. The legumes are very prone to loss of leaf and, because the leaf dries more rapidly than the stem, they need to be handled with care during the later stages of drying. This is one of the major problems with hay-making in conditions of high insolation. Shepherd (1961) demonstrated differences between varieties of white clover in their susceptibility to mechanical losses during hay-making, but there has been no attempt to breed legumes which are the more resistant to mechanical loss.

Some advantages of haymaking a a method of conservation are:

(i) Hay is a dry feed that can be easily transported and fed to small groups of sheep.

(ii) Only a low level of mechanization is needed and hay can be made in uneven fields with little difficulty.

(iii) Hay has a long shelf life and quantities can be taken from the store at any time.

(iv) Batches of hay of differing nutritive value can easily be kept separate and used according to the particular requirements of the sheep at different times.

Major disadvantages are:

(i) The extreme sensitivity of the method to weather conditions and the consequent difficulty when the conservation programme is being planned in forecasting the feeding value of the product and the levels of loss that will occur.

(ii) The high losses of nutrients that can occur through mechanical damage during haymaking.

(iii) The difficulty in completely mechanizing the feeding of hay.

Silage

The level of production of ruminant animals fed silage is commonly lower than that of those fed on fresh or dried forage from the same origin. Although ensiling has little effect on digestibility and on the efficiency with which digested nutrients are utilized, voluntary intake is almost always reduced by ensiling (Wilkins 1974, Wilkinson et al 1976). The reduction in intake is sometimes as much as 50% (Demarquilly 1973) and is most severe when a clostridial fermentation has taken place in the silo. In this case lactic acid produced during the early stages of fermentation is broken down by clostridia to form butyric acid and acetic acid and extensive degradation of the nitrogenous constituents takes place. Wilkins et al (1971) found that a large proportion of the variation in the intake of silage by sheep could be accounted for by measurements of fermentation quality in the silage. There was negative correlation with ammonia nitrogen as a percentage of total nitrogen and a positive correlation with lactic acid as a percentage of total acids. More recently, Barry et al (1978a) and Barry et al (1978b) established close correlations between these factors and the efficiency with which the crude protein in lucerne silage was utilized by sheep. Decarboxylation brought about by clostridia appeared to be of greater consequence than deamination.

Losses of dry matter during ensiling will also tend to be higher if clostridial fermentation has taken place and the silage will have an obnoxious odour. Because of these factors much attention has been paid to the elaboration of methods that will prevent the occurrence of a clostridial fermentation. Broadly speaking clostridial growth will occur if pH does not fall sufficiently during the early stages of ensiling. Two approaches have been made to prevent clostridial fermentations:

(i) Wilting crops prior to ensiling. This concentrates the sugars in the crop and also the growth of clostridia is suppressed in dry conditions.

(ii) The addition of sugar (molasses) or acid at ensiling to increase directly or indirectly acid levels in the silage. In the last decade formic acid has been used extensively in northern Europe as a silage additive and its use has resulted in substantial improvements in silage fermentation quality and animal performance (Saue & Breirem 1969, Wilson & Wilkins 1973a and b).

It is unfortunate that high concentrations of acids in silage can also lead to some lowering in intake (McLeod et al 1970, Wilkinson et al 1976) so that the intake even of well-preserved silage is lower than that of the crop from which the silage is made. This has encouraged research into methods to prevent or severely to restrict fermentation in the silo.

Much of this research has centred around the use of formaldehyde. In addition to restricting fermentation in the silo this chemical will also bind with protein to protect it from digestion in the rumen. It is now clear that the application rate of formaldehyde is critical and Wilkinson et al (1976) concluded that the optimal rate was 3-5 g HCHO/100 g crop CP. When applied at this rate, some acid needs to be added with the formaldehyde in order to prevent clostridial fermentation (Wilson & Wilkins 1977).

A further factor which has a large effect on the intake of silage by sheep is its particle length. Dulphy & Demarquilly (1973) found intakes 44% higher for finely-chopped rather than coarsely-chopped silages. This increase arose in part from better fermentation quality in the finely-chopped silages and in part from an effect of particle length per se.

Much progress has been made in preventing loss during silage making. It is now widely realised that much of this loss may arise from oxidation in the surface layers of the silo. This loss can be completely prevented if the silo is covered with polythene sheeting and the sheeting covered, with for example manure, to prevent damage and consequent air ingress during the storage period. Methods of filling silos are discussed by Raymond et al (1972) and by the Ministry of Agriculture, Fisheries and Food (1977). If the silo is sealed well and clostridial fermentation prevented, the loss of dry matter during ensiling will be below 15%.

Some advantages of silage making as a method of conservation for a sheep enterprise are:

i. A large degree of independence from weather conditions. This allows a conservation programme to be worked out on the basis of the growth of the crop and gives conserved forage with a consistent and predictable digestibility.

ii. Nutrient losses can be low.

iii. A high level of mechanization is possible.

Major disadvantages are:

i. Low levels of intake and low efficiency of crude protein utilization.

ii. Silage is a heavy feed to man-handle (as may be required with small groups of sheep).

iii. Silage has a low shelf life. Once a silo is opened, feeding should be continuous rather than intermittent and, in extreme situations, losses of dry matter after the silo has been opened may reach 30% (Honig 1975).

iv. The high moisture content of silage will lead to high levels of urine output and consequently a need in housed conditions for more bedding or the use of slatted floors.

v. Listeriosis may be associated with silage feeding, particularly if the fermentation quality of the silage is low (see Khan et al 1970).

CONSERVED FORAGES FOR SHEEP IN DIFFERENT PHYSIOLOGICAL CONDITIONS

Consideration is given here to the use of conserved forages in feeding the ewe at different stages of the reproductive cycle and in finishing lambs.

The requirements of ewes for metabolizable energy (ME) and digestible crude protein (DCP) are illustrated in Figure 2. These are based on data from the Meat and Livestock Commission (undated), the Ministry of Agriculture, Fisheries and Food (1975) and the Agricultural Development and Advisory Service (1976) and refer to ewes weighing 70 kg at the start of pregnancy and bearing twin lambs. Changes in the requirements for ME and DCP closely match one another and the total demand for nutrients is of course highest during late pregnancy and lactation. In Figure 3 these requirements are interpreted in terms of the minimum quality of forage which would be needed to satisfy these requirements. Sheep were considered to be able to consume feed to a standard quantity of indigestible organic matter (0.57 kg indigestible organic matter for a dry ewe weighing 70 kg) (Agricultural Development and Advisory Service 1976). The maximum intake of indigestible organic matter per kg liveweight was considered to vary through the year and was taken as 80% in the fifth month of pregnancy, 110, 130, 130 and 110% in the first four months of lactation and as 100% for the rest of the year. These intakes of indigestible organic matter would be reached with fresh or dried feeds, but for a well-preserved silage might be some 15% lower. The lower intake of silage gives a need to satisfy requirements at any physiological stage for material with a ME some 0.5 MJ/kg higher than that of hay and a DCP some 5 g/kg DM higher.

There is little difficulty in producing conserved forage to satisfy the requirements of ewes that are dry and non-pregnant or in the first three months of pregnancy. The requirement for feed of particularly high quality in early lactation reflects both the large demand for nutrients at that time and a rather lower voluntary intake than in mid-lactation. This requirement can be satisfied by grazed grass but, of the conserved forages, it is likely that only dried grass would be adequate without the use of supplements.

Surprisingly few critical experiments have been carried out with ewes fed hay or silage. Ferguson & Black (1974) fed hays varying in ME from 7.0 to 10.2 MJ/kg DM to pregnant Scottish Halfbred ewes. Cereal-based concentrates were given at rates from 0 to 30 kg per ewe. Hay intake was positively correlated with metabolizable energy concentratin. They concluded that the quality of hay required to satisfy nutrient requirements for satisfactory performance during pregnancy increased from 7.8 MJ ME/kg DM from 56-40 days prior to lambing to 9.8 MJ ME/kg DM from 22-0 days prior to lambing; the figures are in close accord with those given in Figure 3.

Figure 2

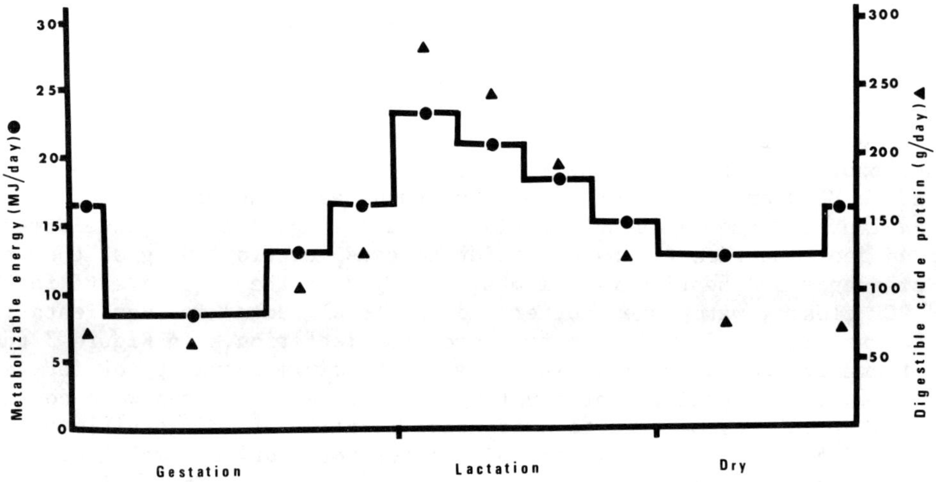

Metabolizable energy (MJ/day)●
Digestible crude protein (g/day)▲
30
25
20
15
10
5
0
300
250
200
150
100
50
Gestation
Lactation
Dry

Figure 3

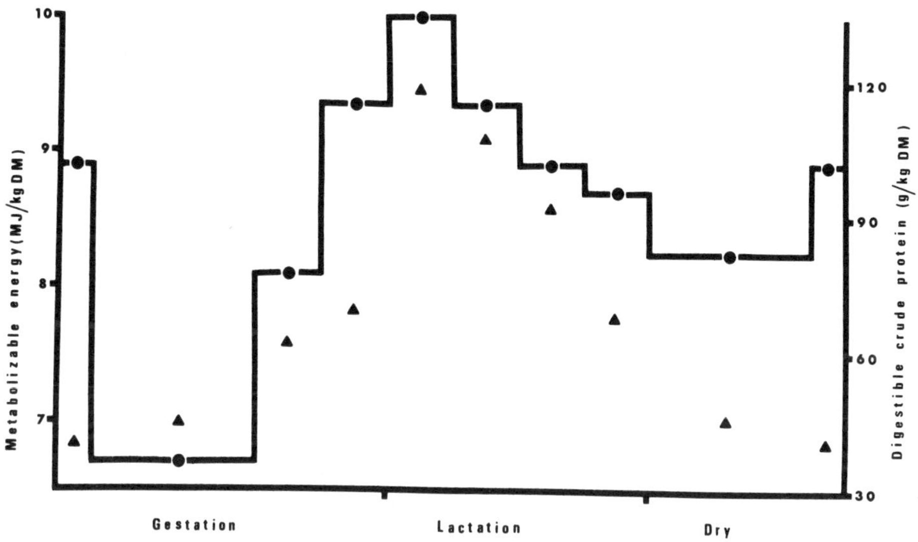

10
9
8
7
Metabolizable energy(MJ/kgDM)
120
90
60
30
Digestible crude protein (g/kg DM)
Gestation
Lactation
Dry

Unpublished data from Drayton Experimental Husbandry Farm also show that the nutrient requirements in the last month of pregnancy of ewes carrying twin lambs can be satisfied by hay with a ME concentration of 10 MJ/kg DM.

Nedkvitne (1969) clearly showed that silages of poor fermentation quality (high ammonia-N content) without supplementation could not satisfy the requirements of ewes in late pregnancy and Treacher & Wilkins (1973), working with sole diets of well-preserved grass silages of high digestibility fed to ewes bearing twin lambs, found depressions in lamb birth weight. In two years of experiments at Liscombe Experimental Husbandry Farm the supplementation of high-quality silage with cereals did not increase the birthweights of twin lambs (Liscombe Experimental Husbandry Farm 1976). Work in France, in which ewes were fed maize silage, is summarised by Tissier & Molenat (1975). They noted much higher intakes of finely-chopped rather than long silages and obtained the best results with silages of 25-35% dry matter content. In late pregnancy the maize was generally supplemented with hay protein concentrate and minerals.

Conserved forages are also used in fattening lambs. Table 2 has been calculated from the Agricultural Development and Advisory Service (1976) for lambs weighing 30 kg and consuming 0.25 kg indigestible organic matter per day, a limit considered appropriate to fresh forages and hay. The growth rates obtained from lambs ranging from 6 to 12 months of age in some recent experiments in Great Britain and Ireland are given in Table 3. The highest gains were from well-preserved lucerne and maize silages and, for both of these feeds and for the hays, rates of liveweight gain were above those indicated from Table 2. A direct comparison between hay and silage was made in the work of Sheehan & Fitzgerald (1976) and performance was higher for sheep given hay, in line with higher dry matter intakes. It is not clear however whether the silage used in this work had been made with a precision-chop or a flail harvester.

Liveweight gain is likely to be increased by the provision of supplementary feed. Wilkins (1974) showed however that the depression in silage voluntary intake was much greater for silages with high than low voluntary intakes when fed alone. The magnitude in the response in liveweight gain to supplementary feeding will therefore vary. For instance the provision of a barley/soya/urea concentrate at a daily rate of 16 g OM/kg $LW^{0.75}$ increased the liveweight gain of sheep given sorghum silage from -19 to +178 g/day but resulted in an increase only from +161 to +231 g/day for sheep fed maize silage (Wilkins & Cook unpublished data).

Nolan (1974) noted a high fat content in the carcasses of sheep fed silage, despite their low growth rates. This probably resulted from the inefficient use of protein and thus the limited quantity of absorbed amino acid while on silage diets. A similar effect has been noted in experiments at the Grassland Research Institute (Lonsdale 1977) and stresses the need to improve the efficiency of protein utilization in silages. It is not established to what extent this effect on carcass fatness is modified when the silage is supplemented with concentrates.

Table 2 Predicted liveweight gains of lambs weighing 30 kg given ad libitum feeds with differing metabolizable energy concentrations (consumption of 0.25 kg indigestible organic matter/day)

ME concentration (MJ/kg DM)	Liveweight gain (g/day)
8	0
9	35
10	115
11	220

Calculated from Agricultural Development and Advisory Service (1976)

Table 3 Liveweight gains (g/day) for 6 to 12 month old lambs given silage or hay ad libitum without supplements

	ME concentration[1] (MJ/kg DM)	Liveweight gain (g/day)
Hay (Sheehan & Fitzgerald 1976)	7.79	45
	8.20	83
Grass silage		
Sheehan & Fitzgerald (1976)	8.1	-56
	8.4	-14
Nolan (1974)	8.3	50
Wilkinson & Tetlow (unpublished)	9.8	68
Lucerne silage (Barry et al unpublished)		
Flail harvester, no additives	9.3	62
Flail harvester, formic acid at 6 l/t	9.5	97
Precision chop, no additives	9.5	105
Precision chop, formic acid at 6 l/t	9.5	212
Maize silage (Wilkins & Cook unpublished)	10.0	161
Sorghum silage (Wilkins & Cook unpublished)	7.8	-19
Barley silage (Wilkins et al 1970)	9.0	-30

[1]Derived from in vivo digestibility of OM, DM or of OM in the DM, using equations from Ministry of Agriculture, Fisheries and Food (1975)

CONSERVATION STRATEGY IN RELATION TO NUTRIENT REQUIREMENTS

During much of the winter ewes do not require feeds of high nutritive value. High-quality feeds are needed only in late pregnancy and early lactation and perhaps for fatterning lambs. In general there is an inverse relationship between the yield of grass and other forage crops and their quality. For example in recent experiments at the Grassland Research Institute, the total seasonal dry matter yield with 3-weekly cutting was only 50% that obtained with cutting at 9-weekly intervals. Mean ME concentrations were 10.14 and 8.76 MJ/kg DM respectively. In many situations it may therefore be the best policy to cut grass of low ME concentration for conservation and to supplement this with concentrates at critical periods during the year. It is easier to take a more flexible approach with hay than with sheep, because different lots of hay can conveniently be stored separately whereas losses are often high if several small silos (from cuts differing in quality) are made. Thus with hay, batches of high quality can be used in the periods of maximum nutrient demand.

It is important to reduce dry matter losses during conservation in all situations and in sheep production this is probably more important than endeavouring to make conserved forage with extremely high ME concentrations.

REFERENCES

Barry, T.N.;Cook, J.E.;Wilkins, R.J. (1978) The influence of formic acid and formaldehyde additives and type of harvesting machine on the utilization of nitrogen in lucerne silages. 1. The voluntary intake and nitrogen retention of young sheep consuming the silages with and without intra-peritoneal supplements of DL-methionine. Journal of Agricultural Science, UK 91,701-715.

Barry, T.N.;Mundell, D.C.;Wilkinson, R.J.;Beever, D.E. (1978) The influence of formic acid and formaldehyde additives and type of harvesting machine on the utilization of nitrogen in lucerne silages. 2. Changes in amino-acid composition during ensiling and their influence on nutritive value. Journal of Agricultural Science, UK 91,717-725.

Boaz, T.;Forbes, J.M.;Kilkenny, J.B.;Mudd, C.H.;Robinson, J.J.;Russel, A.J.F.; Treacher, T.T. (1973) Feeding the ewe. Technical Report, Sheep Improvement Service, Meat and Livestock Commission No.2. 52pp.

Demarquilly, C. (1973) Composition chimique, caractéristiques fermentaires, digestibilité et quantité ingérée des ensilages de fourrages: modifications par rapport au fourrage vert initial. Annales de Zootechnie 22,1-35.

Dulphy, J.P.;Demarquilly, C. (1973) Influence de la machine de récolte et de la finesse de hachage sur la valeur alimentaire des ensilages. Annales de Zootechnie 22,199-217.

Ferguson, J.A.;Black, W.J.M. (1974) Hay quality and the pregnant ewe. Edinburgh School of Agriculture Annual Report 1974 41-43.

Honig, H. (1975) Umsetzungen und Verluste bei der Nachgärung. Wirtschaftseigene Futter 21,25-32.

Khan, M.A.;Seaman, A.;Woodbine, M. (1970) Review of listeriosis: a new zoonotic challenge to mankind? Report of the School of Agriculture, University of Nottingham 1969-1970 103-110.

Klinner, W.E. (1975) Research and development in crop mowing and conditioning. ARC Research Review 1,51-58.

Klinner, W.E.;Holden, M.R. (1977) Advances with chemical preservatives for hay. Paper presented at ASAE/CICR International Grain and Forage Harvesting Conference, Ames, Iowa, September 1977.

Klinner, W.E.;Shepperson, G. (1975) The state of haymaking technology - a review. Journal of the British Grassland Society 30,259-266.

Lonsdale, C.R. (1977) The efficiency of utilization of digested nutrients by calves. Annual Report, Grassland Research Institute, Hurley, 1976 92.

McLeod, D.S.;Wilkins, R.J.;Raymond, W.F. (1970) The voluntary intake by sheep and cattle of silages differing in free-acid content. Journal of Agricultural Science, UK 75,311-319.

National Institute of Agricultural Engineering (1977) Annual Report, National Institute of Agricultural Engineering 1976-1977.

Nedkvitne, J.J. (1969) Hay and silage in the winter feeding of sheep. In: Spedding, C.R.W. (Ed) Grassland in sheep and goat production, EAAP Report No.2 from Study Commission on Animal Nutrition p40-51.

Nellist, M.E. (1976) Drying hay and grain. Agricultural Engineer 31,55-60.

Raymond, W.F.;Shepperson, G.;Waltham, R. (1972) Forage conservation and feeding. Ipswich; Farming Press Ltd. SBN 85236-022-3.

Roy, M.G.;Wilkins, R.J. (1969) The effects of weather conditions on the moisture loss from cut grass in the field. Meteorological Office Agricultural Memorandum No 273.

Saue, O.;Breirem, K. (1969) Comparison of formic acid silage with other silages and dried grassland products in feeding experiments. In: Crop conservation and grassland. Proceedings of the 3rd General Meeting of the European Grassland Federation, Braunschweig, 1969, pp282-284.

Sheehan, W.;Fitzgerald, J.J. (1976) Lambs do better on hay than on silage. Farm and Food Research 7,85-86.

Shepherd, W. (1961) The susceptibility of hay species to mechanical damage. II. Effects of moisture content and variety. Australian Journal of Agricultural Research 12,783-796.

Shepperson, G. (1960) Effect of time of cutting and method of making on the feed value of hay. In: Proceedings of the 8th International Grassland Congress, Reading, Berks, July 1960. Hurley, Berks; British Grassland Society, Grassland Research Institute, pp704-708.

Spatz, G.;Eimern, J. van; Lawrinowicz, R. (1970) Der Trocknungsverlauf von Heu im Freiland. Bayerisches Landwirtschaftliches Jahrbuch 47,446-464.

Spedding, C.R.W. (1967) The grazing sheep in relation to its environment. In: Report of the Proceedings and Invited Papers, 9th International Congress of Animal Production, Edinburgh 1966, pp174-187.

Tayler, J.C.;Wilkins, R.J. (1976) Conserved forage - complement or competitor to concentrates. In: Swan, H.;Broster, W.H. (Eds) Principles of cattle production. London, UK; Butterworth & Co Ltd, pp343-364.

Tetlow, R.M.;Wilkins, R.J.;Evans, W.E. (1975) Pre-harvest desiccation with formic acid. Annual Report, Grassland Research Institute, Hurley, 1974 61-62.

Thaine, R.;Harris, C.E. (1973) Formic acid as a desiccant for grass leaves. Journal of Agricultural Science, UK 80,349-351.

Tissier, M.;Molénat, G. (1975) L'ensilage de mais dans l'alimentation des brebis. Bulletin Technique, Centre de Recherches Zootechniques et Vétérinaires de Theix No 20,37-42.

Treacher, T.T.;Wilkins, R.J. (1973) Comparison of the voluntary intake by pregnant ewes of grass conserved by dehydration or ensiling with and without the addition of formaldehyde. Annual Report, Grassland Research Institute, Hurley, 1972 116.

UK, Ministry of Agriculture, Fisheries and Food (1975) Energy allowances and feeding systems for ruminants. Technical Bulletin 33. London, UK; HMSO, viii+79pp. ISBN 0-11-240894-X.

UK, Ministry of Agriculture, Fisheries and Food (1977) Silage. Bulletin, Ministry of Agriculture, Fisheries and Food No 37, 10th edn. London, UK; HMSO, 82pp. ISBN 0-11-241337-4.

UK, Ministry of Agriculture, Fisheries and Food, Agricultural Development and Advisory Service (1976) Feeding in-lamb ewes. Liscombe Sheep Bulletin No 1.

UK, Ministry of Agriculture, Fisheries and Food, Agricultural Development and Advisory Service (1976a) Nutrient allowances and composition of feedingstuffs for ruminants. ADAS Advisory Paper No 11, 2nd edn., i+38pp.

Wieneke, F. (1972) Verfahrenstechnik der Halmfutterproduktion. Gottingen; F. Wieneke, 607pp.

Wilkins, R.J. (1974) The nutritive value of silages. In: Swan, H.;Lewis, D. (Eds) Nutrition conference for feed manufcturers: 8. London, UK; Butterworth & Co. Ltd, pp167-189. ISBN 0-408-70658-9.

Wilkins, R.J.;Hutchinson, K.J.;Wilson, R.F.;Harris, C.E. (1971) The voluntary intake of silage by sheep. 1. Interrelationships between silage composition and intake. Journal of Agricultural Science, UK 77,531-537.

Wilkins, R.J.;Osbourn, D.F.;Tayler, J.C. (1970) The feeding value of silages made from whole-crop barley. Journal of the British Grassland Society 25, 37-43.

Wilkinson, J.M.;Wilson, R.F.;Barry, T.N. (1976) Factors affecting the nutritive value of silage. Outlook on Agriculture 9,3-8.

Wilson, R.F.;Wilkins, R.J. (1973a) Formic acid as a silage additive for wet crops of cocksfoot and lucerne. Journal of Agricultural Science, UK 80,225-231.

Wilson, R.F.;Wilkins, R.J. (1973b) Formic acid as a silage additive. 1. Effects of formic acid on fermentation in laboratory silos. Journal of Agricultural Science, UK 81,117-124.

Wilson, R.F.;Wilkins, R.J. (1977) The effects of mixtures of formalin and acid on silage fermentation. Proceedings of the XIII International Grassland Congress, Leipzig 1977 (In press).

RECENT ADVANCES IN THE EPIDEMIOLOGY OF SHEEP ENDOPARASITES

PROFESSOR J. ARMOUR

Department of Veterinary Parasitology, University of Glasgow

Over the past decade considerable advances have been made in the epidemiology of ruminant helminthiases. Although a plethora of factors are now known to influence the transmission and survival of helminths, these can conveniently be grouped under 4 key headings. Two of these relate to seasonal changes in the numbers of free-living stages in the environment (usually the pasture) and two to changes in the composition of parasitic populations within the host. While the latter are often seasonal in nature, other factors also have a modifying effect on host/parasite relationships.

SEASONAL FLUCTUATIONS OF FREE-LIVING STAGES ON PASTURE

It is accepted that in most areas of the world, the numbers of free-living stages of parasitic helminths usually to be found on the pasture fluctuate in relation to season. These fluctuations are usually regular in temperate or mediterranean zones with distinct summer and winter periods and in semi-arid or arid zones with alternating periods of rainfall and dry conditions; the fluctuations are less distinct in humid or semi-humid tropical areas. Two factors which have a major influence on seasonal changes in the numbers of pasture larval stages are, (a) the longevity of the existing infective stages and, (b) the seasonal development of new populations of those infective stages.

Longevity of Infective Stages

The survival of infective stages varies according to the species of helminths, the nature of the infective stage, e.g. egg, larva or cyst and the existing climatic conditions. In most of the trichostrongyles and strongyles of economic importance, the third larval stage is the infective one and in temperate zones they are capable of survival in significant numbers for periods of up to one year. They survive particularly well under conditions of moderate temperatures (10-18°C) and high humidity but are also capable of resisting periods of sub-zero temperatures. At higher temperatures (<20°C) and where the relative humidity is low the mortality of most species of larvae increases sharply. Thus in areas such as Western Europe with relatively mild winters, there is a regular pattern of larvae successfully overwintering on the pasture only to succumb to rising temperatures and dry weather in the following spring. Obviously, where susceptible sheep are grazed throughout the winter, they will be continuously infected until mortality of the overwintered infection on the herbage occurs; where sheep are housed during the winter, the date on which they are turned out in the spring will influence the level of infections acquired from the overwintering infection on the herbage. In countries with a prolonged cold winter and a good snow cover such as Norway, Sweden and Switzerland, survival of infective larval stages on pasture can be surprisingly good; this probably represents an adaptation for survival by the helminths.

If the infective larval stage is enclosed within an egg, such as occurs with Nematodirus battus in sheep, the winter survival rates are particularly high.

The infective metacercarial cysts of the trematode Fasciola hepatica appear to have similar requirements for survival to those of nematode infective larvae. Thus, studies in temperate zones such as the Netherlands, Scotland, Denmark and France have clearly shown that metacercariae, which appear on the pasture in autumn, survive the winter only to succumb in the following summer.

Few studies have been completed on the longevity of cestode infective stages apart from those by Gemmell (1976) in New Zealand. His results indicate that cestode eggs are as capable of survival during winter periods as their nematode and trematode cousins.

The important fact is that infective helminth larvae, eggs or cysts are capable of surviving for a sufficient period through conditions adverse to the development of free-living stages and so infect the next batch of susceptible young lambs; the latter are usually born when climatic conditions suitable to both the growth of pasture and development of free-living stages of helminths prevail.

Seasonal Development of Infective Stages

For many years, helminthologists thought that nematode eggs in the faeces of sheep would hatch to the first larval stage and develop through the second instar to the third and infective stage in one or two weeks. It is now known that the number of helminth generations per annum is low and in some instances may be only one or two. The latter was first demonstrated by Gibson, Michel and their colleagues at Weybridge, England, who found that due to temperature limitations, the eggs of the common ruminant nematodes took several weeks to develop during spring, but that during summer the development period became shorter and so infective larvae accumulated on the pasture within a short period, usually during the months of July and August. Thereafter the period of development lengthened as temperatures fell, and by late autumn and throughout the winter no development took place. Since these results, there have been articles from many parts of the world on the seasonal development of strongyle and trichostrongyle infective larvae with a limited number of annual generations (reviewed by Michel, 1976, and Armour, 1978).

There have been fewer studies on trematodes, although in temperate zones such as Western Europe it has been shown that cercariae produced by Lymnaea truncatula snails infected in the so-called summer infection during May, June and early July, are released on to pasture over a limited period of 6-8 weeks in August, September and October. From studies on the size distribution of L. truncatula in West Scotland, it was concluded that the infection was produced by one generation of snails. Although extra generations will occur in warmer climates, the concept of a limited number of generations per annum remains the same whether it is nematode larvae or *Lymnaea* snails.

The epidemiological significance of the limited number of annual generations of helminths and the concentrated production of infective stages during a specific part of the grazing year is that the latter usually occurs at a time when susceptible young stock are consuming increasing quantities of herbage and, in the case of suckling animals, are becoming less dependent on the mothers' milk supply; it also coincides in many instances with one of the most traumatic and stressful episodes in an animal's life, the post-weaning period.

CHANGES IN HELMINTH POPULATIONS WITHIN THE HOST

An analysis of helminth populations in livestock over a period of several years in many different countries has shown that both the numbers and the structure of these populations fluctuate and that in some instances these fluctuations are seasonal. Two principal factors are responsible for these variations, (a) the phenomenon of arrested larval development or hypobiosis, and (b) the immune status of the host.

Hypobiosis (Arrested or Inhibited Larval Development)

This phenomenon, which is seen primarily in nematodes, occurs when the normal rate of development of parasitic larvae to the sexually mature adult temporarily ceases; in most but not all nematodes it affects the fourth larval stage. It is known that nematode larvae may become arrested in development within the host as a manifestation of acquired immunity. There is increasing recognition however that larval development may arrest as a result of prior experience of certain climatic or seasonal influences, a phenomenon usually referred to now as hypobiosis. In several European countries it has been shown that larvae of *Ostertagia* spp, *Cooperia* spp, *Haemonchus contortus* and *Dictyocaulus* spp, ingested by grazing livestock in late autumn or winter, do not develop to the adult stage but remain as hypobiotic or low metabolising larvae for varying periods which may extend to several months and then resume their development. The stimulus for the onset of hypobiosis is not definitely known but it is independent of the number of days grazed and the numbers of worms established. Although the precise stimulus for the onset of hypobiosis is unknown, it can be induced experimentally with *Ostertagia* spp if they are subjected to adverse environmental conditions such as cold or declining temperatures for a period of 4-6 weeks. This capacity to arrest development in response to cold stimuli appears to exist in only certain strains of larvae and it is likely that other strains will respond to other adverse stimuli.

In other areas of the world, such as Australia, Israel and Nigeria, hypobiosis has been shown to occur at a time when dry conditions are prevalent and apparently in response to rising temperatures and declining humidity.

If the precise cause of hypobiosis is not known, even less is known regarding the stimulus which triggers larvae to recommence their development. It has been shown however that both natural and experimentally induced hypobiotic larvae of various species may resume development spontaneously, synchronously or asynchronously, or following the removal of the existing adult population by an anthelmintic or administration of large doses of steroid. Resumption of development may also coincide with parturition, although in some instances this may be coincidental.

Whatever the mechanisms involved, the phenomenon of larval hypobiosis is important epidemiologically, because it ensures the presence of at least some mature adults at the end of a period when reinfection of stock is unlikely, either due to housing during adverse conditions or due to mortality of infective stages as in arid grazing conditions, and so ensure pasture contamination with the onset of favourable developmental conditions. The latter is doubly important as it usually coincides with the season when most young stock are born.

Until recently, hypobiotic larvae were particularly recalcitrant to anthelmintic therapy, but the recent advent of some of the new benzimidazoles with activity against these larvae may result in this phenomenon playing a less important epidemiological role.

Immune Status of the Host

In host-parasite relationships, the development of immunity is usually associated with previous exposure to the parasite and not to age *per se.*

With most nematodes, particularly intestinal ones, immunity is acquired relatively slowly and it often requires a full grazing season before any significant host immunity is present. An obvious exception to this are the lungworms of sheep and cattle where immunity is acquired relatively quickly; this has enabled the development of live X-ray attentuated larval vaccines for these species. Fluctuations in the immune status of older sheep, generally considered to be highly immune, can occur for various reasons, the most common and important being the relaxation in immune status which occurs around parturition, the so-called peri-parturient relaxation in immunity (PPRI). This increased susceptibility is characterised by an elevated nematode faecal egg count, the responsible adult worm population being recruited mainly from freshly ingested larval populations or the maturation of larvae previously arrested in development as a result of acquired immunity or a coincident release of hypobiotic larvae. The epidemiological importance of the phenomenon is that it ensures contamination of the grazing areas with helminth eggs at a time when susceptible young stock are present. The aetiology of the PPRI has been linked to the circulating levels of the lactogenic hormone prolactin, which is known to have a depressive effect on the cellular component of the immune system and, significantly, the removal of the suckling young from the dam results in a fall in circulating prolactin and a restoration of immunity.

Host immune status may be altered by several other factors, and these may conveniently be divided into those affecting susceptibility to existing infection and those affecting susceptibility to fresh infections. Increased susceptibility to the effects of existing infection occurs during late pregnancy and lactation as mentioned above. Dietary changes particularly if there is a deficiency in protein, may also increase susceptibility; mineral or trace element deficiencies, especially those of phosphorus or cobalt, are important as some trichostrongyle infections impair utilisation of these elements and a dietary deficiency of the same mineral could result in an exacerbated pathogenic effect.

Increased susceptibility to the acquisition of new infection is also influenced by late pregnancy and lactation, while inter-current infections with other microbiological and protozoal agents have been shown to impair the host immune response to gut nematodes. The administration of long-acting steroids is also known to lower the host immune response and in man there are documented instances of severe parasitism following the administration of these drugs. Anthelmintic therapy can affect the local immunity, in some instances enhancing and in others temporarily suppressing the response; but since this is currently under investigation, it would be unwise to alter the present approach to the regular use of anthelmintics as a prophylactic for animal helminthiasis.

Finally, it has recently been demonstrated that in immune sheep, where the acquired immunity to a gut nematode is dependent on or accompanied by a hypersensitivity reaction, reinfection resulted in an exacerbated local hypersensitive reaction in the gut; the latter was reflected in a loss of essential protein through the mucosa and impaired productivity of wool (Anderson 1973). It seems likely that similar reactions occur in other ruminants. All of these factors impairing immunity to nematodes are important epidemiologically, because in most instances they permit an accumulation of adult worms in hosts previously considered to be immune, allowing unexpected contamination of the grazing environment to occur.

The factors which affect the immunity of sheep to cestode and trematode infections are still unclear. An acquired immunity to certain infections with intermediate stages of cestodes is known to develop e.g. with Cysticercus ovis (Rickard & Bell 1971) but it is not known if this immunity is affected by similar factors to those affecting immunity to nematodes. With infections of trematodes such as F. hepatica it seems unlikely that a significant immunity develops, since clinical fascioliasis is regularly seen in adult sheep. Indeed, the possibility exists that F. hepatica infections are themselves immuno-suppressive during certain stages of their parasitic development (Tiggele & Over 1977). It has also been shown that F. hepatica migrated more rapidly through the liver of pregnant sheep, suggesting that a similar phenomenon to the PPRI against nematodes may operate (Sinclair 1972).

In summary, the past decade has seen a marked increase in epidemiological knowledge of sheep helminthiases. In particular, precise data are now available on the longevity of infective stages in the environment of the most economically important helminths, and the timing of seasonal increases of fresh infections on the herbage is well documented. The existence of hypobiosis and its epidemiological importance has also been demonstrated.

Several factors have been shown to affect the immune status of sheep in relation to helminth infections, and in particular the relaxation of immunity from 4 weeks prior to parturition until weaning has been thoroughly investigated and its aetiology established. A knowledge of the local variations of the above epidemiological factors should enable any veterinary investigator to formulate a successful control programme.

REFERENCES

Anderson, N. (1973) Trichostrongylid infections of sheep in a winter rainfall region. II. Epizootiological studies in the western district of Victoria, 1967-68. Australian Journal of Agricultural Research 24,599-611.

Armour, J. (1978) Recent advances in the epidemiology of endoparasites. In 'Perspectives in the control of parasitic disease in animals in Europe'. Papers given at a symposium in London, October 1977. Association of Veterinarians in Industry pp27-36.

Armour, J. (1979) Veterinary Parasitology (in press)

Gemmell, M.A. (1976) Factors regulating tapeworm populations: estimations of the build-up and dispersion patterns of eggs after the introduction of dogs infected with Taenia hydatigena. Research in Veterinary Science 21,220-222.

Michel, J.F. (1976) The epidemiology and control of some nematode infections in grazing animals. Advances in Parasitology 14,355-397.

Rickard, M.D.;Bell, K.J. (1971) Successful vaccination of lambs against infection with Taenia ovis using antigens produced during in vitro cultivation of the larval stages. Research in Veterinary Science 12, 401-402.

Sinclair, K.B. (1972) The pathogenicity of Fasciola hepatica in pregnant sheep. British Veterinary Journal 128,249-259.

Tiggele, L.J. Van;Over, H.J. (1976) Serological diagnosis of fascioliasis. Veterinary Parasitology 1,239-248.

Urquhart, G.M.;Armour, J. (Editors) (1973) Helminth diseases of cattle, sheep and horses in Europe. Proceedings of a symposium held at the University of Glasgow Veterinary School, Spring 1973. Robert MacLehose and Co, Glasgow.

EFFECTS OF PARASITISM ON HOST METABOLISM

DR A.R. SYKES

Department of Animal Science, Lincoln College, Canterbury, New Zealand

DR R.L. COOP

Animal Diseases Research Association, Moredun Institute, 408 Gilmerton Road, Edinburgh

The costs to national livestock industries which would result from failure of control programmes against the many ecto- and endo-parasites which cause animal losses do not need to be emphasised. What is perhaps more difficult to accept, and certainly more difficult to recognise, is the extent to which the metabolism of an animal may adapt to a particular parasitic challenge or worm burden without it showing clinical manifestations of disease but still suffer a significant reduction in the efficiency of nutrient utilisation.

The scope of this review is limited to helminth infections of the liver and gasto-intestinal tract and recent research on their influence on host metabolism.

One of the major problems is to define the parasitic state of a particular animal which, with a continuous intake of larvae, is dynamic and is the result of many complex, interacting variables. These include rate of larval intake and condition of larvae, the particular species of parasite involved and the breed, age and nutritional status of the host.

Rate of larval intake is crucial in determining the influence of infection on the metabolism of the host. At one extreme a rapid intake of many thousands of larvae can result in sudden death without apparent effect on performance or body composition, for example acute fascioliasis, nematodiriasis and haemonchosis in young lambs. At the other extreme, low or gradually increasing larval intakes may enable the host to adapt to the infection due to the development of resistance coupled with minor metabolic adjustment. Such responses appear to be enhanced by good nutrition (Spedding 1970), increasing age of the host and previous worm experience. Acquisition of resistance to the development of a mature worm population may not necessarily imply complete resistance to metabolic consequences of infection; Barger & Southcott (1975) recently demonstrated a 16% reduction in wool growth in sheep with a high degree of resistance to *Trichostrongylus colubriformis* infection induced by previous experience.

Between these extremes exist a range of larval intakes which cause severe metabolic disturbance with consequent loss of performance, without the host necessarily displaying clinical symptoms. Such relatively stable host-parasite relationships, which incidentally are an essential prerequisite for reliable metabolic studies, have recently been achieved by low intakes of larvae at a rate and pattern commonly experienced by sheep at pasture.

Species of parasite will influence the metabolic consequences of any infection through (i) the life cycle and damaging activity of the particular species of parasite, (ii) the site specificity in the host and (iii) general systemic effects. The common parasites of sheep in Britain are classified according to site in the host and their activities (Table 1).

LIFE CYCLE AND ACTIVITY

The gastro-intestinal Trichostrongyles have similar life cycles. Infective third stage larvae are ingested with herbage, penetrate the predilected site and as adult worms inhabit the lumen of the alimentary tract, live in close association with the mucosa or burrow under the surface epithelium where they can cause extensive damage. The particular effect on host metabolism may vary with the phase of development of the parasite. In *Ostertagia circumcincta* infections considerable abomasal damage is associated with emergence and development of fourth stage larvae 8 to 16 days after infection, whereas in *Trichostrongylus colubriformis* infections the main intestinal damage results from activity of adult worms and is present from 3 to 4 weeks after infection. In fascioliasis one may anticipate two phases; disruption of normal hepatic functions as young immature flukes migrate through the liver parenchyma and, subsequently, interference with haematopoiesis due to haematophagic activity of adult flukes in the bile ducts.

The pathogenicity of a particular species will also depend on its feeding mechanism. Endo-parasites can generally be classified into those which directly ingest blood and those which feed on tissue exudate some of which may also cause haemorrhage into the gastro-intestinal tract. Consequently, many of the parasites listed in Table 1 are associated with anaemia and kinetic data on iron metabolism and red cell turnover are available for some of these infections.

Haemonchus contortus pierces the abomasal mucosa with its buccal lancet, causing blood loss and anaemia. This may be associated with rapid loss of blood in acute infections or, in persistent infections, with depletion of iron stores or reduction of erythropoeisis associated with protein loss. *Bunostomum* has mouth parts adapted for lacerating intestinal tissue and in common with other hook worms causes anaemia similar to that associated with haemonchosis.

Anaemia in fascioliasis may arise from direct haemorrhage of the liver parenchyma due to migration of large numbers of immature fluke or from haematophagic activities of adult fluke in the bile ducts. Kinetic studies have shown extensive turn-over of erythrocytes and plasma protein, and blood losses of 0.5 to 1.0 ml per fluke per day (see review by Jennings 1976).

Several of the small and large intestinal worms *(Oesophagostomum, Chabertia, Trichuris)* have been associated with intestinal bleeding during some stage of their development and can result in mild anaemia (Ross et al 1969, Bremner 1970, Herd 1971).

Table 1 Some important helminths of sheep, according to site in host and feeding activities

Site	Species	Direct blood feeders	Causing blood loss
Abomasum	Haemonchus contortus	*	
	Ostertagia circumcincta		
	Trichostrongylus axei		
Duodenum and Proximal Small Intestine	Nematodirus battus		
	Trichostrongylus colubriformis		
	Trichostrongylus vitrinus		
	Cooperia curticei		
	Strongyloides papillosus		
	Capillaria longipes		
	Bunostomum trigonocephalum	*	
Caecum colon and distal ileum	Oesophagostomum venulosum		*
	Oesophagostomum columbianum		*
	Chabertia ovina		*
	Trichuris ovis		*
Liver	Fasciola hepatica	*	

Mild anaemia has been reported in Ostertagia circumcincta, Trichostrongylus axei and Trichostrongylus colubriformis infections, none of which feed directly on blood or produce extensive haemorrhage. Reduced food intake, protein leakage into the gastrointestinal tract and impaired protein metabolism are a feature of these infections and probably contribute to reduced erythropoiesis (see Berry & Dargie 1976).

SITE SPECIFIC EFFECTS

Abomasal Parasitism

O. circumcincta, H. contortus and T. axei cause sufficient damage to parietal cells to impair secretion and consequently elevate abomasal pH from normal values of 2-3 to 6-7. Evidence for reduction in pepsinogen secretion by chief cells is less clear but, since its conversion to the active proteolytic enzyme pepsin requires a pH below 4.0, impaired proteolytic digestion may be anticipated. In Figure 1 evidence for such an effect on protein digestibility is present in 4 month-old sheep with a continuous daily intake of 4000 larvae of O. circumcinta. The leakage of pepsinogen into plasma as a result of damage to cell junctional complexes has allowed the development of an assay for detection of abomasal damage, although false negative values may be obtained with certain abomasal parasites such as T. axei. Plasma pepsinogen concentrations show rapid increases in the early stages of the infection coincident with the 25% reduction in N digestibility (Figure 1). Sustained abomasal damage was suggested by the maintenance of high pepsinogen levels and by abomasal pH of up to 6.0 on slaughter at the end of the experiment. The return of N digestibility towards normality during the course of the experiment therefore suggests either that some of the initial reduction in digestibility was due to other causes, e.g. increased gut motility, which resolved during the course of the experiment, or that a compensatory increase in either pancreatic proteolytic digestion or caecal degradation occurred. The latter would have resulted in an increased proportion of N being absorbed as ammonia rather than as amino acid and an increase in the proportion of apparently digested N excreted in the urine. This appeared not to be the case and it seems more likely that increased pancreatic proteolytic digestion occurred, an hypothesis which is consistent with findings of normal proteolytic digestion in humans suffering from pernicious anaemia who do not have normal gastric function. Further support for the conclusion that adaptation was associated with amino acid rather than ammonia absorption is provided in the carcass analysis (Table 2) which showed that little further reduction in protein deposition occurred in infected sheep (ALI) than took place in pair-fed controls (PF), relative to animals fed ad libitum.

The increase in abomasal pH in such infections may have implications for pH sensitive phsyiological processes in more distal regions of the tract. Increases in pH in the proximal small intestine to values above 6.5, at which precipitation of calcium phosphate occurs (Smith 1969), have been recorded in abomasal parasitism (Mapes & Coop 1970) and interference with calcium and phosphorus absorption could theoretically be anticipated. Table 2 shows quite clearly a specific reduction in skeletal mineral deposition in animals with abomasal parasitism by comparison with pair-

Figure 1

Effect of abomasal parasitism induced with 4000 Ostertagia circumcincta larvae/day on serum pepsinogen concentration and nitrogen utilisation.

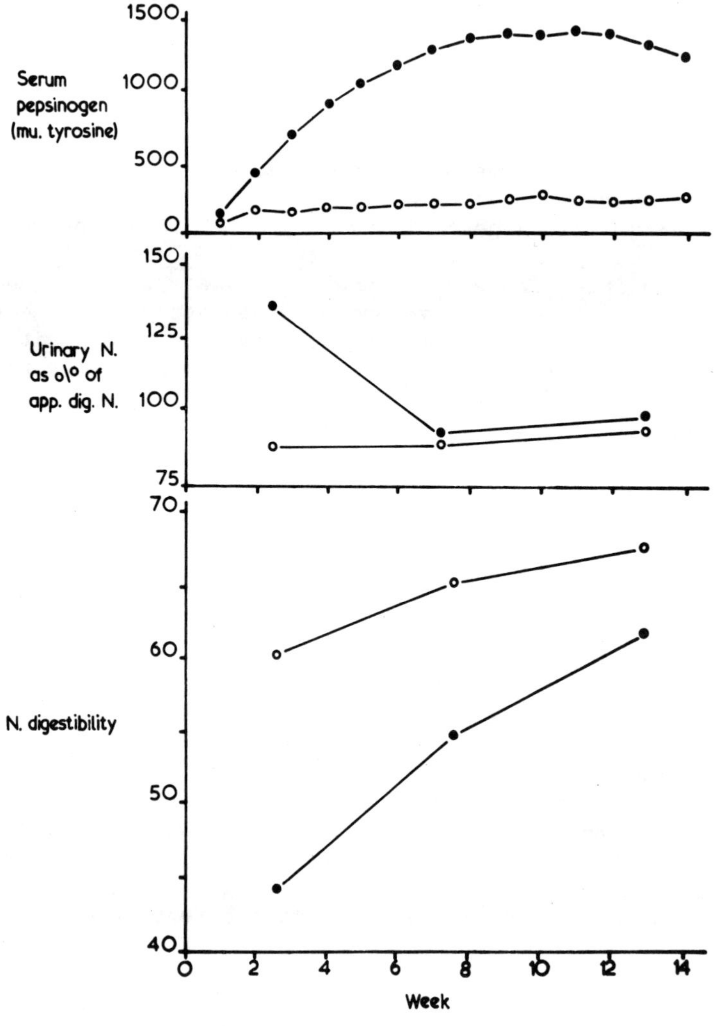

fed controls (PF). The characteristic serum hypophosphataemia and hypocalcaemia which occur in Ca and P deficiency do not occur in abomasal parasitism, and Sykes et al (1977) speculated that more general effects of induced protein or energy deficiencies or even hormonal changes may therefore be more probable explanations of poor skeletal growth.

Table 2 The effect of chronic infection (4000 larvae/day for 100 days) with Ostertagia circumcincta on the body composition of growing lambs. Values are expressed as a percentage of those obtaining in controls fed ad libitum

	ALC[1]	AL1[2]	PF[3]
Liveweight gain	100	48	80
Fat deposition	100	58	84
Protein deposition	100	49	66
Ca deposition	100	36	75

ALC[1] = Controls fed ad libitum

AL1[2] = Infected sheep fed ad libitum

PF[3] = Controls pair-fed to AL1

Intestinal Parasitism

It is difficult to predict the effect which extensive villus atrophy, mucosal thickening and stunting of micro-villi in infections with Trichostrongylus colubriformis (Barker 1975 a and b, Coop & Angus 1975), or simple stunting of micro-villi in Nematodirus battus infections (Coop et al 1973), have on the physiological function of the site. Reductions in absorption of amino acid, fat and minerals may be anticipated, although there may be considerable capacity for compensatory absorption from distal regions of the tract which would moderate the effect on the animal. Amino acids can be absorbed efficiently in the ileum and, in rats, calcium absorption in distal portions of the intestine increases during calcium deficiency.

Evidence for reduced apparent digestibility of protein is equivocal, although the majority of studies have failed to demonstrate an effect. Direct studies of protein metabolism between the abomasum and terminal ileum (Symons & Jones 1970, Barger et al 1973) have similarly failed to demonstrate reduced protein absorption. Nevertheless, it is a fact that in relation to N absorption urinary N excretion is elevated in intestinal parasitism of sheep (Roseby 1973, Sykes & Coop 1976) and other species, and severe reductions in protein deposition in the body have been demonstrated (Table 3).

Table 3 The effect of chronic infection (2,500 larvae/day for 100 days) with Trichostrongylus colubriformis on the body composition of growing lambs

	ALC[1]	AL1[2]	PF[3]
Liveweight gain	100	38	74
Fat deposition	100	50	72
Protein deposition	100	23	83
Ca deposition	100	1	110

ALC[1] = Controls fed ad libitum

AL1[2] = Infected sheep fed ad libitum

PF[3] = Controls pair-fed to AL1

Furthermore, urea production in sheep suffering from intestinal parasitism is increased (Roseby & Leng 1974) which is consistent with increased amino acid degradation. Whether this occurs in the body or alimentary tract is not yet clear. Studies of acute parasitic states have demonstrated reduced muscle and increased liver protein synthesis (Symons & Jones 1971 and 1972, Symons et al 1974). This may be attributable to priority synthesis of serum proteins in the liver to compensate for increased endogenous losses in the digestive tract which are a feature of all forms of parasitism. Possibly more significant may be the losses into the tract resulting from increased intestinal cell sloughing (Symons 1965) though estimates of the quantitative significance are not available. Endogenous secretions are considered to be efficiently reabsorbed as amino acid (Fauconneau & Michel 1970) and may not contribute to faecal excretion, although the resulting increase in protein cycling between the body and gastro-intestinal tract must impair the overall efficiency of amino acid use. Its quantitative significance has not yet been tested.

Poor Ca and P retention is a feature of intestinal parasitism though the possible magnitude of the effect has only recently been established (Reveron et al 1974, Sykes & Coop 1976, Sykes et al 1975). Complete cessation of skeletal mineral deposition (Table 3) can occur in animals continuously infected with T. colubriformis larvae at the rate of only 2,500/day. Bone pathology showed bone growth to be markedly reduced. Mineral osteoporosis, i.e. a reduction in deposition of bone matrix and in its degree of mineralization, together with concurrent hypophosphataemia and normocalcaemia (Figure 2) strongly suggested a specific impairment of Ca and P absorption. The other intestinal trichostrongyle, T. vitrinus has similar effects on mineral metabolism. (Coop & Sykes, unpublished data).

Large Intestinal Parasitism

There are little data from which the significance for nutritional phsyiology of infection in this site can be determined. Infection with Oesophagostomum columbianum reduces feed conversion efficiency (Bawden 1969) while, in calves, O. radiatum increases plasma protein leakage (Bremner 1969).

Liver

This organ appears to have enormous capacity for regeneration after damage or surgical removal of tissue. Regeneration has been accomplished within a few weeks of removal of up to 70 or 80% of the liver and function appears to follow the same pattern (Bucher 1963). In chronic infection with F. hepatica a two-fold increase in weight of liver tissue can occur which appears to be due to an increase in functional tissue (Table 4). This is an important concept in understanding the effect of chronic fascioliasis on the metabolism of the host.

Loss of liver functional capacity may not necessarily occur if the rate and capacity for synthesis of new functional tissue is not exceeded by cell necrosis. Although theoretically the accumulation of large numbers of mature flukes in the bile duct would be considered likely to obstruct biliary flow and therefore interfere with digestion, bilirubinaemia is not a marked feature of the disease (Sinclair 1967).

There are reports of reductions in N and DM digestibility in acute infections but this has never been clearly established. Our own data, from sheep ingesting 10 metacercariae of F. hepatica per day for 150 days, resulting in severe damage to the ventral lobe of the liver and accumulations of up to 250 mature flukes, have shown no evidence for impaired dry matter or N digestibilities, a finding substantiated by a lack of effect on carcass fat or protein deposition (Table 4). Although there are reports of depression of serum Ca and P concentrations as a result of fascioliasis, which might imply interference with absorption, a comparison of the body Ca and P deposition in infected and pair-fed animals in Table 4 suggests there to be on specific effect on mineral metabolism.

Table 4 The effect of chronic infection (10 metacercaria/day for 22 weeks) with Fasciola hepatica on the body composition of growing lambs during a 40 week period

	ALC[1]	AL1[2]	PF[3]
Liveweight gain	100	76	76
Fat deposition	100	58	69
Protein deposition	100	80	90
Ca deposition	100	73	77

ALC[1] = Controls fed ad libitum

AL1[2] = Infected sheep ad libitum

PF[3] = Controls pair fed to AL1

GENERAL EFFECTS OF PARASITISM AT DIFFERENT SITES

Food Intake

Reductions in food intake are a common and consistent feature of parasitism of the abomasum, intestine and liver. The severity may range from 15-20% in chronic subclinical infections (Tables 2, 3 and 4) to virtual complete anorexia in acute infections. This clearly has consequences for growth of skeleton and muscle and fat deposition. The effect of reductions in food intake of only 15-20% can be seen in comparisons between animals fed ad libitum (ALC) and those pair-fed (PF) to infected sheep, in Tables 1, 2 and 3. Reduction in growth of the skeleton may be particularly severe.

Food Utilization

Poor food conversion in parasitised animals has been recognised for many years (Andrews et al 1944). The precise nature and extent of this change in efficiency in subclinical parasitism has only recently become apparent. It has required the use of pair-fed animals, since a reduction in feed intake necessarily implies a reduction in gross efficiency of food utilization because of the large, constant component of energy required for maintenance. Though there are reports of reduced DM digestibility in abomasal (Spedding 1954, Sykes & Coop 1977) and in intestinal parasitism (Steel 1974) these have been small (1 or 2 percentage digestibility units) and this does not appear to be an important factor in reducing feed utilization. Our own data (Table 5) have shown very large reductions of 30-50% in gross efficiency of utilization of metabolizable energy (ME) for body energy deposition in abomasal and intestinal parasitism, though not in fascioliasis.

Table 5 Gross efficiency of utilization of metabolizable energy for growth in sheep with abomasal intestinal and liver parasitism (see Tables 1, 2 and 3 respectively).

	ALC[1]	AL1[2]	PF[3]
Abomasal parasitism (4000 Ostertagia circumcincta/ day	19	14	20
Intestinal parasitism (2500 Trichostrongylus colubriformis/ day)	26	13	24
Liver parsitism 10 metacercariae of Fasciola hepatica/ day; 230 mature flukes in bile ducts	20	15	17

ALC[1] = Controls fed ad libitum

AL1[2] = Infected sheep fed ad libitum

PF[3] = Controls pair-fed to AL1

We are unable to assess the precise cause of this change; it could result from a parasite-stimulated increase in metabolic rate or a change in efficiency of utilization of end products of digestion. The former seems more likely and indeed would in part be a consequence of the generalised increase in leakage and turnover of plasma proteins (Mulligan 1972) and in intestinal cell-turnover (Symons 1965) which occur in these states. Such increases in protein synthesis have an energy cost. The reduction in efficiency of utilization of ME seen in abomasal parasitism (Table 4) could be accounted for by an increase in protein synthesis of 20-30 g/day, assuming the cost of protein synthesis to be 30 kJ/g (Kielanowski 1965). It is not possible to judge whether total endogenous protein losses increase to this extent; increases in albumin losses of up to 6 g/day were recorded in abomasal parasitism by Holmes & MacLean (1971).

The possibility that hormonally induced changes in metabolic rate are involved cannot be excluded. An elevation in plasma cortisol and a reduction in thyroxine (T_4) concentrations was observed in acute intestinal parasitism (Prichard et al 1974) and Sykes & Coop (unpublished data) have evidence for changes in thyroid hormone metabolism in several chronic parasitic states. These aspects of metabolism require further investigation.

REFERENCES

Andrews, J.S.;Kauffman, W.;Davis, R.E. (1944) Effects of the intestinal nematode, Trichostrongylus colubriformis, on the nutrition of lambs. American Journal of Veterinary Research 5,22-29.

Barger, I.A.;Southcott, W.H. (1975) Trichostrongylosis and wool growth. 3. The wool growth response of resistant grazing sheep to larval challenge. Australian Journal of Experimental Agriculture 15,167-172.

Barger, I.A.;Southcott, W.H.;Williams, V.J. (1973) Trichostrongylosis and wool growth. 2. The wool growth response of infected sheep to parental and duodenal cystine and cysteine supplement. Australian Journal of Experimental Agriculture 13,351-359.

Barker, I.K. (1975a) Intestinal pathology associated with Trichostrongylus colubriformis infection in sheep: histology. Parasitology 70,165-171.

Barker, I.K. (1975b) Intestinal pathology associated with T. colubriformis infection in sheep: vascular permeability and ultrastructure of the mucosa. Parasitology 70,173-180.

Bawden, R.J. (1969) Relationships between Oesophagostomum columbianum infection and the nutritional status of sheep. I. Effects on growth and feed utilization. II. Haematological changes. Australian Journal of Agricultural Research 20,589-599;601-606.

Berry, C.I.;Dargie, J.D. (1976) The role of host nutrition in the pathogenesis of ovine fascioliasis. Veterinary Parasitology 2,317-332.

Bremner, K.C. (1969) Pathogenetic factors in experimental bovine oesophagostomosis. Experimental Parasitology 25,382-394.

Bremner, K.C. (1970) Pathogenetic factors in experimental bovine oesophagostomosis. V. Intestinal bleeding as cause of anaemia. Experimental Parasitology 27,236-245.

Bucher, N.L.R. (1963) Regeneration of mammalian liver. International Review of Cytology 15,245-300.

Coop, R.L.;Angus, K.W. (1975) The effect of continuous doses of Trichostrongylus colubriformis larvae on the intestinal mucosa of sheep and on liver vitamin A concentration. Parasitology 70,1-9.

Coop, R.L.;Angus, K.W.;Mapes, C.J. (1973) The effect of large doses of Nematodirus battus on the histology and biochemistry of the small intestine of lambs. International Journal of Parasitology 3,No.3 349-361.

Fauconneau, G.;Michel, M.C. (1970) In Munro, H.N. (Ed) Mammalian Protein metabolism IV. Academic Press, New York & London.

Herd, R.P. (1971) The parasitic life cycle of Chabertia ovina (Fabricius, 1788) in sheep. International Journal of Parasitology 1,189-199.

Holmes, P.H.;McLean, J.M. (1971) The pathophysiology of ovine ostertagiasis: a study of the changes in plasma protein metabolism following single infections. Research in Veterinary Science 12,265-271.

Jennings, F.W. (1976) The anaemias of parasitic infections. In: Soulsby, E.J.L. (Ed) "Pathophysiology of parasitic infection", pp41-67. Academic Press, London & New York.

Kielanowski, J. (1965) Estimates of the energy cost of protein deposition in growing animals. Publications of the European Association of Animal Production No.11, pp13-20.

Mapes, C.J.;Coop, R.L. (1970) The interaction of infections of Haemonchus contortus and Nematodirus battus in lambs. Journal of Comparative Pathology 80,123-136.

Mulligan, W. (1972) The effect of helminthic infection on the protein metabolism of the host. Proceedings of the Nutrition Society 31,47-51.

Prichard, R.K.;Hennessey, D.R.;Griffiths, D.A. (1974) Endocrine responses of sheep to infection with Trichostrongylus colubriformis. Research in Veterinary Science 17,182-187.

Reveron, A.E.;Topps, J.H.;Gelman, A.L. (1974) Mineral metabolism and skeletal development of lambs affected by Trichostrongylus colubriformis. Research in Veterinary Science 16,310-319.

Roseby, F.B. (1973) Effects of Trichostrongylus colubriformis (Nematoda) on the nutrition and metabolism of sheep. I. Feed intake, digestion and utilization. Australian Journal of Agricultural Research 24,947-953.

Roseby, F.B.;Leng, R.A. (1974) Effects of Trichostrongylus colubriformis (nematoda) on the nutrition and metabolism of sheep. II. Metabolism of urea. Australian Journal of Agricultural Research 25,363-367.

Ross, J.G.;Dow, C.;Purcell, D.A. (1969) A study of Chabertia ovina infections in lambs. British Veterinary Journal 125,136-144.

Sinclair, K.B. (1967) Pathogenesis of Fasciola and other liver-flukes. Helminthological Abstracts 36,115-134.

Smith, R.H. (1969) Absorption of major minerals in the small and large intestines of the ruminant. Proceedings of the Nutrition Society 28,151-160.

Spedding, C.R.W. (1954) Effect of a sub-clinical worm burden on the digestive efficiency of sheep. Journal of Comparative Pathology 64,5-14.

Spedding, C.R.W. (1970) Management in sheep production and grazing. Bailliere, Tindall & Cassels 167pp.

Steel, J.W. (1974) Pathophysiology of gastrointestinal nematode infections in the ruminant. Proceedings of the Australian Society of Animal Production 10,139-147.

Sykes, A.R.;Coop, R.L. (1976) Intake and utilization of food by growing lambs with parasitic damage to the small intestine caused by daily dosing with Trichostrongylus colubriformis larvae. Journal of Agricultural Science 86,507-515.

Sykes, A.R.;Coop, R.L. (1977) Intake and utilization of food by growing sheep with abomasal damage caused by daily dosing with Ostertagia circumcincta larvae. Journal of Agricultural Science 88,671-677.

Sykes, A.R.;Coop, R.L.;Angus, K.W. (1975) Experimental production of osteoporosis in growing lambs by continuous dosing with Trichostrongylus colubriformis larvae. Journal of Comparative Pathology 85,549-559.

Sykes, A.R.;Coop, R.L.;Angus, K.W. (1977) The influence of chronic Ostertagia circumcincta infection on the skeleton of growing sheep. Journal of Comparative Pathology 87,521-529.

Symons, L.E.A. (1965) Kinetics of the epithelial cells, and morphology of villi and crypts in the jejunum of the rat infected by nematode Nippostrongylus brasiliensis. Gastroenterology 49,158-168.

Symons, L.E.A.;Jones, W.O. (1970) Nematospiroides dubuis, Nippostrongylus brasiliensis, and Trichostrongylus colubriformis: Protein digestion in infected mammals. Experimental Parasitology 27,496-506.

Symons, L.E.A.;Jones, W.O. (1971) Protein metabolism. I. Incorporation of ^{14}C-L-Leucine into skeletal muscle and liver proteins of mice and guinea pigs infected with Nematospiroides dubuis and Trichostrongylus colubriformis. Experimental Parasitology 29,230-241.

Symons, L.E.A.;Jones, W.O. (1972) Protein metabolism. 2. Protein turnover, synthesis, and muscle growth in suckling, young, and adult mammals infected with Nematospiroides dubuis or Trichostrongylus colubriformis. Experimental Parasitology 32, No.3 335-342.

Symons, L.E.A.;Jones, W.O.;Steel, J.W. (1974) Protein metabolism. III. Relationship between albumin metabolism and elevated liver protein synthesis in the guinea pig infected with Trichostrongylus colubriformis. Experimental Parasitology 35, No.3 492-502.

GROWTH AND DEVELOPMENT OF LAMBS

DR J.H.D. PRESCOTT

Department of Animal Husbandry, Edinburgh School of Agriculture, West Mains Road, Edinburgh

INTRODUCTION

Sheep production is dominated by the overhead costs of maintaining the breeding ewe and there is advantage in spreading this cost over as great an output of lamb carcass as possible. Knowledge of the influence of genotype, nutrition and stage of maturity on the growth of lambs must be applied to obtain efficient conversion of feed by the growing lamb into saleable meat. This focuses specific attention on the production of muscle together with an optimum amount of fat, appropriately distributed, and a minimum of bone.

However, selection of sheep of the most appropriate body size and composition for meat production must be undertaken with due regard to those features of body form that fit the sheep for efficient function as a grazer, breeder and survivor under the wide variety of conditions in which sheep are kept. In many parts of the world this requires a capacity to survive feed deficiency and to recover normal body weight and conditions when more liberal feed supplies are available. In other circumstances more abundant provision of grazing, forage crops or by-product feeds and judicious use of cereals may enable lambs to grow at a rate close to their genetic potential up to slaughter weight. The latter is the concern of this paper.

The adaptation of the body structure for the efficient physiological function essential to survival is associated with a notable constancy in relative proportions of skeletal and muscular development in different regions of the body as sheep grow larger. The exceptions are of minor practical significance (Berg & Butterfield 1976). The selection of short legged, thick bodied sheep is not justified in terms of muscle content or muscle proportion in their carcasses, (Kirton & Pickering 1967, Boccard et al 1961). Furthermore, there is evidence that in the South Country Cheviot extremely blocky body conformation may be associated with the incidence of dystocia (Whitelaw & Watchorn 1975).

Features of major concern in the growth of the meat producing lamb are rate of muscle growth and the relative rates of fat deposition and bone growth. The growth of bone in relation to muscle varies appreciably between different genotypes, but is little affected by sex or plane of nutrition. The relative rate of deposition of fat on the other hand is markedly influenced by genotype, sex and nutrition. Furthermore since lambs are commonly selected for slaughter on the basis of fatness assessed by appraisal of subcutaneous fat cover (e.g. Russel et al 1969), the relative development of fat on the surface of the carcass, between the muscles and within the body cavity is also of practical importance.

LIVEWEIGHT GROWTH

Growth is most conveniently measured as liveweight change and the potential growth curve of the lamb under optimum environmental conditions is typically sigmoidal. Growth rate accelerates up to puberty and slows down progressively as maturity is approached. Well grown lambs are typically slaughtered at about 50% of mature adult weight under UK conditions (UK/MLC 1975).

Liveweight, when measured under standard conditions can provide a useful index of muscle growth. The relationship between liveweight and the muscle or protein content of the body can be distorted by variation in gut fill and in the weight, condition and moisture content of the fleece. Muscle comprises a remarkably constant proportion of the fleece free, empty bodyweight even in association with considerable variation in fatness. The proportion of viscera, inedible offal and bone in liveweight varies inversely with the fat content of the body. McClelland et al (1976) in a study with lambs of four breeds, slaughtered over a range of liveweight from 14 to 80 kg, observed that total muscle as a percent of fleece-free, empty body weight was remarkably close to 28.5% in all breeds, of both sexes and at a range of weights between 40 and 70% mature body weight (Table 1)

Table 1 Developmental changes in dissected tissue content with increasing age, body weight and stage of maturity

Age (days)	100	172	208	236
Liveweight (Kg)	25	32	40	47
[1]FFEB (Kg)	19	25	31	37
% Mature Weight	40	50	62	73
Weights (Kg)				
Carcass	9.5	13.3	17.5	21.2
Muscle	5.4	7.2	8.9	10.4
Fat	2.1	3.9	5.7	7.8
Bone (excl. hd and ft)	2.0	2.1	2.8	3.0
[1]Percentage FFEB				
Carcass	48.3	51.3	54.9	56.5
Muscle	27.9	29.0	28.7	28.7
Fat	9.9	13.5	16.8	19.5
Bone	10.4	8.6	8.9	8.3
Percentage Carcass				
Muscle	57.9	56.8	52.4	51.2
Fat	20.0	25.6	30.2	34.0
Bone	21.7	17.0	16.3	14.7

[1]Fleece-Free, Empty Bodyweight

(from McClelland et al 1976)

Reid et al (1968) observed that empty body weight provided a useful prediction of the body protein content of sheep that had sustained positive growth to slaughter. However the coefficient of variation of the prediction was almost four times as great with respect to body fat content.

Chemical Component	Prediction Equation	Coefficient of Variation
Protein Y =	0.80 X - 0.50	4.6%
Fat Y =	1.99 X - 2.18	16.1%

(i) Prediction equation:
Log chemical component (Y) (kg) = b Log empty body weight (X) (kg) - a

(ii) Coefficient of variation:

$$\% = (Sy.x/\overline{Y}) \times 100$$

DEVELOPMENT IN RELATION TO AGE

Body composition changes as sheep grow older and heavier and attain a higher proportion of their mature body size. Carcass weight comprises an increasing proportion of liveweight due to the relatively slower growth of viscera, skin and fleece, head and feet than that of the carcass tissues and most notably fat. Within the carcass the proportion of muscle and bone decreases as the proportion of fat increases. The ratio of muscle to bone increases (Table 1).

These relationships are clearly shown in the study carried out by Fourie et al (1970) under New Zealand conditions. Their study involved male and female sheep of Southdown, Romney and Southdown x Romney cross, raised under conventional grazing conditions and slaughtered at intervals from birth to maturity.

Figure 1 shows the progressive changes in weight of body tissues with age. Figure 2 shows these changes in terms of proportions of bone, muscle and fat in the carcass.

RELATIVE GROWTH

The predominant changes as the animal grows are the increase in body size and the concurrent increase in the proportion of fat in the body. The fat content of the body, whether determined by dissection or chemical analysis, is the most variable feature of carcass composition and at constant carcass weight may be markedly influenced by both the genotype of the sheep and their nutrition.

Our understanding of the separate but interrelated effects of the weight of the structural tissues of the body (muscle and bone) and of fatness on the composition of the live sheep and its carcass may be clarified by relating the growth of individual organs and tissues to one another or to the body as a whole.

Figures 1 and 2

Muscle Wt(kg)	4.4	7.4	8.0	9.8	14.5
M % L.Wt.	30	31	28	29	27
Carcass %	51	52	51	53	57

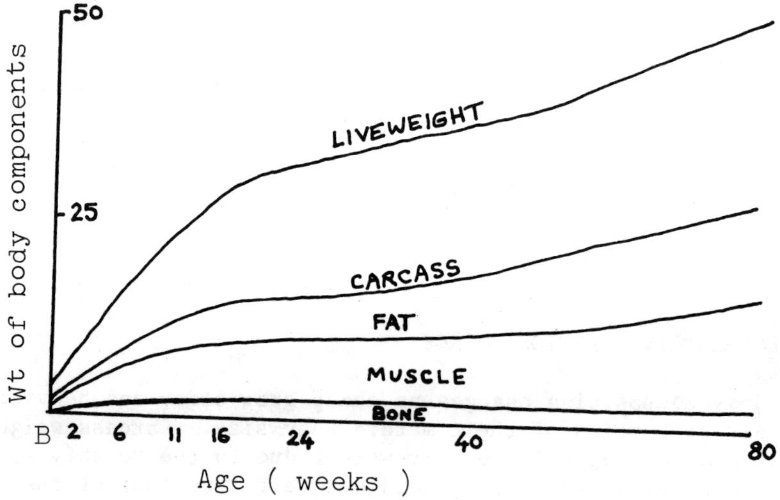

Fig. 1. Growth and development with age.
(from Fourie et al 1970)

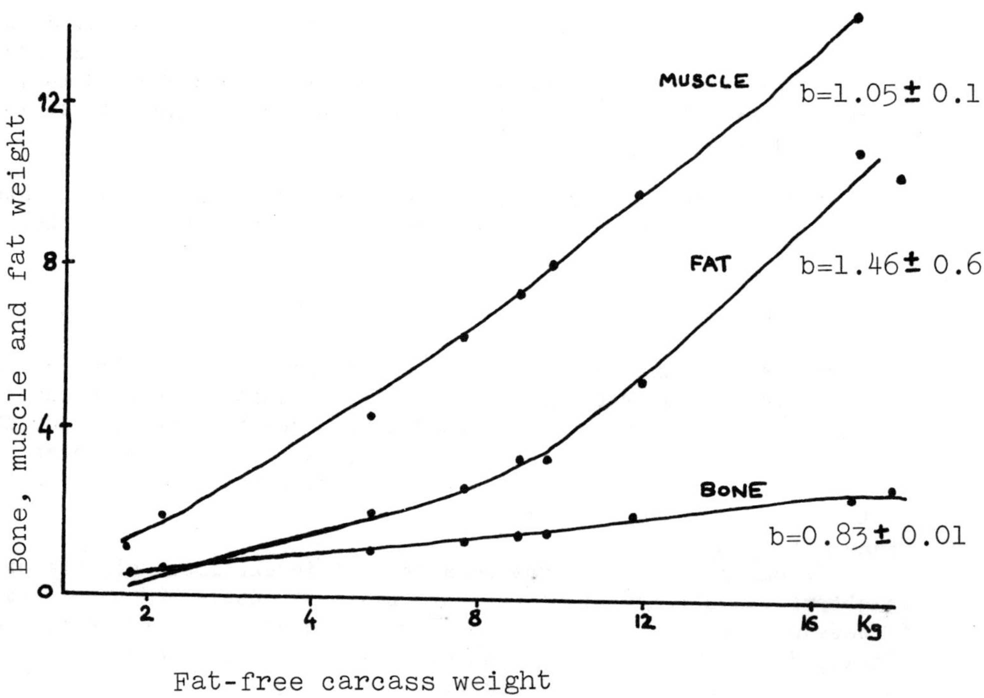

Fig.2. Relative growth of bone, muscle and fat.
(from Fourie et al 1970)

An appropriate quantitiative description of the growth of parts and tissues relative to each other or to the body as a whole is provided by the allometric equation of Huxley (1932):

$$Y = aX^b$$

which describes the curvilinear relationship between the growth of most tissues. Transformed by logs, it conforms to the simple linear regression of:

$$\text{Log } Y = \text{Log } a + b \text{ Log } X$$

In this equation Y = the weight of the organ or tissue and X = the weight of another part or of the body as a whole, a = the intercept of the log regression line on the Y axis and b = the relative growth coefficient.

The relative growth coefficient b x 100 provides a convenient means of expressing the percentage change in the part under study relative to a one per cent change in another part or in the body as a whole.

The general importance of this is that when the growth of a part or tissue relative to body weight conforms closely to this relationship (with a low standard error of the coefficient and the regression) that feature of body composition is dependent upon body weight and largely unaffected by the time sheep take to attain that weight.

GROWTH OF BONE, MUSCLE AND FAT IN RELATION TO BODY SIZE

The relationship between growth of bone, muscle and fat in relation to fat-free carcass weight is shown in Figure 3 for the sheep of Fourie et al (1970). These relationships are quantified in Table 2, in which the relative growth coefficients relating tissue growth to starved liveweight, carcass weight and fat-free carcass weight are presented.

Table 2 Allometric Growth Coefficients Relating Growth of Body and Carcass Components to Starved Liveweight, Carcass Weight and Fat-Free Carcass Weight

Components	Relative to Starved Liveweight[1]	Carcass Weight[2]	Fat-Free Carcass Wt.[2]
Bone	0.77 ± 0.02	0.77 ± 0.02	0.83 ± 0.01
Muscle	1.00 ± 0.01	0.97 ± 0.01	1.05 ± 0.01
Fat	1.45 ± 0.05	1.37 ± 0.04	1.46 ± 0.06

[1] total body components
[2] carcass components (Romney males: from Fourie et al 1970.)

It may be noted that each 1% increase in starved liveweight was associated with a similar 1% change in total muscle, while bone increased relatively more slowly (0.77%) and fat much more rapidly (1.45%). The coefficients were modified by the effect of fat when carcass weight was taken as the reference base. They were more comparable with those based on starved liveweight when the reference base was fat-free carcass weight.

In all these relationships it should be noted that the standard errors of the b coefficients were as low as $\pm$ 0.01 for muscle and $\pm$ 0.01 - 0.02 for bone, but $\pm$ 0.04 - 0.06 for fat. This suggests that muscle and bone weights were varying in predictable proportion to the structural body size of the animal, while the amount of fat was much more variable in animals of similar size.

The change in muscle:bone (M:B) ratio as the sheep grew is closely related to the increase in size of the animal and largely independent of its fatness. In this study of Fourie et al (1970), M:B ratio increased from 3:1 at birth, with 1.5 kg muscle + bone in the carcass, to 4.6:1 at 5.4 kg muscle + bone and 5.0:1 at 8 kg muscle + bone.

The relatively faster rate of increase in fat than in either muscle or bone is reflected in the progressive increase in fat percentage in the carcass and the decrease in muscle and bone percentage as the sheep grow heavier. The ratio of muscle:fat rapidly changed from 6:1 at birth to 2:1 at conventional slaughter weights, e.g. 50-60% mature muscle or liveweight. At this stage each unit increase in percent fat is associated with a reduction of approximately 0.75% in muscle and 0.25% in fat (Timon 1963).

It may be noted however that the progressive increase in fatness was associated with a decrease in bone and offal such that muscle developed in constant proportion to starved liveweight and muscle weight remained a relatively constant proportion of starved liveweight.

FAT DISTRIBUTION

Within the rapidly increasing fat content of the sheep carcass it is appropriate to identify the relative rate of development of subcutaneous, intermuscular and perinephric fat (Figure 4).

At birth intermuscular fat comprised the highest proportion of the small amount of fat in the lamb carcass. Then, during post-natal life in these Romney and Southdown sheep, the relative rate of increase was more rapid in subcutaneous (b = 1.7) than intermuscular (b = 1.3) or perinephric fat (b = 1.2). The proportion of subcutaneous fat however did not exceed that of intermuscular fat until the total fat content in the carcass was over 7 kg (Fourie et al 1970).

SUMMARY

In practical terms the most important changes in body composition as sheep grow older, heavier and more mature are the increase in

Figures 3 and 4

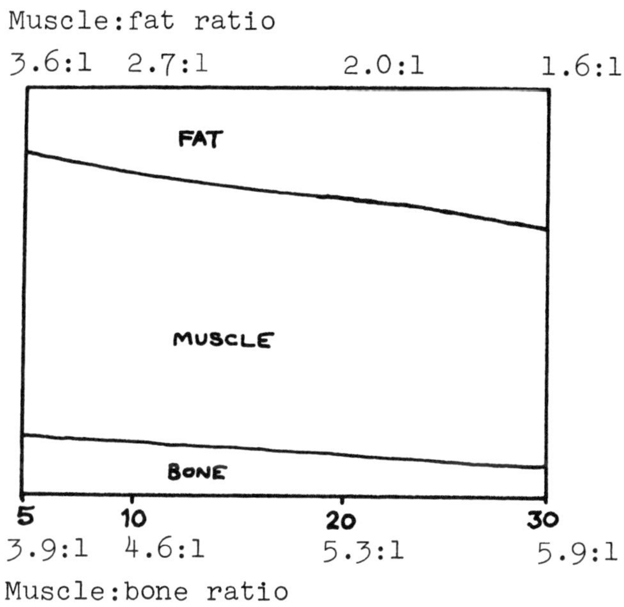

Fig.3. Percentage composition of the carcass. (from Fourie et al 1970)

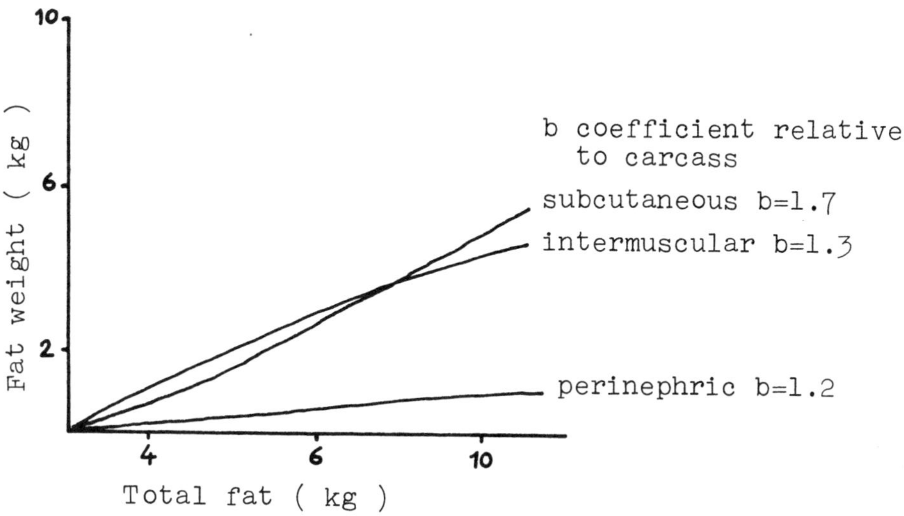

Fig.4. Relative growth of fat depots. (from Fourie et al 1970)

muscle:bone ratio that is closely correlated with increase in structural body size and the markedly more variable deposition of fat.

POTENTIAL LIVEWEIGHT AND MUSCLE GROWTH

There is an upper limit to the daily rate of liveweight and muscle gain that a lamb can achieve. Furthermore, there is a close relationship between potential growth rate at any stage of postnatal life and mature body size. Birth weight and weights at successively later ages are each correlated one with another and with mature weight (Taylor 1968). The closeness of these relationships implies that there is only limited scope for changing the shape of the growth curve by selection and that a large part of genetic variation in weight for age is removed when mature weight is held constant (Taylor 1968).

The relationship between size, growth rate, body composition and feed efficiency is of fundamental concern in the production of meat. In this respect the criteria for slaughter, on a weight, age, fatness or percentage of mature weight basis, have a crucial influence on differences in growth rate, body composition and feed efficiency between genotypes of different body size.

COMPARISON AT EQUAL WEIGHTS

Potentially larger animals grow faster to attain any particular weight. They therefore require to be maintained for fewer days than slower growing and smaller animals. Furthermore, since at the same weight these larger animals have attained a lower proportion of their mature size, their gains to that weight generally contain a lower proportion of fat. They therefore require less energy above maintenance to produce each unit of gain.

When slaughtered at the same weight as a genetically smaller animal, the larger animal is likely to have converted feed into gain more efficiently.

COMPARISON AT EQUAL AGE OR EQUAL FATNESS

In practical systems larger and potentially fast growing animals are heavier at all stages of growth, i.e. birth, weaning and at the start of the finishing period. They are generally fed to reach a similar slaughter condition (that is subcutaneous fat cover) to the smaller animal but they are expected to do this at a heavier weight.

Though their mean growth rate is higher, their maintenance requirement is also higher because throughout the growing period they are supporting a heavier mean liveweight. Furthermore, since the aim is to slaughter them at a similar carcass fat content to the potentially smaller, slower growing animal the gross efficiency of energy conversion into carcass gain should be similar. The larger animals may be expected to reach slaughter condition at a heavier weight, but at a similar proportion of their mature muscle (or liveweight) and at this point both body composition and efficiency of feed conversion into carcass gain may be similar.

For this reason when comparing genotypes it is instructive to make the comparisons of growth rate, body composition with due allowance for differences in mature size. This involves comparing growth rate and feed conversion over the same segment of the growth curve in relation to maturity and comparing body composition at equivalent stages of maturity.

In this connection it is important that the definition of mature size should either be based specifically on mature muscle weight or on mature liveweight of the different breeds and sexes under similar environmental conditions, avoiding problems due to variation in fill and fleece weight.

A study recently reported by McClelland et al (1976) of ABRO exemplifies this approach. They compared a range of contrasting breed types, Soay, Finnish Landrace, Southdown and Oxford, which were fed ad lib on a standard diet and serially slaughtered at 40, 50, 60 and 70% of their estimated mature weight, which was defined under standard conditions. The results are summarised in Table 3.

Differences in liveweight at each stage of maturity were reflected in differences in carcass weight and the actual weights of the tissues. Total muscle ranged from 3.3 kg in the Soay to 13.0 kg in the Oxford, and fat from 1.2 to 9.2 kg.

When body composition was expressed as a percentage of fleece-free, empty bodyweight (FFEB) there were only small differences between the Finn, Southdown or Oxford in the percentage of carcass, fat or bone. There was however a marked difference between these three breeds and the Soay which was notable for its much lower fat percentage, higher bone content and lower percentage of carcass in the FFEB.

The markedly lower fat content and higher bone content of this breed was also evident when tissue composition was expressed as a percent of carcass weight. Smaller differences between the other three breeds in percentage of muscle in the carcass were also evident, reflecting differences between these breeds in both fatness and muscle:bone ratio at the same degree of maturity.

The study showed that among domesticated breeds apparent differences in body composition are much reduced when due allowance is made for mature adult size and comparisons are made at a similar stage of maturity. Clearly however breeds such as the Soay, adapted to a feral existence, do deviate markedly from the norm and are markedly less fat at the same stage of maturity. Furthermore, it is noteworthy that the deviation of the Soay from the mean was 40% for dissected fat in FFEB but was 30% for chemical fat content of the FFEB.

POTENTIAL FOR MUSCLE GROWTH

The definition of the muscle growth potential in terms of rate of gain per day and mature muscle weight is a feature of particular interest with reference to different genotypes and sexes.

Table 3 Body and Carcass Composition of Contrasting Genotypes at Equal Stages of Maturity

		Genotype			
		Soay	Finn	Southdown	Oxford
Estimated Mature Liveweight	(Kg)	25	63	60	110
Birth Weight	(Kg)	1.3	2.5	2.7	4.2
[1]Slaughter Age	(days)	162	169	186	198
[1]Slaughter L. Wt.	(Kg)	14.2	35.9	34.0	60.3
Component Weights at 56% Mature Liveweight					
Muscle	(Kg)	3.3	7.9	7.7	13.0
Fat	(Kg)	1.2	4.7	4.6	9.2
Bone	(Kg)	1.3	2.4	2.0	4.2
Components as Percent Fleece-Free, Empty Body Weight					
Carcass	(%)	47.7	53.3	53.3	56.8
Muscle	(%)	28.3	28.5	29.0	28.3
Fat	(%)	8.3	16.5	15.9	18.9
Bone	(%)	10.3	8.8	7.8	9.3
Components as Percent Carcass Weight					
Muscle	(%)	59.7	53.8	54.7	50.1
Fat	(%)	16.5	30.8	29.5	33.1
Bone	(%)	21.9	16.6	14.8	16.4
Muscle:Bone Ratio		2.54:1	3.29:1	3.85:1	3.10:1

[1]mean of four slaughter weights 40, 50, 62 and 73% mature weight = 56% maturity.

The growth curve of muscle in lambs over a range from 40 to 80% maturity has been characterised by Jackson et al (1973) in relation to estimated mature muscle weight. The study involved serial slaughter of lambs receiving four levels of metabolisable energy intake over the post-weaning period, allowances being adjusted to provide for the increasing maintenance requirements of the lambs as they grew heavier.

The estimated mature muscle weights and daily muscle growth potential (over the range 40-80% mature weight) were 15 and 19 kg, and 35 and over 50g/day, for Cheviot and Suffolk x Border Leicester - Cheviot lambs respectively.

Muscle growth potential was shown to decline markedly with advancing maturity above conventional slaughter weights, e.g. in Cheviots from 50g/day at 45% maturity to 20g/day at 75% maturity (Weddell, personal communication).

Rate of muscle growth was also affected by level of energy intake at any specific stage of maturity. The limits in terms of response to increasing energy intake by faster muscle gain differed between the two genotypes. (Jackson et al 1973)

In this particular study the Suffolk-cross lambs showed a linear response to increased energy allowance up to the highest level that was fed. The Cheviots on the other hand showed little response to raising energy intake over a mean intake of 2.6 Mcal ME/lamb/day.

When compared on the basis of a similar amount of gain (6 kg muscle) above 40% muscle maturity, the difference between the genotypes was appreciably greater than when the comparison was made over a similar stage of muscle maturity (40-80%). Comparison over a similar stage of maturity involved the smaller Cheviots gaining just over 6 kg and the larger Suffolk-crosses 7.6 kg of muscle.

Differences in the gross efficiency of energy conversion into muscle growth were also evident in terms of Mcals ME/kg muscle gain.

On these levels of feeding, which did not allow the Suffolk crosses to express their full growth potential, they were markedly more efficient even when the comparison was made over a greater total gain but comparable stage of maturity.

In the case of the Cheviots the gross efficiency of energy conversion was reduced at the highest level of energy intake. This was associated with the higher content of fat in the daily gains of lambs as energy supply exceeded the potential of the lamb to respond in terms of extra muscle growth and the extra energy was diverted into fat.

MUSCLE GROWTH, FATNESS AND FEED CONVERSION

The gross efficiency of feed energy conversion into muscle is affected by both the daily rate of gain in muscle weight and the amount of fat associated with each unit of muscle gain. The former influences the partition of feed utilisation between maintenance and liveweight gain, and the latter the energy cost of each unit of gain, which varies according to the ratio of muscle to fat. The energy cost of growing each kg of fat is about 3½ times that of growing an equivalent amount of muscle. This difference largely reflects the fact that muscle contains 75% water and fat only 10%.

Furthermore, as the lamb grows each unit weight of muscle growth is likely to be associated with a minimum amount of fat deposition. Thus in young pigs (20-100 kg) each kilo of muscle growth is associated with a minimum of about 0.22kg fat, or 1:1 on a chemical fat:protein basis (Whittemore 1976). Black (1974) in studies with male, milk-fed lambs of from 3 to 38 kg indicates that the chemical fat:protein ratio in the weight gains attained with energy intakes of 4 MJ ME/lamb/day were about 0.5:1.

Jackson et al took as a starting point lambs of about 33 kg liveweight containing a specified 10 kg of muscle. This was associated with about 3 kg of dissected fat in the Suffolk crosses and close to 4 kg in the Cheviots.

The lowest level of energy allowance above maintenance (of 0.5 Mcal ME/lamb/day) supported liveweight gains of 90g and 80g/lamb/day in the Suffolk crosses and the Cheviots respectively and the relative growth of fat to lean was 1.7:1 in both genotypes.

It can thus be calculated that the growth of an extra kilo of muscle above 10 kg was associated with 0.5 kg fat in Suffolk crosses and 0.7 kg in Cheviots.

With a higher level of energy allowance above maintenance, 1.5 Mcal/lamb/day supporting liveweight gains of 180 and 140g, relative growth of fat:lean increased to 2.3:1. In this situation the growth of an extra kilo of muscle would be associated with 0.7 and 0.9 kg fat in the Suffolk crosses and Cheviots respectively.

Jackson et al have related the fat content of the two genotypes at defined muscle weights both to level of feeding and to the level of fatness that is acceptable in the local market. This is shown in Table 4.

Jackson et al derived separate regressions relating fat depth (mm) to fat:muscle ratio. The relative rate of deposition of fat in subcutaneous and other regions of the body has been observed to vary significantly between different genotypes, for example between Blackface and Finnish Landrace (McClelland & Russel 1972), Suffolk-crosses and Clun Forest (Prescott & Hinks 1967), but not between Southdown and Romney (Fourie et al 1970), though the latter found differences in relative growth of total fat and muscle both between breeds and sexes.

Table 4 Fat Content of Lamb Carcasses at Defined Muscleweights and Energy Allowances in Relation to Acceptable Carcass Fatness

Genotype	"Estimated"[1] Slaughter Weight (Kg)	Defined Muscle Weight (Kg)	Daily Energy Allowance (Meals/day) 2.1	2.6	Max. Acceptable[2] Level of Fatness (Kg)
Cheviot	33	10	4.8	5.6	5.9
Cheviot	40	12	7.2	8.5	7.0
Suffolk X	40	12	4.7	5.5	5.6
Suffolk X	47	14	6.6	7.8	6.7

[1]assuming muscle 30% liveweight

[2]6mm fat cover over eye muscle at 12th rib

(from Jackson et al 1973)

MODELLING GROWTH

Quantitative information of this type is required to characterise the growth potential of different genotypes and predict the consequences of variation in nutrition. The approach has been developed to the stage of practical application in pig production (Whittemore 1977) and Black (1974) and his colleagues in Australia have described a similar approach to growth modelling in lambs. Characterising the influence of energy intake on the partition of metabolisable energy between metabolic activity, protein synthesis and fat deposition is central to this development and is illustrated in Figure 5. Extension of its usefulness would need to incorporate ad libitum feeding in many practical situations.

A current study at Edinburgh (Thorgeirsson, Purser and Sales, personal communication) involves the analysis of growth responses of Blackface lambs from three different selection lines. These are being fed a high concentrate diet at three levels of intake; ad lib, 90% and 75% of ad lib. Lambs are being slaughtered at the start of the feeding period (25 kg) and at 30, 35, 41 and 46 kg liveweight, with a further group on ad lib feeding being taken to two years of age. The aim is to define the muscle growth potential of the different selection lines and relate this to voluntary feed intake. The influence of different levels of energy intake on muscle growth, body fat content and feed conversion efficiency will be evaluated in relation to age, weight and degree of maturity at slaughter.

Jackson's system of growth analysis enables different genotypes to be characterised in terms of muscle growth potential (and relative feed intake, when fed ad lib) and quantifies the response to level of energy intake in a way that enables the implications for more practical situations to be deduced.

Figure 5

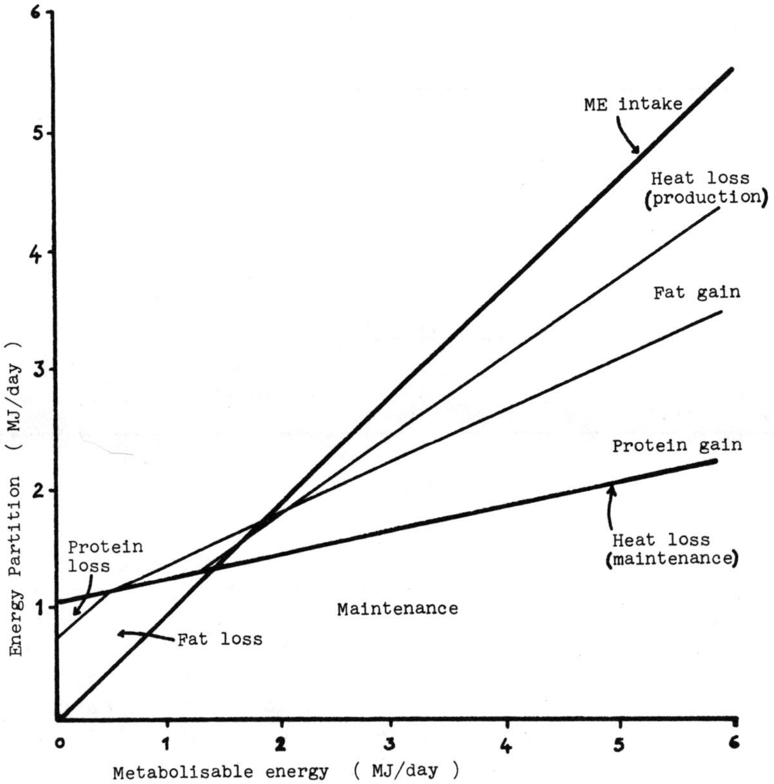

Fig.5. Effect of ME intake on energy partition for milk-fed male lambs, 3 - 38 kg L.Wt. (from Black 1974)

PRACTICAL IMPLICATIONS

In practice genotypes of different size, growth potential and feeding behaviour must be matched to different feed situations and marketing opportunities (Speedy & Jackson 1975, UK/MLC 1975).

Where lamb production is geared to seasonal market prices, fast liveweight gain to finished weight is most important. Under UK conditions fast growth and early finishing at a light carcass weight is at a premium for early lamb, which must be fed concentrates in order to reach the market before the main supply of grassfed lamb is available. Small, early-maturing breeds such as the Dorset Down when mated to milky crossbred ewes fit well in this situation.

Under the most liberal concentrate feeding conditions the Dorset Down crosses may be expected to reach slaughter condition about a week earlier than Suffolk crosses, 4 to 5 kg lighter and producing 2 kg less carcass weight. In this connection it must be recognised that variation in growth potential within a breed such as the Dorset Down is likely to be greater than the mean difference between this breed and others such as the Oxford and Suffolk (Table 5(a)).

For the temperate grassland flock a combination of good growth potential with some flexibility in time of finishing is required. Suffolks are deservedly popular combining fast growth with an ability to produce a finished carcass in the early or mid-season. It can also be taken on more slowly to heavier weights for late lamb production without becoming excessivley fat.

	Suffolk-cross	Clun
Carcass weight (kg)		
August	18.2	17.7
November	23.2	24.1
Fat % Best Neck (6-12th rib)		
August	28	36
November	41	48

(Prescott & Hinks 1967)

For systems in which early growth is not of major importance as with the winter finishing of older and heavier lambs, there may be particular advantage in a breed that can be taken to heavy weights without becoming excessively fat. Recently, the Texel breed has been introduced to the UK, and appears to hold promise in this respect (Table 5(b)).

In two major comparisons of this and a range of other breeds as fat lamb sires, it clearly emerges as slower growing than the Suffolk, but at the same liveweight produces a carcass containing much less fat and a higher content of lean.

Table 5 Breed Comparisons with Growth and Carcass Evaluation

(a) Feeding Concentrates ad lib. Slaughter at 16-22 weeks (Mean 20).

	Dorset Down cross	Suffolk cross	Oxford cross	
Mean Daily Gain (kg)				
Birth-Slaughter	0.28	0.29	0.31	*
12 wks - Slaughter	0.33	0.35	0.38	*
Daily Feed Intake (kg)	1.63	1.69	1.74	NS
Feed Conversion				
12 wks - Slaughter	5.01	4.85	4.62	*
Carcass Weight (kg)	21.6	21.8	22.2	NS
Carcass Fatness (%)	34.8	30.5	32.4	*

(Prescott & Wilson, unpublished)

(b)

(i) Grazing. Slaughtered at 35 or 40 kg (Smith ABRO)

	Age at Slaughter (days)	Muscle Weight (kg)
Oxford	141	8.6
Suffolk	141	8.6
Dorset	162	8.5
Texel	166	9.4

(ii) Grazing. Slaughtered at 36 or 45 kg (O'Ferrall, Irish Institute)

	Age at Slaughter (days)	Carcass Fat (%)	Muscle:Bone ratio	Carcass Lean %
Oxford	204	26.6	3.32	56.1
Suffolk	206	27.4	3.39	55.9
Dorset	247	29.8	3.59	54.7
Texel	212	22.7	3.58	59.9

(McClelland 1975)

The implications of this are that the Texel probably has a lower muscle growth potential than the Suffolk hence its slower growth. Its voluntary intake of feed is however so regulated that its relative intake above maintenance requirement supports this muscle growth with only limited additional energy being available for fat deposition.

Furthermore genetic variation in relative intake (that is intake in relation to maintenance) may under practical management conditions be more important than has been appreciated to date (Taylor et al 1962, Taylor & Young 1964, Graham 1969, Davies et al 1978). Differences between genotypes in eating behaviour and sensitivity to stress may be particularly significant under field conditions in which feed supply fluctuates and flock behaviour is subject to a variety of disruptive stimuli.

It is of interest that the difference between the Suffolk and Texel was less under conditions of the Irish trial in which the sheep were taking over 220 days to reach 40 kg than in the ABRO trial in which these weights were being achieved soon after 160 days. McClelland (1975) has noted that there were no undue difficulties in lambing Texel cross lambs in either of these studies.

CONCLUSION

A better understanding of growth and development in sheep should stimulate an awareness of the need to scale animals for size (if possible in relation to mature adult liveweight) whenever growth rate, body composition and feed efficiency are being compared. This finds practical application in the selection of optimum slaughter weight for different genotypes (UK/MLC 1975).

A clear distinction must be made between muscle growth potential (which is reflected in liveweight gain) and carcass leanness on which the dominant influence is the relative rate of fat deposition. Fat content is the major variable in carcass composition.

The rate of fat deposition relative to muscle growth will largely depend upon the balance between intake of energy above maintenance and the potential for utilisation of this energy in muscle synthesis. It is evident that there may be important variation between genotypes in this respect; possibly this is exemplified by the difference between the Suffolk and Texel. Certainly it is a feature that has a major influence on the meat production potential of our sheep.

ACKNOWLEDGEMENT

Particular reference has been made to the work (1970-73) in the Edinburgh School of Agriculture of the late Dr. T.H. Jackson and his colleague J.R. Weddell.

I am grateful to Mr. Weddell for permission to refer to unpublished information from his postgraduate studies at the University of Edinburgh.

REFERENCES

Berg, R.T.;Butterfield, R.M. (1976) New concepts of cattle growth. Sydney University Press, Australia.

Black, J.L. (1974) Manipulation of body composition through nutrition. Proceedings of the Australian Society of Animal Production 10,211-218.

Boccard, R.;Dumont, B.L.;Le Guelt, P.;Arnoux, J. (1961) Étude de la production de la viande chez les ovins. IV. Relation entre la forme et la composition du membre postérieur. Annales de Zootechnie 10,155-160.

Davies, D.A.R.;Owen, J.B.;Read, J.L. (1978) Artificial rearing as a system for evaluating male lambs. Animal Production 26,31-38.

Fourie, P.D.;Kirton, A.H.;Jury, K.E. (1970) Growth and development of sheep. II. Effect of breed and sex on the growth and carcass composition of the Southdown and Romney and their cross. New Zealand Journal of Agricultural Research 13,753-770.

Graham, N.McC. (1969) The influence of body weight (fatness) on the energetic efficiency of adult sheep. Australian Journal of Agricultural Research 20,375-385.

Huxley, J.S. (1932) Problems of relative growth. Methuen, London.

Jackson, T.H.;Weddell, J.R.;Mansour, Y.A. (see Jackson, T.H. (1974) European Association of Animal Production 25th Annual Meeting, Copenhagen, 1974. Sheep and Goat Commission).

Kirton, A.H.;Pickering, F.S. (1967) Factors associated with differences in carcass conformation in lamb. New Zealand Journal of Agricultural Research 10,183-200.

McClelland, T.H. (1975) The Texel breed of sheep. Scottish Agricultural Colleges, Technical Note No.CN 8.

McClelland, T.H.;Russel, A.J.F. (1972) The distribution of body fat in Scottish Blackface and Finnish Landrace lamb . Animal Production 15,301-306.

McClelland, T.H.;Bonaiti, B.;Taylor, St.C.S. (1976) Breed differences in body composition of equally mature sheep. Animal Production 23,281-293.

Prescott, J.H.D.;Hinks, C.E. (1967) University of Newcastle-upon-Tyne, Department of Agriculture Marketing Report No.7.

Reid, J.T. et al. (1968) Peculiarities in body composition of animals. In 'Body composition in animals and man. Proceedings of a symposium, Columbia, May 1967' National Academy of Sciences/National Research Council Publication No.1598,pp19-44.

Russel, A.J.F.;Doney, J.M.;Gunn, R.G. (1969) Subjective assessment of body fat in live sheep. Journal of Agricultural Science, UK 72,451-454.

Speedy, A.W.;Jackson, T.H. (1975) Scottish Farmer, August 23rd.

Taylor, St.C.S. (1968) Genetic variation in growth and development of cattle. In: Lodge, G.A.;Lamming, G.E. (Eds) 'Growth and development of mammals'. Butterworths, London pp267-290.

Taylor, St.C.S.;Watson, J.H.;Young, G.B. (1962) Variation in efficiency of food utilization in twin calves. I. Individual variation and age trends in efficiency. Journal of Agricultural Science, UK 59,287-299.

Taylor, St.C.S.;Young, G.B. (1964) Variation in growth and efficiency in twin calves. Journal of Agricultural Science UK 62,225-236.

Timon, V.M. (1963) The measurement and inheritance of lamb carcass traits. PhD. Thesis, University of Durham.

UK, Meat and Livestock Commission (1975) Planned crossbreeding and lamb carcass weights. The Commission, Bletchley, UK.

Whitelaw, A.;Watchorn, P. (1975) An investigation into dystocia in the South Country Cheviot flock. Veterinary Record 97,No.25/26 489-492.

Whittemore, C.T. (1976) A study of growth responses to nutrient input by modelling. Proceedings of the Nutrition Society 35,383-391.

Whittemore, C.T. (1977) The prediction of growth responses in pigs. In: Haresign, W. & Lewis, D. (Eds) 'Recent advances in animal nutrition - 1977'. 11th Conference for Feed Manufacturers, University of Nottingham. Butterworths, London, UK pp158-166.

SHEEP CARCASS AND EATING QUALITY

A. CUTHBERTSON and A.J. KEMPSTER

Meat and Livestock Commission, P.O. Box 44, Queensway House, Milton Keynes

INTRODUCTION

Carcass and eating quality may be defined as the sum of the various characteristics desired by those who purchase meat. The consumer is the final arbiter by the meat she is willing to buy in the butcher's shop but this definition of quality also includes the characteristics which the wholesaler and retailer require as likely to give them greatest profit.

We must recognise that quality considerations may not be the same for wholesalers, retailers and consumers. It may, for example, be better for a retailer to purchase carcasses giving meat not considered best quality by the consumer, if by buying and selling more cheaply the differential is such that he can make a greater profit.

So, we can think of the optimum carcass and eating quality ultimately in terms of what the consumer is willing to pay most for. If this is not compatible at present with least cost production, processing and marketing, we must look for improvement in the efficiency of these factors. If quality and efficiency are not compatible even in the long term with improved production, processing technology and better marketing (and they may not be) it is necessary to strive for that quality cost ratio which is most acceptable to the consumer.

Taking these points a little further we can see that there is a distinction between improvements in carcass and meat quality and production characteristics. The former may be seen as movement towards an ideal when those purchasing the carcasses and meat from them will be satisfied. In contrast, for production characteristics it will always be desirable to improve efficiency. In reality, production, processing and marketing costs are interwoven with quality and price considerations and are difficult to disentangle.

In this paper, we shall try to define ideal carcass and meat quality and consider how far we are from this at present, the constraints on improvement and the ways in which better quality may be achieved. There is, of course, no single ideal; individual consumer requirements will vary about this, but we may think of an ideal optimal range with anything outside this undesirable. Alternatively, we may think of different optima for a series of markets.

CARCASS AND MEAT QUALITY CHARACTERISTICS

The main characteristics influencing carcass value are as follows:

(i) Carcass weight
(ii) Carcass lean, fat and bone content

(iii) The distribution of the weights of lean, fat and bone between the different parts of the carcass (tissue weight distribution)
(iv) Carcass conformatin (or shape) and muscle thickness
(v) Meat quality (this presents quite a different problem and is considered in a separate section of the paper)

Carcass weight or the related liveweight is obviously of major importance to the producer since this represents his output. The requirements of the meat trade show quite a wide range of variation. This is because wholesalers and retailers in certain areas have become accustomed over generations to the breeds and crosses traditionally produced in their areas and to the marketing pattern in terms of the time of year that the lambs are slaughtered. With the increasing movement of meat about the country, the widening catchment areas of large wholesalers and the influence of the Continental market, meat trade preferences in carcass weight are becoming more clearly defined.

Local influences are also evident in the considerable variation in levels of fatness. Besides the movement towards standardisation, in this case there is increasing pressure from the consumer for leaner meat. Today few people want fat: not only is it said to kill one with heart disease but it can be bought more cheaply as margarine or butter. Lamb fat is also often considered more unpalatable than beef or pork fat. Some evidence of the extent to which fat is rejected on the plate by the consumer is given in Table 1 from a survey conducted among 1500 consumers by the Meat Research Institute (Rhodes 1976).

Table 1 Percentage of U.K. consumers who reject visible fat on the plate. (Age groups are divided at 16 years) (Rhodes 1976)

Age/sex group	Number sampled	Fat type		
		Beef	Pork	Lamb
Boys	122	67	55	62
Girls	116	64	67	68
Men	634	34	38	45
Women	623	42	42	50
Overall	1495	42	43	50

Variation in lean weight distribution through the carcass is potentially very important because of the wide price differentials between different cuts of lamb. However, as will be indicated later, there is little variation to exploit in this characteristic even between extreme types of lamb.

When considering the importance of conformation and muscle thickness, one has to distinguish between the value of these characteristics as predictors of carcass composition (examined later in the paper) and the commercial value they have in their own right. It is difficult to assess their value per se; joints from carcasses with better conformation and muscle thickness are possibly less wasteful to cut and more attractive to the consumer but there is little scientific evidence of this. Examination of these characteristics is an important area for future research.

Before going on to examine the variation in carcass characteristics, it is necessary to spend a little time on methods of assessment of carcass composition.

ASSESSMENT OF CARCASS COMPOSITION

Carcass composition can be measured directly using three main techniques or combinations of them.

(i) Chemical analysis - the carcass is minced and a representative sample chemically analysed for fat, moisture, protein and ash.

(ii) Dissection - the carcass is jointed either in a commercial way or anatomically and the joints separated into fatty tissue, lean, bone and waste.

(iii) Commercial trimming - the carcass is separated into wholesale or retail joints and these trimmed of fat to a commercial specification.

Ideally, the technique used should be accurate, repeatable and relevant both in biological and commercial terms. None of the methods meets this requirement exactly and it is often necessary to use a combination of them. The major problem associated with chemical analysis, in addition to the extra cost due to loss in value, is that it is difficult to evaluate market requirements in terms of chemical components. Dissection does permit the measurement of the carcass in terms which are used by the housewife and butcher but there may be problems in standardizing techniques between operators and from year to year. Furthermore, separable muscle always contains intramuscular fat and this can lead to an overestimate of the muscle in fat animals since their muscle contains more intramuscular fat than the muscles of lean animals. These limitations are much more serious for commercial trimming where it is very difficult to standardize the level of subcutaneous and intermuscular fat left on the joints.

The method chosen will depend very much on the work in question. Most workers are now satisfied with measuring a single side (half carcass) since the increase in precision achieved by the use of the second side in predicting total carcass composition is so small in relation to the virtually doubled cost. Side dissection is used as the baseline for all

MLC lamb carcass evaluation work. Analysis of the fat content of the dissectable lean is used only as a standardization check on the dissection technique.

Full dissection of the carcass is too expensive or impossible to operate in many circumstances and simpler methods have to be used to predict carcass composition. It is also obviously necessary to predict carcass composition when information is required for live animals. Selection of the most suitable predictor for a given application depends on a number of factors:

(i) the cost of taking the predicting measurements

(ii) the precision with which the measurements predict carcass composition

(iii) the stability of prediction relationships between types of lamb differing for example in breed, sex or the feeding regimen on which they were grown

(iv) and physical constraints as for example in commercial carcass classification where it is necessary to take the measurements on the hot intact carcass under fast-moving slaughter line conditions.

The problem of selecting the best predictor for a given application can only be solved satisfactorily with reference to the particular circumstances, the resources and labour force available and the required sensitivity of the work in question. The best predictors are generally the most expensive and it is usually necessary to compromise between cost and precision. The various methods will be discussed briefly and consideration given to some of the problems of carcass assessment encountered in the course of operating MLC schemes.

Live animal evaluation

The development of accurate methods for predicting carcass composition in live animals has major significance for breeding schemes and for experiments where it is necessary to follow changes in carcass composition during growth or identify slaughter points defined in terms of carcass composition. Among the objective methods suitable for use on live animals, ultrasonic techniques have the greatest practical value at present. Although they are used extensively in pig breeding and increasingly in cattle breeding and research, there is little published information on their use on sheep.

The MLC has been carrying out an evaluation of the Scanogram ultrasonic machine, developed by Professor J.R. Stouffer and manufactured by Ithaco in the USA, for use on lamb. This is based on a modified linear scanning technique. The probe is moved mechanically on a track and sound waves are sent into the body at right angles to the surface. As the probe moves, the signals reflected from tissue interfaces are displayed on a cathode ray screen as a cross-sectional picture which is recorded by a polaroid camera. A cross-sectional picture of the M. longissimus and overlying fat is thus obtained. Preliminary results have been published by Kempster et al (1977) and some are shown in Table 2. The best

Table 2 Prediction of carcass tissue percentages from ultrasonic measurements taken using the Scanogram: residual standard deviations (Kempster et al 1977)

The evaluation was carried out on 51 castrate male lambs by nine sire breeds out of Scotch Halfbred ewes. Analysis was carried out on a pooled within sire breed basis.

	% subcutaneous fat	% lean
Standard deviation	2.64	3.02
Prediction from:		
Live weight at scanning (W)	2.42	2.90
W + fat depth[1]	2.20	2.73
W + fat area [1]	1.93	2.47
W + eye muscle area[1]	2.07	2.55
W + fat area + eye muscle area	1.86	2.39

[1]Scans were taken of the cross section of the eye muscle and overlying fat at the 12th rib. Fat depth ws measured at 3.5cm from the dorsal mid line; fat area was measured from the dorsal mid line to a point at right angles from the skin to the dorsolateral corner of the eye muscle.

combination of measurements tested accounted for 40% of the variation in percentage carcass lean (residual standard deviation = 2.39). It is relevant to compare this level of precision with that obtained with cattle. Our experience indicates that the residual standard deviation for predicting percentage lean in cattle, with the same overall variation and using similar combinations of measurements to those used here for lamb, would be approximately 1.9 (60% of the variation explained).

Whether the level of precision is adequate for use in experiments and breeding schemes is difficult to answer without considering particular circumstances. We think the technique has a role in experiments where it is necessary to identify slaughter points at a given level of fatness and possibly in selection programmes within nucleus flocks of the more important meat sire breeds. The Meat and Livestock Commission are now carrying out tests on other sophisticated ultrasonic machines in collaboration with the Meat Research Institute and Animal Breeding Research Organisation.

Other techniques of estimating carcass composition in the live animal include X-rays, dilution methods, density and the measurement of naturally occurring K^{40} and electromagnetism. These have been discussed briefly by Cuthbertson (1978). None of these methods has consistently yielded sufficiently close correlations with total leanness for its wide application, particularly in view of the skill and cost required.

Carcass evaluation

Kempster et al (1976) carried out a detailed evaluation of different predictors for use in particular circumstances using 420 lambs covering a range of breed types. The main results are shown in Tables 3 and 4. Table 3 shows the results obtained for some simple meaurements and subjective scores for predicting the percentage of lean in the carcass.

Table 3 Prediction of percentage lean in carcass from simple measurements and subjective scores: residual standard deviations (Kempster et al 1976)

Evaluation was carried out on 420 lambs comprising seven breed-type groups. Overall results and results pooled with breed-type groups (different intercepts but common slopes) are given below.

	Overall	Within breed-type group
Standard deviation of % lean	3.96	3.70
Prediction from:		
Carcass weight	3.78	3.45
Subcutaneous fat score[1]	3.17	2.85
Internal fat score[2]	3.35	3.17
Conformation score[3]	3.57	3.33
Circumference of buttocks	3.92	3.66
Side length	3.93	3.66
Carcass weight/length ratio	3.61	3.39

[1]Subjective assessment of subcutaneous fat development on a seven-point scale.

[2]Subjective assessment of KKCF development on a seven-point scale.

[3]Subjective assessment of fat and muscle thickness in relation to the size of skeleton (seven-point scale).

Table 4 provides some further information where the carcass is actually cut across and where sample joints may be separated. The results will now be considered with reference to their possible use in different circumstances.

Table 4 Prediction of percentage lean in carcass from cut surface measurements and sample joint dissections: residual standard deviations (Kempster et al 1976)

Results are given in the same manner as Table 3.

	Overall	Within breed-type group
Prediction from:		
Best single fat depth[1]	3.00	2.80
Fat depth/muscle depth ratio[1]	3.20	3.06
Eye muscle area[1]	3.94	3.69
KKCF % in carcass	3.27	2.98
Percentage lean in leg	2.44	2.15
chump	1.91	1.87
loin (1)	1.94	1.81
breast	1.92	1.82
best end neck	1.61	1.53
middle neck (2)	2.88	2.75
shoulder (3)	1.67	1.62
neck	3.56	3.37
Best pair of sample joints (1+3)[2]	1.11	1.03
Combination of sample joints used in MLC breed evaluation (2+3)[3]	1.58	1.49

[1]Measurements taken on the cut surface at the 12th rib.

[2]Pair of sample joints giving the most precise prediction.

[3]Combination of joints considered best in terms of cost and precision under MLC conditions.

Table 6 Commercial lamb survey. Least squares means by breed-type group and sampling time.[1] (Kempster & Cuthbertson 1977)

	Carcass weight (kg)	Composition of carcass (%)				% lean distributed in higher-priced cuts
		Lean	Bone	Subcutaneous fat	Total fat	
Breed-type[2]						
Welsh Mountain	12.8	54.7	14.8	12.7	29.0	57.3
Blackfaced Mountain	15.8	57.1	16.3	11.2	25.1	56.0
Longwool crosses (1)	18.6	57.2	15.8	12.2	25.8	55.7
Suffolk crosses (2)	18.4	55.5	16.7	11.7	26.3	56.2
Intermediate (3)	16.6	55.9	14.9	13.1	27.9	56.3
Southdown crosses (4)	15.3	53.3	14.6	14.5	30.8	57.2
Lowland Longwool	22.0	52.6	14.2	15.6	32.0	54.7
F ratio for differences between means	62.9	11.7	18.9	12.4	15.1	15.4

Sampling time[3]						
First	16.5	55.7	15.8	12.4	27.0	56.3
Second	17.1	54.8	15.2	13.6	28.7	56.3
Third	17.6	55.0	15.0	13.0	28.7	55.9
F ratio for differences between means	4.7	2.3NS	9.6	4.7	5.4	2.7NS
Residual s.d. after fitting the model	2.58	3.60	1.56	2.92	4.66	1.54

[1]The least squares model included the effects of breed-type group, sampling time and their interaction.

[2]Breed-type group (1) British Longwool crosses on hill ewes i.e. Greyface, 'Mule' and Masham

(2) Lambs by Suffolk rams out of Scottish Halfbred ewes

(3) Intermediate-type lambs by Dorset Down or Hampshire Down rams out of ewes of intermediate size e.g. Clun Forest and Welsh Halfbred

(4) Smaller-type lambs by Southdown rams

[3]Three occasions corresponding with the beginning (first), middle (second) and end (third) of the normal slaughtering season for each group.

So far as commercial carcass classification is concerned, the need is for a simple system of identifying differences in fatness or leanness between carcasses. The predictor must be capable of identifying such differences on intact carcasses on fast-moving killing lines. The subcutaneous fat score was the best single predictor amongst those tested for use in such circumstances. Addition of either the internal fat score or the circumference of buttocks produced a useful increase in precision and these predictors may be of value for use in classification as a supplement to the subcutaneous fat score. Individual linear carcass measurements taken on the intact or on the split carcass had no worthwhile predictive value.

The importance of conformation in the prediction of lamb carcass composition has been the subject of a number of studies over the years. The general finding is that conformation does not provide any guidance on lean content. Conformation is normally defined as the thickness of muscle and fat in relation to the size of the skeleton and it is, therefore, confounded with the fatness. The important question from the point of view of carcass classification is whether conformation improves the prediction of leanness for carcasses of similar fatness. In various studies by the MLC, including the one referred to above, the addition of conformation score to subcutaneous fat score in multiple regression did not improve the precision of lean prediction and neither did weight to length ratio which is the conventional objective measure of conformation. Existing evidence indicates therefore that subcutaneous fat score should be the principle index of carcass leanness for use in commercial classification. However, this approach is not without some problems. The first is that the results indicate the relationship between the fat score and percentage lean is affected by breed type. Secondly, there are problems associated with maintaining standards within and between classifiers, particularly between different times of the year and geographical regions where there are wide differences in the mean level of fatness. It is important therefore for a suitable objective predictor to be found.

There is a need for carcass evaluation in connection with large scale breeding schemes which will involve progeny or sib testing within breeds and in connection with breed comparisons. On the basis of various MLC studies, it is doubtful whether the precision achieved with carcass measurements on the intact, split or quartered carcass would be adequate for use in such testing, and it would probably be necessary to use sample joint dissection. Among the sample joints studied by MLC, the shoulder and best end of neck joints gave consistently good predictions across groups and were also cheap to dissect in relation to their precision. A combination of percentage lean in either of these joints plus the percentage of kidney knob and channel fat probably represents the best compromise between precision and cost so far as within group studies are concerned. In the case of comparisons between breeds the stability of the prediction equations between breeds is particularly relevant. The prediction equations for the shoulder joint were very stable from one group of lambs to another and it is recommended for use in circumstances where a sub-sampling technique cannot be used to estimate the prediction equations for use in individual groups.

VARIATION IN THE CARCASS CHARACTERISTICS OF COMMERCIAL BRITISH LAMBS

A survey of the carcass characteristics of the main types of British lamb was carried out in 1971/72 as part of the development programme for a national classification scheme (full results were reported by Kempster & Cuthbertson 1977). Carcasses of 420 castrated male lambs were evaluated and their left sides dissected. The sample comprised seven groups representing the main breed types in the national population. Within these groups, lambs were selected from three regions on each of three occasions corresponding with the beginning, middle and end of the normal slaughtering season for each group. Overall means, standard deviations and coefficients of variation for the main characteristics are shown in Table 5. These show the considerable variation that exists both in weight and carcass fatness. Eye muscle area was related as expected to carcass weight ($r = 0.64$) and equally variable. Lean weight distribution showed little variation.

Table 5 Commercial lamb survey. Means, standard deviations and coefficients of variation computed over all data ignoring subgroups (Kempster and Cuthbertson 1977)

	Mean	s.d.	CV (%)
Carcass weight (kg)	16.6	3.60	21.8
Composition of carcass (%)			
Lean	55.5	3.95	7.1
Bone	15.5	1.80	11.6
Subcutaneous fat	12.6	3.19	25.2
Total fat	27.6	5.21	18.9
Total lean distributed in higher priced cuts[1] (%)	56.3	1.74	3.1
Eye muscle area (cm^2)	10.8	1.90	17.6

[1] Higher-priced joints are leg, chump, loin and best end neck.

Table 6 sets out the means for different breed-type groups and sampling times and the variation within breed-type groups at the same sampling time (residual s.d. after fitting the model). There were major differences in carcass weight between breed-type groups; Welsh Mountain lambs were the lightest, weighing 2.5kg less than the next heaviest group, Southdown crosses, and 9.2kg less than Lowland Longwools which were by far the heaviest group. There were also important differences in level of fatness with a range of 4.4% in subcutaneous fat content at 7.1% in total fat. Lambs slaughtered in the middle of the season were heavier and significantly fatter than those slaughtered towards the beginning of the season.

It is possible to examine the variation in fatness in the national population further using results from MLC's classification scheme (details of the scheme are given later). In the scheme, a visual assessment of subcutaneous fat cover is carried out on a five-point scale. The fat scale has an objective basis in that the points are standardised against ranges of subcutaneous fat percentage. In Table 7, the distribution of survey carcasses between fat classes (determined by their actual subcutaneous fat percentage) for each breed-type is compared with the distribution of all carcasses classified in 1977. The classification results are based on 2,190,000 carcasses from 40 abattoirs and embracing some 25% of the total clean sheep kill in Great Britain. Considered as a whole, the survey carcasses were fatter than those classified in 1977, with 18% more in fat classes 4 and 5. The difference was largely a reflection of the fatness of the Southdown cross and Lowland Longwool lambs which made up a much larger proportion of the survey sample than they do in the population of classified lamps. The Suffolk cross group most clearly represented the classified population in terms of the distribution of carcasses between fat classes. The commercial significance of the proportion of carcasses in the higher fat classes can be seen by considering the fat trimming necessary to make the cuts from these carcasses acceptable to the consumer. Unpublished results from MLC breed comparison trials, which involve a commercial trimming technique, indicate that some 11% of the weight of fat class 4 carcasses and 15% of the weight of fat class 5 carcasses are likely to be trimmed before retail sale.

Taking the results for classified carcasses we can make some estimates of national waste fat production using a similar method to that employed by Harrington & Kempster (1977) for cattle. Table 8 gives our best estimate of excess fat production in lamb carcasses of different fat classes i.e. the fat in excess of that directly eaten by the consumer. Some fat is naturally produced in lambs converting feed into lean meat with optimum efficiency and if we want an estimate of true waste from our current lamb population, it is necessary to remove this obligatory fat from these figures.

There is relatively little information available on levels of fatness at optimum efficiency of conversion of feed into lean meat for lamb. On the limited evidence available, fat class 2 is probably close to the optimum. On this basis, waste fat is the excess fat produced above fat class 2. If we assume that the weight x fatness distribution in the national slaughter population is the same as the sample of carcasses classified in 1977, the waste fat production is 12,000t. Taking 150 MJ metabolizable energy at £0.85 as the cost of producing 1kg fat, the cost of annual waste fat production is £10 million. This is equivalent to 90p/carcass slaughtered.

Table 7 A comparison of the distribution of survey carcasses between fat classes (determined from actual subcutaneous fat percentages) with the distribution between fat classes of carcasses classified in 1977 (modified from Kempster & Cuthbertson 1977)

Fat class	1	2	3	4	5
Subcutaneous fat in carcass (%)	<6.0	6.0-9.9	10.0-13.9	14.0-17.9	≥18.0
Survey results					
Welsh Mountain	-	13.9	55.7	25.3	5.1
Blackfaced Mountain	1.3	31.6	48.7	18.4	-
Longwool crosses	1.3	19.5	50.6	26.0	2.6
Suffolk crosses	3.1	18.5	66.1	9.2	3.1
Intermediate	2.0	7.8	51.0	35.3	3.9
Southdown crosses	-	15.1	32.1	32.1	20.7
Lowland Longwool	-	2.9	34.3	45.7	17.1
Overall	1.1	17.2	50.0	25.5	6.2
1977 classifications	3.1[1]	27.8[1]	54.9	12.7	1.5

[1] 4.5% of carcasses were not classified for fatness because they had poor conformation. It has been assumed that these were distributed evenly between fat classes 1 and 2.

Table 8 Estimate of national waste fat production

Weights of components are expressed as percentages of carcass weight unless otherwise stated.

	Fat class				
	1	2	3	4	5
Total dissectable fat[1]	14.3	20.3	26.6	32.7	38.9
Fat trim in trade	2.0	3.7	6.7	11.2	15.0
Plate waste if the consumer eats 10 parts lean to 1 part fat	5.8	10.7	14.3	16.3	19.1
Fat produced in excess of consumer requirements	7.8	14.4	21.0	27.5	34.1
Excess fat above fat class 2, i.e. waste fat if fat class 2 is assumed to correspond with optimum efficiency of converting feed into lean meat	-	-	6.6	13.1	19.1
National waste fat production (t) if we assume that the national weight x fatness distribution is the same as that in the sample of classified carcasses	-		7460	3640	650
				11,750t	

[1]Includes kidney knob and channel fat but not other depots in the body cavity.

THE ROLE OF CLASSIFICATION AND GRADING

It is clear from these estimates that there is a need for the industry to reduce levels of fatness and improve the quality of its product as quickly as possible if lamb is to retain its position relative to other foods. Government organisations, research workers and to some extent the meat industry have been making noises about this for many years. But, while there has been considerable success in reducing the level of fat in pig carcasses, little headway has been made with sheep (or for that matter with cattle). A major influence on pigs is that most are sold directly from producer to abattoir on a grade and deadweight basis with fatter carcasses being penalised and leaner carcasses given premiums. In contrast only 35% of lambs are sold on a deadweight basis and perhaps a third to half of these paid for on the basis of grade. The remaining 65% of lambs are sold in live auctions where neither the buyer nor seller really knows what is under the skin.

In order for the producer to provide what the consumer wants, it is necessary to have standard methods of describing the important commercial characteristics and for this information to be fed back down the marketing chain to the producer. Given clear price differentials by the wholesaler for different types of carcasses, the producer can exploit present knowledge of animal breeding and management to provide what is required. Without such price differentials, there can be no incentive to improve carcass characteristics unless the improvement is directly related to improved production efficiency.

MLC SHEEP CARCASS CLASSIFICATION SCHEME

In Great Britain, where trade requirements for carcasses are so diverse, it is particularly important to draw the distinctioin between classification and grading. The word "classification" covers the description of carcasses by the characteristics which are of prime importance to meat traders. These characteristics are described separately without attributing relative importance or cash value differences to them. The buyer creates grades by combining the classification data to form a summarized assessment to which prices are attached. It is therefore for the individual trader to decide what is good and what is poor in relation to his own business. However, if he can describe his requirements in terms of a generally accepted decriptive classification scheme, understanding should be increased and result in greater efficiency in trading.

The scheme describes carcasses by 4 key characteristics; weight, category, fatness and conformation.

<u>Weight</u> This is defined as the cold carcass weight obtained after carcasses have been subjected to a clearly defined carcass dressing procedure. Currently, the specification includes kidney knob and channel fat in the weight. It is important that producers do not underestimate the importance of having this weight authenticated by an independent third party.

Category Five age/sex groups are currently distinguished: lamb, hogget, sheep, ram and ewe. There may be a desire to change their definition, or at least some of the names, once the Government's Fat Sheep Guarantee Scheme, which uses these categories, is phased out.

Fat class This is a vital characteristic since, amongst carcasses of similar weight, it has a marked effect on the quantity of meat for sale. In describing fatness in the scheme, it was decided that this should be done subjectively until a more accurate objective technique could be found which could be applied to fast moving hot carcasses on a killing line. The evidence given earlier indicates that classification for fatness can be achieved by subjective appraisal with sufficient accuracy for practical purposes. The fat classification involves assessing the degree of subcutaneous fat development on a scale of 5 points where 1 = very lean to 5 = very fat, leaving classes 2, 3 and 4 to describe the fatness of the majority of carcasses. Although the classification is carried out by eye, there is an objective base in that each point on the scale is linked to a particular percentage range of subcutaneous fat in the carcass, which is set as shown in Table 7. A recent modification to the scheme is the option for the wholesaler to have fat class 3 divided into two halves: 3L (low fat) and 3H (high fat).

In addition to describing the degree of subcutaneous fat development, carcasses which have more than 5% of their weight as kidney knob and channel fat (judged subjectively) are identified by the letter K.

Conformation class An important feature of the scheme is the separate description of carcass fatness and conformation. This means, for example, that the value which traders place on conformation can be assessed amongst carcasses of similar weight and fatness. Four conformation classes are being distinguished in the scheme. Conformation is defined as the overall shape of the carcass and is represented by the thickness of lean + fat in relation to the size of the skeleton. In drawing up the conformation classes, it was decided that there should be a wide average conformation class, a class (E) identifying carcasses with very good conformation, while the small percentage of carcasses with poor or very poor conformation would be identified by the letters C and Z respectively. The conformation assessment was included in the scheme principally as an index of muscle thickness and not of lean content.

The five fat and four conformation classes are used to produce the ten class combinations shown in Table 9. The intention is to keep the scheme as simple and as workable as possible while allowing traders to draw the distinctions which they consider important. It can be seen from the Table that conformation is not distinguished amongst very lean and very fat carcasses and that fatness is not distinguished amongst carcasses of poor or very poor conformation. Such distinctions are not thought to be important since, for example, most of the poor or very poor conformation carcasses have low fat levels.

An attempt has been made to keep the nomenclature of the class combinations simple. Thus carcasses of average conformation in fat class 2, 3 and 4 (i.e. the majority of carcasses) are described by the fat class number only. If they have extra conformation the fat class number is followed by the letter E.

Some of the results obtained from the operation of the experimental scheme are given in Table 9.

Table 9 Fat x conformation grid used in the MLC classification scheme and the percentage of classified carcasses falling into each cell in 1977. Based on a total of 2,190,000 carcasses.

<table>
<tr><th rowspan="2">CONFORMATION CLASS</th><th></th><th colspan="5">FAT CLASS</th></tr>
<tr><th></th><th>1
leanest</th><th>2</th><th>3</th><th>4</th><th>5
fattest</th></tr>
<tr><td></td><td>E
(extra)</td><td rowspan="2">0.8</td><td>1.0</td><td>8.6</td><td>3.9</td><td rowspan="2">1.5</td></tr>
<tr><td></td><td>Average</td><td>24.6</td><td>46.3</td><td>8.8</td></tr>
<tr><td></td><td>C
(poor)</td><td colspan="5">3.9</td></tr>
<tr><td></td><td>Z
(very poor)</td><td colspan="5">0.6</td></tr>
</table>

BREED DIFFERENCES IN CARCASS COMPOSITION

Of the possible ways of improving carcass lean content, breed substitution probably offers the quickest response if there is to be no major reduction in carcass weights. MLC are currently comparing a range of terminal sires for use on different dams and preliminary results for the first year are given in Table 10. Comparison is being made at equal subcutaneous fat percentage in carcass. This basis is considered most satisfactory for practical purposes since fatness is the main factor influencing slaughter point and also closely relates to the overall efficiency of converting feed into lean meat. The results indicate that the ranking of breeds for carcass weight produced at equal fatness is very similar to their ranking for mature size. However, there are still sizeable differences in lean content due to differences in lean to bone ratio and fat distribution between breeds.

There is much interest in the Texel breed at present, due to the high quality carcass it is claimed to produce. Recent research findings in this country and Eire indicate that the claims are well founded (Table 11). The results together indicate that Texel cross carcasses in comparison with those of Suffolk crosses had a 5% larger eye muscle area at the same weight but there was little difference in lean weight distribution. Their lean to bone ratio was 8% higher and this together with a lower fat content at equal carcass weight gave them an advantage of 7% in carcass lean meat content.

Table 11 also shows the relative performance of the breeds in performance characteristics. The results indicate that the Suffolk crosses were superior to the Texel crosses in terms of daily rates of live weight and carcass weight gain and would thus be a better choice, ceteris paribus, for a producer marketing lambs simply on a weight basis. However, the Texel crosses were superior in terms of daily lean meat gain. Thus, for a producer selling lambs to a market which pays a premium for leaner lambs, the Texel could be a better choice. The implications of this are that if the producer is to change from Suffolk sires to Texel sires there must be a financial incentive from the meat trade for him to do so. Without a premium for leanness, it would not be worth his while.

MEAT QUALITY

In this section, consideration will be given not only to eating quality as such, but also to some of the factors influencing meat buying and eating habits.

The consumer's choice of a particular piece of meat in the shop is affected by its visual appeal, convenience and suitability for a particular purpose, and expected value for money. Some aspects of these will be considered in turn.

Visual appeal embraces a number of characteristics and depends, to some extent, on the type of cut and on whether and how it is packaged. The consumer is generally looking for meat with only the minimum covering of fat since most consumers, and particularly the younger ones, show resistance to eating fat. This was illustrated in Table 1. The fat which is present should not be markedly yellow as this will lead to consumer resistance, particularly when displayed alongside cuts with whiter fat. There is a recessive gene which produces such yellow fat, with an incidence of probably less than 1%, but there is no reason to suppose that such meat is nutritionally inferior.

When fresh meat is being purchased the colour and appearance of the lean meat should convey freshness, although lighting in display cabinets can be used to modify the colour of the meat. In packaged meat or, even, uncovered meat displayed on trays in shops, the appearance of drip is unattractive. Meat which drips is clearly undesirable and represents a

Table 10 Preliminary results from the first year of MLC's Ram Breed Comparison Trial

Results are based on a total of 582 carcasses, half of which were fully dissected and half of which had carcass tissue percentages predicted from sample joint dissection.

The ram breeds were evaluated in four flocks, two Scotch Halfbred and two Scottish Blackface. Results below are least squares means from a model including the effects of sex and flock with regression to equal carcass subcutaneous fat content.

Sire Breed	Number of lambs	Carcass weight (kg)	Composition of carcass		Eye muscle depth[2] (mm)	Conformation[1] Score
			% lean	% of total fat		
Southdown	73	15.6	56.2	26.0	25.9	3.26
Dorset Down	60	17.6	56.0	26.6	27.3	3.19
N.C. Cheviot	56	18.5	56.1	26.1	26.4	3.09
Hampshire Down	73	17.7	56.1	26.0	26.5	3.19
Ile de France	75	18.8	56.8	26.3	28.4	3.28
Suffolk Down	65	19.8	56.3	25.8	27.4	3.25
Border Leicester	69	19.9	56.0	26.0	26.8	2.91
Oxford Down	58	20.4	55.7	26.4	27.3	3.03
Wensleydale	53	20.3	56.6	25.5	27.1	2.90

[1] 5 point where 5 is best conformation

[2] At the 12th rib

Table 11 Comparison of the carcass characteristics of progeny from Texel and Suffolk sires.[1] Texel is given relative to Suffolk (100%).

	Trials carried out by ABRO in collaboration with MLC[2]	Trials in Eire (Timon 1974)[3]
Comparison is made at equal live weights (kg)	37.5	37.0
Carcass composition (%)		
Lean	107	107
Fat	88	83
Bone	96	102
Lean/bone ratio	112	106
Conformation score	99	NA
Percentage lean distributed in higher-priced cuts	99	NA
Eye muscle area	103	107
Daily live weight gain	94	97
Daily carcass weight gain	96	96
Daily lean weight gain	103	102

[1]Different dam breeds were used but the analyses were carried out so that dam breed effects were the same for progeny of the two ram breeds.

[2]Based on two years results for 117 Texel progeny and 87 Suffolk progeny. Carcasses were fully dissected.

[3]Based on 469 Texel progeny and 828 Suffolk progeny. Composition was predicted from sample joint dissection.

loss of meat to the trader and to the consumer. The extent to which meat may drip is dependent on its pH and temperature after slaughter. In animals where physiological and biochemical factors cause the pH to fall rapidly while the carcass is still warm, the ability of the meat to hold its moisture is impaired due to the denaturation of muscle proteins. In such carcasses, the quicker the temperature is reduced the better, but if it is reduced too quickly there is a risk of cold shortening which toughens the meat. This will be discussed further below.

Convenience aspects of meat are important, particularly as more and more housewives are involved in some form of paid activity outside the home (in Britain about 45%) and have less time to spend on cooking. There is also a change of attitude, with many young people having neither the skills nor the desire to spend time on cooking requiring significant preparation time. Admittedly, promotional agencies are reacting to the social changes by emphasising to school children the merits of good nutrition, meat as a food and simple methods of preparation. A further factor is the forecast expansion in freezer ownership. By 1985, it is predicted that about half the households in Britain will own deep freezers compared with about one third now (Bansback 1977). This paves the way for expanding the frozen lamb market with the merchandising of frozen retail cuts.

Studies conducted by the Department of Agricultural Marketing at the University of Newcastle-upon-Tyne are beginning to throw more light on consumer attitudes to different types of meat. There are indications that some cuts of lamb are not looked upon favourably by consumers. In particular, lamb chops are considered to provide poor value for money. These attitudes may account in part for the continuing decline in lamb consumption per head in Britain. One of the reasons for lamb being regarded in this way is doubtless its fatness. Many find lamb fat distasteful, particularly when cold, and this is primarily due to the high degree of saturation or hardness of the fat and the relatively high temperature at which it starts to solidify. Rapid solidification on cooling results in a tendency for the fat to stick to the palate. There is evidence that the degree of saturation is less in lambs which have been grown quickly on high concentrate diets or raised at low temperatures, and that it may be less in ram lambs than in castrates. The characteristic flavour of lamb originates mainly from the fat, but this can lead to adverse consumer reaction due to the presence of certain unattractive volatile components which arise from certain types of feed.

The obvious solution to the fat problem is to prevent sheep becoming too fat in the first place. Research work, particularly on beef and pigs, has failed to show that the eating quality of lean meat from fatter carcasses is any better than that from leaner carcasses (with the exception of beef of very low fat) of otherwise similar characteristics. The implication is that one could afford to produce more lambs at lower fat levels than those now found in Britain without impairing the eating quality of the lean meat. Apart from reducing the amount of fat a retailer has to trim in order to make the meat more saleable to the consumer, there may be production advantages in slaughtering animals at lower levels of fatness. As indicated earlier, it is likely that the optimum point at which to slaughter, in terms of the point where the feed cost per unit of lean meat produced is lowest, lies around Fat Class 2 in the MLC Sheep Carcass Classification Scheme. Many producers take their lambs on to higher fat levels because the lambs are gaining weight for which they are paid and unless there is an alternative use for the forage it may pay him to do this, particularly if the wholesale market does not discriminate to any signficant extent against this product. However, with meat wholesalers in Britain looking for more opportunities to export lamb, where fat is not wanted, there is likely to be more discrimination against fat lamb in British plants, particularly those with approval to export.

Apart from the way in which meat is cooked, which will be referred to below, the factor which probably has the most significant effect on the eating quality of meat is its tenderness. The age of an animal has an effect on the tenderness of the meat, since there are changes in the amount and structure of connective tissue as it grows which lead to tougher meat. However, these changes are only likely to affect consumer acceptability of meat from animals beyond the lamb/hogget age range. The more important effect on tenderness amongst younger animals seems likely to be due to the cold shortening effect referred to earlier. This is due to the behaviour of the myofibrillar proteins under conditions where the temperature of any part of the meat falls below some 10°C in the first 10 hours after slaughter. Under these conditions the actin and myosin filaments become inter-locked to a greater extent than would occur under less rapid cooling conditions. Even after ageing, such cold shortened meat, as it is called, will be tougher. Cold shortening has been a particular problem in New Zealand where it is customary to freeze lambs as soon as possible after slaughter thereby avoiding the need to have conventional chillrooms to bring down the temperature more slowly and reducing evaporative weight loss. To overcome the cold shortening problem, the application of electrical stimulation to the carcass shortly after slaughter has been investigated. This involves applying a high voltage to the carcass on the killing line either before or immediately after dressing. The effect is to cause very rapid post-mortem glycolysis so that the ultimate pH is attained in perhaps 5 hours compared with 18 - 24 hours under conventional procedures. The rate of glycolysis is such that low temperatures can be applied to the carcass shortly after stimulation without running the risk of cold shortening. The New Zealanders are now starting to introduce electrical stimulation into their meat works and are thought to be using some 3000 volts. Work which has been in progress at the Meat Research Institute in Britain indicates that much lower voltages can achieve the same effect, but there is still a good deal of development work necessary to establish optimum procedures. Nevertheless, the technique is very promising.

Reference has already been made to the influence of cooking on the eating quality of meat. From various studies which have been conducted, it is clear that there is a great deal of variation in the way in which a given cut is cooked, which will affect its eating characteristics. There appears to be a need to educate housewives, and particularly school children, who are tomorrow's housewives, not only to know more about where the various cuts come from in a carcass but, more importantly, to understand how best to cook them. There is much to be said for including with meat purchases instructions on alternative ways of preparing and cooking the particular cuts which have been purchased. It has been suggested that, in spite of the existence of many cookery books and recipes in magazines, many housewives do not refer to these when deciding how to cook particular cuts. If this were confirmed by further study of the issue, incorporating an assessment of the extent to which housewives are prepared to put into practice the recipes which they read about, it would indicate that a new approach is needed in communicating cookery techniques.

REFERENCES

Bansback, R.J. (1977) United Kingdom meat consumption in 1979, 1982 and 1985. In: Meat demand and price forecasting. Proceedings of Meat and Livestock Commission Symposium, Wye, 1977.

Cuthbertson, A. (1978) Carcase evaluation of cattle, sheep and pigs. World Review of Nutrition and Dietetics 28,210-235.

Harrington, G.;Kempster, A.J. (1977) An examination of the fatness of commercial British beef carcasses and estimates of waste fat production. Animal Production 24,152.

Kempster, A.J.;Avis, P.R.D.;Cuthbertson, A.;Harrington, G. (1976) Prediction of the lean content of lamb carcasses of different breed types.
Journal of Agricultural Science, UK 86,23-24.

Kempster, A.J.;Cuthbertson, A. (1977) A survey of the carcass characteristics of the main types of British lamb. Animal Production 25,165-179.

Kempster, A.J.;Cuthbertson, A.;Jones, D.W.;Owen, M.G. (1977) A preliminary evaluation of the 'Scanogram' for predicting the carcass composition of live lambs. Animal Production 24,145-146.

Rhodes, D.N. (1976) Eating quality of meat - the interaction of composition, preference, regulation and marketing. In: Proceedings of 22nd European Meeting of Meat Research Workers, Malmo, Volume I, ppA01-5.

Timon, V.M. (1974) The ealuation of sheep breeds and breeding strategies. In: Proceedings of the working symposium on breed evaluation and crossing experiments with farm animals. Zeist, 1974, pp367-387.

ECONOMIC TECHNIQUES FOR THE ASSESSMENT OF SHEEP SYSTEMS

DR T.J. MAXWELL

Hill Farming Research Organisation, Bush Estate, Penicuik, Midlothian

INTRODUCTION

The technology now available for the improved management of sheep enables increased output almost regardless of the breed, the climate or the vegetation. A major constraint on its application is economic; resources such as land, labour and capital are limited and in general their relative scarcity is reflected in their monetary worth. The possibilities offered by new technology would be infinite if it were not for the fact that its application usually has to be economically viable.

The practical context in which economic assessment is necessary may be considered under three broad headings. These are (i) an assessment of modifications in management and inputs to existing systems, (ii) an assessment of a change in the system of production involving capital investment and (iii) the introduction of a system of production where it did not exist before.

These headings will serve our purposes whether we are concerned with individual farm units or large scale national developments. They are of course by no means definitive and analyses used under one head may appropriately be used under another.

ASSESSMENT OF MODIFICATIONS IN MANAGEMENT AND INPUTS TO EXISTING SYSTEMS

Changes in fertiliser inputs, feed inputs, stock number and land allocated to the system (stocking rate) are the kind of modifications that are considered here. Management changes, for example grazing management, which aim to provide a better distribution of feed and forage in relation to requirements may not cost anything to put into practice but some assessment of their impact in economic terms is desirable. They are also considered.

There are three stages in assessing the impact of change. Firstly, we must assess existing performance; this requires physical performance data, lambs weaned per ewe mated, lamb weaning weights, stocking rate, winter forage requirements, supplementary feed per ewe, etc, and the appropriate prices and costs. We also need comparative data of two kinds; data from previous years for the same enterprise, noting the changes both in the inputs and the outputs, and data which relate to similar enterprises, usually from surveys covering a wide range of management conditions.

The next stage is an estimate of the range of the responses that may be expected from changes in inputs and which of these if any are likely to be both justified and attainable in practice. A decision as to whether changes should be made can then be taken.

This breakdown may appear to be artificial since it is a process which should proceed on a continuous basis but it does provide the context for the particular analysis that is used. We are concerned with the short-term impact of changes in inputs and outputs

In the first instance the gross margin analysis is appropriate for these purposes; it relates to the costs and returns which can be easily identified for an individual enterprise calculated on an annual basis. Gross margin is defined as the difference between gross output and variable costs. Gross output is revenue, i.e. money received from the sale of produce, adjusted for valuation changes less the purchase of livestock and farm produce bought for resale. It represents in the case of sheep, income from the sale of lambs, wool, cast and culled ewes and any production or subsidy grant which directly benefits the enterprise. The main variable costs are feed, seed, fertiliser, sprays, casual labour and contract work.

An example of such an analysis is given in Table 1, which is based on physical data from the Scottish Agricultural College's 'Farm Management Handbook'; the updated prices and costs are my estimates. It refers to a flock of Suffolk x Scottish half-bred ewes, with a Down ram, which lambs from mid December to the end of January and is kept on a cropping farm.

Table 1. Gross Margin Analysis of a flock of low ground breeding ewes - early fat lamb production.

Physical Data (per 100 ewes)		
Rams	3.5	
Lambs sold fat	140	
Ewes: culls	22	
deaths	6	
Wool sold (kg)	300	
Concentrate (kg/ewe)	120	
Summer grazing (ha)	10	
Forage area - Hay (ha)	1.2	
Swedes (ha)	2.0	
Output		£
140 lambs @ £26.4 (22 kg @ £1.20)		3696
22 ewes @ £18		396
300 kg wool @ £1.00		300
		4392
Less: 28 gimmers @ £40		1120
Ram replacement (net)		85
TOTAL		3187
Variable Costs		
Concentrate @ £100 per tonne		1200
Vet., Haulage etc.		170
Forage: Grazing @ £70/ha		700
Hay @ £80/ha		96
Swedes @ £100/ha		200
TOTAL		2366
GROSS MARGIN per 100 ewes		821
GROSS MARGIN per forage hectare		62

Analysing current financial performance on this basis and relating it to past performance is often made difficult by changing prices and costs, particularly if inflation rates are running in excess of 5.0%. It is therefore important to have adequate physical data so that performance may also be examined without attaching financial parameters. It is possible however to adjust for inflation so that the 'real' position over time may be assed.

For example, Table 2 gives some indication of the performance of lowland and upland flocks in Britain, recorded by the Meat and Livestock Commission from 1970-76 with the gross margins adjusted for inflatin.

Table 2 Gross Margins for MLC recorded lowland and upland flocks adjusted for inflation (£)

	1970		1972		1973		1974		1975		1976
Lowland Flocks											
Gross Margin per Ewe	6.0		9.5		10.7		9.9		11.1		17.1
Inflation (%)		8		9		16		24		15	
Recalculated G.M.(1976)	12.8		17.1		17.7		14.1		12.8		17.1
Upland Flocks											
Gross Margin per Ewe	5.4		9.4		10.4		10.1		11.3		19.7
Inflation (%)		8		9		16		24		15	
Recalculated G.M.(1976)	10.5		16.9		17.2		14.4		13.0		19.7

Though the 1976 results show a marked improvement on the 1970 results it is clear that over the six year period they have varied widely. Can we assume real improvement? Examination of both physical and financial data is necessary to confirm trends. Much of the increase in the gross margins in these data is derived from an increase of between 6 and 8% in the number of lambs reared per ewe, combined with a real increase in the value of lamb relative to the cost of inputs. In 1974 and 1975, there was little evidence of improvement in real terms; the cost of inputs, particularly feed inputs, was increasing faster than product prices. Thus the interpretation of financial data has to be done carefully with adequate reference to physical data and rarely taken at its "face value".

Having made a thorough assessment of existing performance against previous performance and survey data from organisations such as the Meat and Livestock Commission, how might improvements be brought about?

Two important factors which influence gross margin per ewe are the number of lambs reared per ewe and the feed costs. These are two aspects which should receive attention. Table 3 shows the influence on the gross margin, of the Table 1 data, of increasing the number of lambs reared while other items remain the same.

Table 3 Variations in Gross Margin with nos. lambs sold (reference to data Table 1).

Lambs sold per 100 ewes	Gross Margin (£) per 100 ewes	Gross Margin (£) per forage hectare
120	293	22
140	821	62
160	1349	102

A 20% increase in lambs weaned can increase gross margin by about 60%; however some of this increase may only be possible through further inputs of feed. Even with a 20% increase in feed costs an increase in gross margin of 35% may still be possible. Obviously this result depends very much on the relative values of lamb and feed and it is not possible to generalise, but the significance of number of lambs produced per ewe has long been recognised as being important in obtaining economic viability in sheep systems. The impact of premating nutrition and the condition and weight of the ewe are also of considerable importance. Appendix 1 gives data drawn from M.L.C. records showing the significant effect of condition of the ewe at mating, for some breeds of ewe, on the number of lambs born.

Land is invariably limiting; the gross margin per unit area is therefore important. The improved use of pastoral resources by increasing stocking rate has also long been recognised as a means of improving output and gross margin per unit area. Table 4 shows that for the early fat lamb flock (ref. Table 1) an increase of about £10 per forage hectare can be achieved for a 20% increase in stocking rate. To achieve this an increase in fertiliser N may be necessary and an appropriate reduction in the financial response will have to be made.

Table 4 Variation in Gross Margin per forage hectare with stocking rate at summer grass

Stocking Rate per ha	8	10	12
Gross Margin per forage ha	52	62	71

Looking at M.L.C. data for gross margins from lowland and upland flocks one is impressed by the improvement in stocking rates (Table 5) in the early years of recording. Perhaps that in itself is significant. This is particularly so in upland flocks.

Table 5 Gross Margins per hectare, MLC recorded lowland and upland flocks (£)

	1970	1972	1973	1974	1975	1976
Lowland						
Gross Margin per ha	52.9	88.9	99.3	95.5	111.7	177.7
Stocking Rate	9.0	9.5	9.5	9.8	10.3	10.4
Upland						
Gross Margin per ha	31.1	78.1	83.2	86.1	93.5	159.0
Stocking Rate	5.8	8.3	8.5	8.5	8.3	8.1

For lowland flocks the variation in stocking rate with given levels of N usage is given in Figure 1; though considerable, it is possible to show a relationship of diminishing return between stocking rate and fertiliser N usage. What matters however is whether output of lamb per unit area increases with increasing N usage. If by increasing stocking rates individual animal performance declines, there will be a point above which further increases will be undesirable. This of course merely draws attention to the economic importance of the interrelationship between individual animal performance, flock output and stocking rate. Figure 2 shows a relationship between stocking rate and gross margin per ha for lowland and upland sheep. It has been noted (Arnold & Bennett 1975, Edelsten 1977) that at high stocking rates profitability and output tend to vary most and therefore those farmers who wish to avoid risk will opt for lower stocking rates.

It is appropriate to draw attention to the dependence of output from intensive grassland systems on fertiliser N. The cost of fertiliser N, relative to lamb prices, has in recent years been favourable but as world demand for N increases, and the real manufacturing costs of N in the UK are accommodated, its retail cost will almost certainly increase. Other means of ensuring high output per unit area will have to be sought.

In assessing the need for change one must be aware of changes that are likely to occur in the cost of inputs and the prices of outputs. This is a notoriously difficult area but one which has important consequences on the long term viability of an enterprise. The supply of feed and fertiliser resources will be much influenced by world trade. Increasingly government agencies attempt to make forecasts so that likely changes in supplies and therefore costs can be taken into account in planning. Similarly information on the supply and demand for food products is monitored closely and, as a consequence of this information, marketing policy can be more soundly based.

Assessment of probable cost and price changes, a gross margin analysis of current performance and comparison with other relevant performance data will provide a basis for making a decision for or against change. For the practising agriculturalist improved profitability will provide the main incentive to change.

In many countries agriculture is 'protected' to a greater or lesser extent from the real economic world for a variety of reasons, not least in an effort to provide cheap food. The use of subsidies, grants or other adjustments can in some circumstances so protect the producer that the incentive for technological change is reduced or lost. Perhaps this emphasises the need to consider different ways in which grants and subsidies might be more successfully applied to achieve improved productivity.

The development of computer simulation models to examine the effects of change in inputs and to predict changes in output is a natural extension of many of the features so far dealt with. Edelsten & Newton (1977) have constructed just such a model of an intensive system of lamb production from temporary grassland. The model is a simplification of computer models

Figure 1

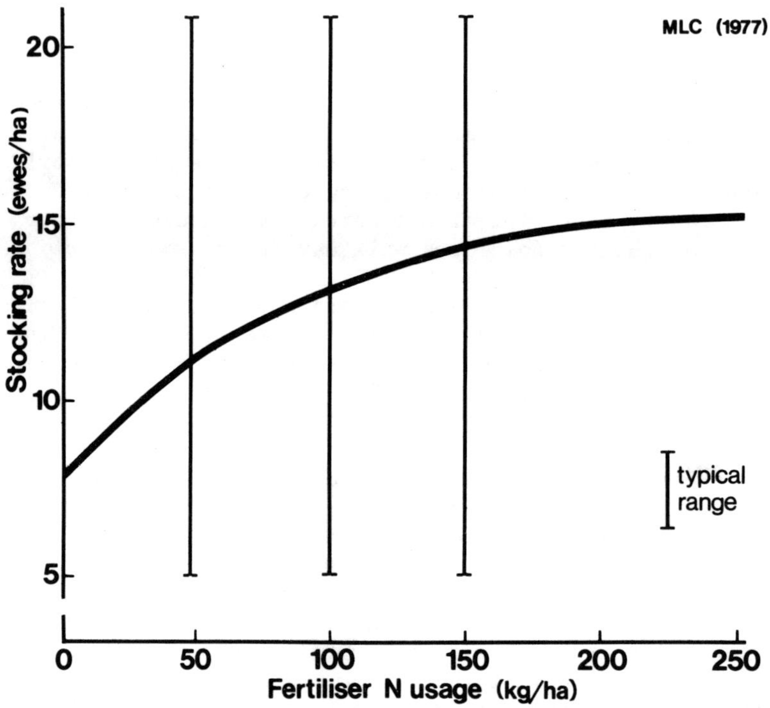
MLC (1977)
Stocking rate (ewes/ha)
20
15
10
5
typical range
0
50
100
150
200
250
Fertiliser N usage (kg/ha)

used for research and has been adapted for interactive use as a management game. Its usefulness in assessing the impact of a technological change is valuable though its application to specific farm situations may be less relevant. Calculation models of the type advocated by Charlton & Street (1975), in which basic performance data may be changed according to experience and not wholly by 'built in' biological relationshiops, may be a more useful approach at the individual farm level.

Some aspects of assessment cannot be covered fully by a simple gross margin analysis. For example labour, except casual labour specific to the sheep enterprise, is regarded as a fixed cost and therefore excluded. The main reason for this is because a permanent allocation of labour to the sheep enterprise is seldom accepted practice, except in the hill and upland areas and extensive pastoral systems. Even in the hills and uplands, production systems have remained economically viable because to a considerable extent labour inputs have been reduced; in Scotland labour input to hill sheep farms has been reduced by 36% since 1947. The efficiency of labour use (man hours per ewe) has improved. Labour costs have however increased very substantially indeed and, though other costs and prices have also changed, output per unit (£) of labour input has changed remarkably little.

Competition for resources of labour, land and working capital between enterprises on the farm is an important feature of farm management. Deployment of these resources has to be done partly in relation to the contribution that a particular enterprise makes to total farm output. It is therefore reasonable to compare the gross margins of different enterprises on the same land resource. Since the area of land is also a major limitation on output, the gross margin per unit area derived from the various enterprises is the most appropriate measure to use for these purposes.

Table 6 is an example of the kind of comparisons that might be made between a beef suckler herd and an upland ewe flock.

Table 6 Comparisons of average gross margins per hectare for an upland beef suckler herd and a ewe flock (£). (MLC Sheep Facts)

	SR/ha	1972	1973	1974	1975	1976	Average
Suckler Herd	1.5	156	172	110	162	230	166
Ewe Flock	8.8	79	97	88	105	158	105

Generally better gross margins were obtained from cattle than sheep even in years when calf prices were low (1974), and over the period returns from cattle were some 37% greater. A comparison of this kind however is not complete unless some assessment of the amount of working capital, the money in breeding stock and feed for the year for each enterprise, is made.

Taking the 1976 data for the UK, Table 7 gives an example of the information that would be helpful in making a critical assessment of how much land, labour and capital might be appropriately deployed to each enterprise.

Table 7 Alternative investment in working capital for livestock enterprises per hectare (£) (MLC 1976)

Item	Dairy Cows	Beef Cows	18-Mth. Beef	Breeding Ewes
Stock Purchase	590	490	119	311
Herd/Flock Replacement	33	30	-	27
Feed and Forage	240	102	321	82
Miscellaneous	33	20	40	15
Total	897	642	480	435
Gross Margin	375	216	269	179
GM as percentage of Working Capital	41.8	33.6	56.0	41.1

Short term capital requirements are of particular importance where capital has to be borrowed and where overdrafts are involved. In these circumstances it is usual to minimise capital outlay and adopt systems that either require little capital or give the best return. Capital is frequently the most limiting resource.

In developing systems of production which involve significant changes in the financial aspects of farm business management, a systematic analysis is essential. Other methods, such as linear programming and Monte Carlo planning, are all useful optimising techniques. They all operate with clearly defined constraints though all the bio-economic relationships that are used are assumed to be linear. It is worth noting that there are few biological relationships that are linear. Non-linear programming methods are available but the mathematical techniques used for the solution of these problems are complex; in some non-linear programming problems there is no technique which can assure a single best solution.

ASSESSMENT OF A CHANGE IN A SYSTEM OF PRODUCTION INVOLVING CAPITAL INVESTMENT

To bring about substantial changes in the output from sheep often requires radical changes in management and investment. These include housing, programmes of land improvement and fencing for controlled grazing and go far beyond the usefulness of a simple gross margin analysis.

During recent years improved systems of hill sheep production have been under test at the Hill Farming Research Organisation. Much of the development work has been concerned with the synthesis and testing of biological knowledge (Eadie & Maxwell 1975) but it was recognised that this could not be done in the absence of an economic framework. The relevance of the systems under development depended on their economic viability and methods of economic analysis necessary to provide an economic framework have also been developed. These have a more general application but hill sheep production provides the background against which they will be described and demonstrated.

It is clear from the development work so far done that substantially increased output from hill sheep farms will ultimately depend upon some form of capital investment in land improvement, stock increases, fencing and possibly road building.

In making investments of this kind it is necessary to know the output objectives that are required to break even. To do this a break-even budget technique, devised by Harkins (1968) and further developed by us, has been found to be helpful.

The break-even equation on which the method is based is written:

$$E_1 . GM_1 = E_2 . GM_2 - Y$$

where E_1 is the number of stock, GM_1 the gross margin per ewe before investment, E_2 the number of stock and GM_2 the gross margin after capital investment. Y is the amount required to service and repay the capital, i.e. the annual capital charge.

Gross margin is used because it reflects the increase or decrease in gross output from a particular enterprise net of the increase or decrease in variable costs. Thus, the technical changes in performance of the enterprise are reflected in economic terms. Changes in fixed costs are not included at this point since they have an impact on other enterprises on the farm and because they tend to vary greatly from hill farm to hill farm. E_1, GM_1 and E_2 are known. The servicing and repayment charges (Y) must be calculated from an amortisation formula.

To calculate Y for a given capital investment (c), the rate of interest (i) and the period of repayment (n) must be stated thus:

$$Y = C \frac{1}{1 - (1 + i)} - n$$

Break-even equation becomes $E_1 = E_2 GM_2 - (E_2 - E_1) V I - Y$

I = annual interest rate on capital invested in stock
V = value per head of extra stock

The break-even equation must also be adjusted to take account of the interest charges on the capital invested in extra stock. This value must be deducted from the gross margin after capital investment (GM_2).

It is now possible to calculate the gross margin per ewe that would be required after capital investment for any given increase in stock numbers. It is rare however that the examination is confined to one particular level of stocking rate increase. It is much more likely that output increases over a range of stocking rate increases require to be examined. The calculations are tedious and a computer programme has been written, the output of which can be graphed as in Figure 3.

The series of curves describes the minimum gross margin per ewe, over

Figure 2

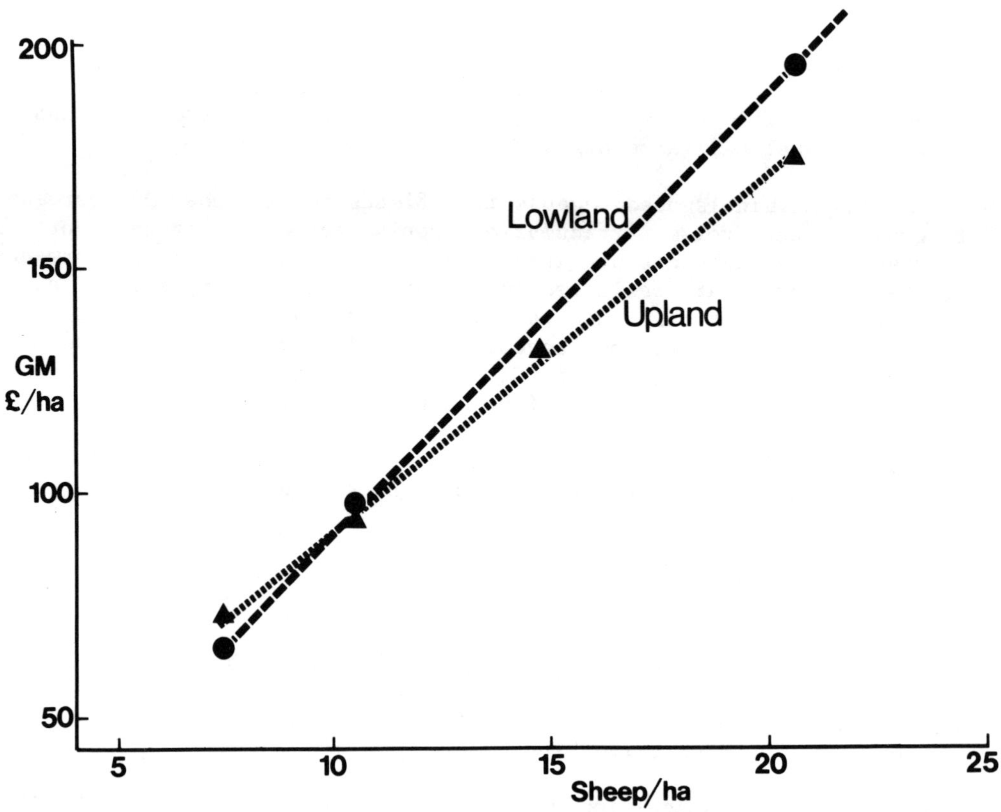

Figure 3

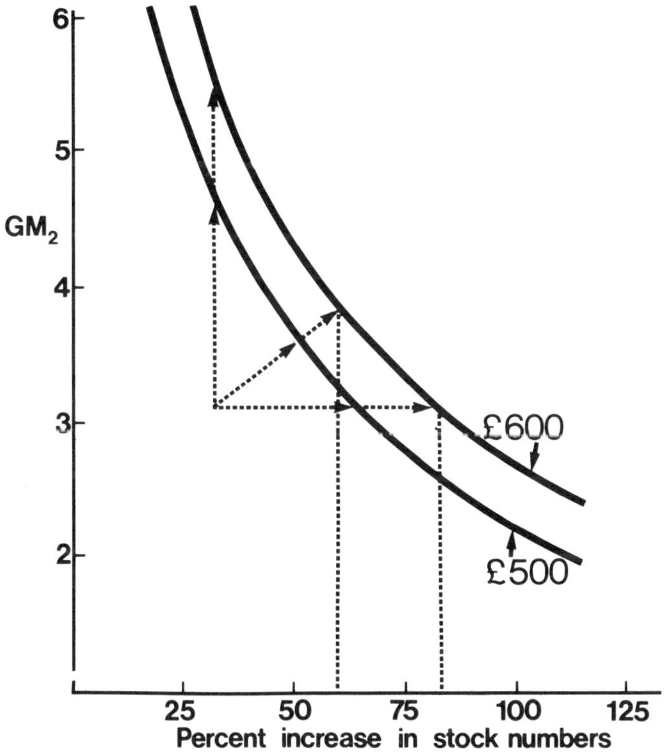

"Break even" lines:

Capital investment = £500 & £600
Original stock no. = 400
Original GM = 3·0

indicate the extent of indebtedness the project will incur. A bank manager will also find it of interest if he is being asked to provide overdraft facilities.

An examination of the cash flows arising from an investment requires a considerable number of calculations, particularly where comparisons between alternative projects are required. In the context of hill sheep systems, these calculations involve different rates of stock number increase, different rates of improvement in animal performance and different timing of capital investment.

A computer programme which can deal with these alternatives has therefore been written (Sibbald & Maxwell 1975). The programme has been designed for use on the multi-access system at the Edinburgh Regional Computing Centre using an ICL 4/75 computer. Access is by teletype, which allows an operator to judge his results, change the input data and "instantly" compute a further set of results.

For convenience of description the process of calculation has been broken down into four sections. The function of each section, the input data necessary and the way they are linked is shown in Figure 4.

The first section deals with physical outputs and stock number adjustments. This part of the programme is so designed that, given the initial number of stock, their age distribution, loss rates and weaning percentages in each year of the project, it will automatically calculate the numbers of ewes entering the next year. If stock numbers are to be increased to an annual target, the programme carries out the necessary calculations. The actual rate of stock increase will depend on a number of factors such as the rate of pasture improvement, current flock weaning percentages and weaning weights.

Stock numbers can be increased by obtaining ewe lambs or by buying in ewe hoggs. In the latter case stock number increases for each year must be specified, although buying in extra stock may bring the hazard of disease.

Stock purchases made early in the life of a project will increase the level of capital investment and may therefore exacerbate negative cash flows; and the effect of purchases on bank overdraft must be considered but early increases in stock numbers will also allow an early potential increase in output from the flock as a whole.

The components of net annual cash flow, as used in liquidity assessment of capital investment, are given in Figure 5. The solid lines represent additions and the broken lines deductions. There are two modifications in calculating net annual cash flow for the purpose of determining net present value (NPV) and internal rate of return (IRR); there are no adjustments to NACF due to interest payments or receipts but a valuation adjustment is made in the last year of the project.

An example of the output from the programme is given in Appendix II.

Figure 4

COMPUTATION

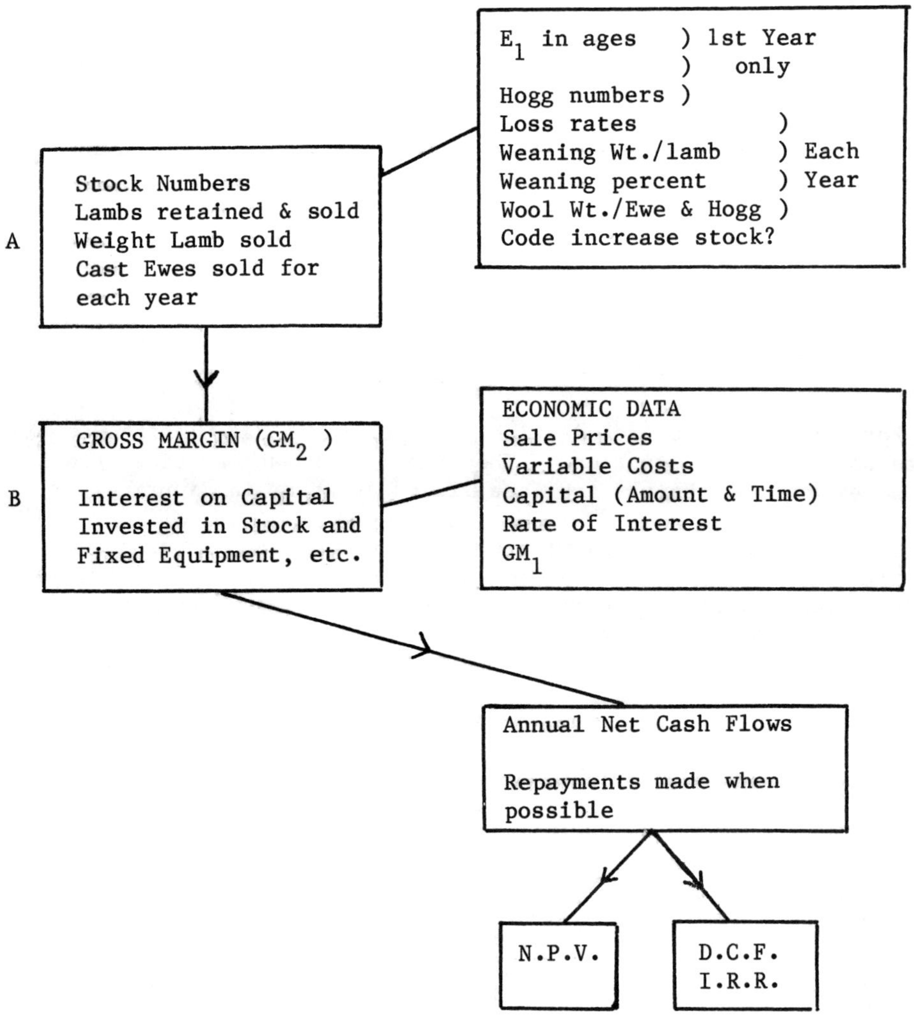

Figure 5

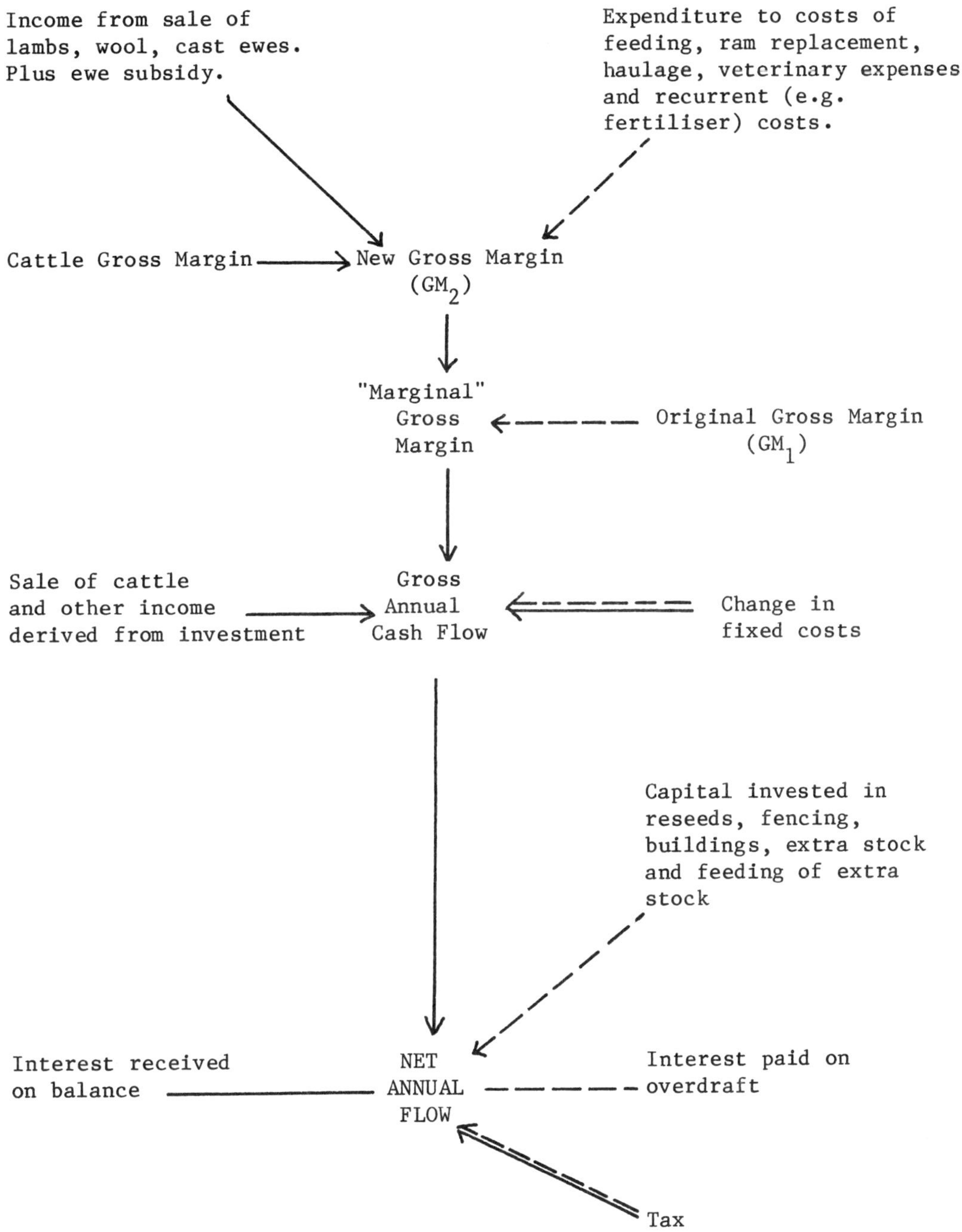
Income from sale of lambs, wool, cast ewes. Plus ewe subsidy.
Expenditure to costs of feeding, ram replacement, haulage, veterinary expenses and recurrent (e.g. fertiliser) costs.
Cattle Gross Margin
New Gross Margin (GM_2)
"Marginal" Gross Margin
Original Gross Margin (GM_1)
Sale of cattle and other income derived from investment
Gross Annual Cash Flow
Change in fixed costs
Capital invested in reseeds, fencing, buildings, extra stock and feeding of extra stock
Interest received on balance
NET ANNUAL FLOW
Interest paid on overdraft
Tax

Using the type of economic analysis described it has been possible to form some general conclusions about the investment of capital in hill sheep farming. These conclusions are made on the basis of our existing biological knowledge and experience gained at HFRO and elsewhere.

It is clear that small amounts of capital invested in land improvement are generally more likely to give rise to cash flows which require only short periods of negative balance on bank overdraft. Large amounts of capital, particularly if invested at one point in time, can produce very difficult cash flow problems. High levels of capital investment must be spread over a number of years and made at times which are closely related to expected increases in animal output.

Net farm incomes can therefore be improved by initially low levels of capital investment. The investment after a few years will give rise to increasing end year balances; further investment can then be made from cash flow.

The return on marginal capital is also important, particularly if capital has to be balanced over long periods. Large businesses such as large farming estates will always have a number of alternative uses for capital and the use of NPV and IRR in assessing and ranking these is obviously necessary.

THE INTRODUCTION OF A SYSTEM OF PRODUCTION WHERE IT DID NOT EXIST BEFORE

The introduction of new enterprises and projects involves the range of economic analyses already referred to.

Within the farm, the introduction of a project which involves the reallocation of land resources is likely to lead to marked changes in output per unit area and output per unit of working capital involved. Justification for change must be well founded on the basis of both of these parameters. Changes in labour deployment and in the seasonal variation of labour requirements must also be closely examined. A new project, which starts on the assumption that no extra labour will be required, but eventually, because of an inadequate earlier analysis of labour requirement, has to support an input of labour, will provide a very different economic return to that originally calculated.

Projects which require long term capital investment will require analysis similar to that carried out under the system described for capital investment. Alternative investment possibilities may exist and comparisons between the NPV's obtained will provide a basis for decision making.

The introduction of a new system of production or a development project may require more fundamental aspects of economic assessment to be examined; this is particularly true of national development schemes of the kind supported by FAO and the World Bank. The question is the same but often it has to be answered in a wider economic context. The major difference that the wider economic analysis requires is in the fixing of prices.

In analysis of the most appropriate use of scarce national resources, a price other than the market price may be a better indication of the value of a commodity or service. It is often called a "shadow price". The need for shadow prices arises because market prices do not always reflect economic values well; they would only do so in perfect market conditions and markets are imperfect because of government intervention, price controls, monopolies, subsidies, and so on.

These considerations are beyond the scope of this paper but are raised to provide an introduction to some of the complex problems often associated with the application of new technologies in developing economies where competition for scarce national resources is severe. The subject is dealt with in a comprehensive but pragmatic manner by Gittinger in his book "The Economic Analysis of Agricultural Projects".

Having overcome the major difficulties associated with valuing goods and services, the subsequent analysis is carried out as described earlier, the choice of project being largely determined by its NPV.

Appendix 1

```
*TITLE
HFRO EXAMPLE - WITH 1 INVESTMENT
*YEARS
 NUMBER       10
 START      1978
*EWE DATA
 START       400
 AGES         88   84   80   76   72
 INCREASE      0
 STOCK NOS
 LOSS %        5.0  4.5  4.0  3.5
 FLEECE WT    1.59
 WOOL PRICE   0.90
 FEED COST    0.70 1.60
 VET COST     0.50
 HAULAGE      0.12
 RAM FEED     0.15
 RAM REPL     0.90
 SUBSIDY
 CAST VALUE   14.0
 VALUE        18.0 17.0 16.0 15.0 14.0
*LAMB DATA
 WEANING %    75.0 83.0 88.0 92.0
 WEIGHT       27.2
 PRICE        0.52
 VALUE       13.00
*HOGG DATA
 START NO.    92
 LOSS %        5.0  4.5
 FLEECE WT    2.04
 FEED I/W     5.50
 VALUE        16.5
*CATTLE DATA
 NUMBER
 PRICE/HD      0
 VALUE/HD      0
*CASH DATA
 FIX COSTS
 BALANCE
 GM1        3115
 INT RATES    15.0 8.00
 TAX RATE     0.33
 CATTLE GM    0
 COSTS FIX
 EQIP/SEED 1625
 BLDG/FNCE  300
 OTHER INC
 RECR COST     0    0    0    0  455    0    0    0    0  455
 INTEREST
 EX INCOME
*PRINT
 INDEX         6R
*FINISH

COMMAND:QUIT
CONSOLE DISCONNECTED  22/08/79 12.20.15
```

Appendix 2

HFRO EXAMPLE - WITH 1 INVESTMENT
Run - 12.08.18 hrs. on 22/08/79
Interest rates (%) - borrowing = 15.0 lending = 8.0
Selected index for output = 6

Total GM1 = $ 3115. GM1/100 = $ 779.

Year	1978	1979	1980	1981	1982	1983	1984	1985	1986	1987
Ewe number	400	399	403	403	402	401	401	401	401	401
Hogg	92	94	88	87	89	90	93	88	87	89
Weaning %	75	83	88	92	92	92	92	92	92	92
Oth income	0	0	0	0	0	0	0	0	0	0
Recur csts	0	0	0	0	455	0	0	0	0	455
Total GM2	3115	3278	3649	3881	3421	3817	3919	3881	3843	3394
GM2 /100	779	821	905	963	851	952	977	968	958	846
Chg Fx-cst	0	0	0	0	0	0	0	0	0	0
Ext income	0	0	0	0	0	0	0	0	0	0
Gross ACF	0	163	534	766	306	702	804	766	728	279
Captl stck	0	0	0	0	0	0	0	0	0	0
EQ/BL	1925	0	0	0	0	0	0	0	0	0
Ex cst	0	0	0	0	0	0	0	0	0	0
Interest	-288	-211	-159	-85	-84	-1	45	90	136	151
Tax (D+I)	0	641	25	-113	-214	-63	-221	-255	-242	-230
Net ACF	-2213	592	400	567	6	637	628	601	621	200
Balance	-2213	-1620	-1220	-652	-646	-8	619	1221	1842	2042

Appendix 2

Data used to calculate net ACF for use with NPV/IRR calculations

Valn chnge	0	0	0	0	0	0	0	0	0	-37
Tax (dep)	0	546	-43	-166	-242	-90	-221	-255	-242	-230
Net ACF	0	708	490	600	62	611	582	510	484	12

Discounted value of all capital invested = $ 1781
I.R.R. yield = $ 1781
I.R.R. rate of interest = 18.05%

Net present value = $ 221
Rate of interest = 15.00%

Command:Quit
Console disconnected 22/08/79 12.16.18

REFERENCES

Arnold, G.W.;Bennet, D. (1975) The problem of finding the optimum solution. In 'Study of agricultural systems' (Ed) Dalton, G.E. London: Applied Science Publishers.

Charlton, P.J.;Street, P.R. (1975) The practical application of bio-economic models. In 'Study of agricultural systems' (Ed) Dalton, G.E. London: Applied Science Publishers.

Eadie, J.;Maxwell, T.J. (1975) Systems research in hill sheep farming. In 'Study of agricultural systems' (Ed) Dalton, G.E. London: Applied Science Publishers.

Edelsten, P.R. (1977) Bio-economic models of sheep systems. PhD Thesis, University of Reading.

Edelsten, P.R.;Newton, J.E. (1977) A simulation model of lowland sheep systems. Agricultural Systems 2,17-32.

Gittinger, J.P. (1972) The analysis of agricultural projects. Baltimore and London: The John Hopkins University Press.

Harkins, J. (1968) Assessing new capital investment on hill sheep farms. Scottish Journal of Agriculture 47,196.

Hawkins, C.J.;Pearce, D.W. (1971) Capital investment appraisal. London: Macmillan.

Meat and Livestock Commission (1978) Sheep Improvement Services. Data Summaries on Upland and Lowland Sheep Production. Data Sheet 6.

Meat and Livestock Commission. Sheep Facts. (In preparation.)

Sibbald, A.R.;Maxwell, T.J. (1975) Program for the appraisal of investments in hill sheep production systems. Internal Report, Hill Farming Research Organisation.

INTENSIVE SYSTEMS ON LOWLAND GRASS

J.E. NEWTON

Grassland Research Institute, Hurley, Maidenhead, Berks

INTRODUCTION

In 1972 lowland grass supplied 56% of the lamb carcasses produced in Britain (MLC - Sheep Improvement 1972). Our research at Hurley in the past 5 years has been aimed at increasing the lamb carcass output per ha of lowland grass.

EWE SIZE AND REPRODUCTIVE PERFORMANCE

Sheep eat in proportion to their weight and so it is logical to consider the productivity of small ewes kept at double the stocking rate of large ewes. Stocking rate is the number of ewes kept per ha per annum. Should insufficient conservation (silage) have been made on the sheep area when assessing productivity, a calculation has been made of the shortfall and the necessary adjustment made to the stocking rate.

The similarity in biological output between the large and small ewes is shown in Table 1. Had the small ewe grown a slightly more valuable fleece, the financial output would have been almost identical. Similar numbers of lambs were sold per ha (17.5 x 1.99 = 34.8 Masham, 35 x 0.91 = 31.8 Welsh) but the Suffolk x Welsh lambs were sold at a slightly lighter weight than the Suffolk x Masham (carcass weight Masham twins 14.5kg, Welsh singles 14.2kg).

Table 1 Lamb production from temporary grassland (1975)

	Large ewe	Small ewe
Ewe size	70kg	32kg
Ewe breed	Masham	Welsh Mountain
Date of lambing	March	March
Selling %	199	91
Stocking rate (per ha)	17.5	35
Sward	PRG S24	PRG S24
Nitrogen kg/ha	250	250
Grazing management	8 paddocks with forward creep	4 paddocks
Carcass output (kg meat per ha)	459	448
G.M. per ewe	£10.22	£4.66
G.M. per ha	£179	£165

(from Newton et al 1977)

THE EFFECT OF GRASS SPECIES

The amount of grass grown in a field is partly dependent on soil type, rainfall and fertilizer level but it can also be affected by grass species and utilization. An experiment was carried out to test whether Italian ryegrass RvP would increase the productivity and profitability of our lowland sheep system compared with perennial ryegrass S24. Italian ryegrass RvP was known to have as high a dry matter yield and to be eaten more readily than S24; it should therefore produce more carcass meat per ha. The results of the experiment are shown in Table 2.

Table 2 RvP v. S24 - carcass output kg meat/ha

	L (14)	M (17)	H (20)
		1973	
RvP	440	400	419
S24	440	479	452
		1974	
RvP	357	307	281
S24	400	441	445

In the first year the carcass output was similar on the low stocking rate for the two grass species but higher on the perennial ryegrass at the medium and high stocking rates. By the second year the advantage in favour of the perennial ryegrass had increased. Table 3 shows the amount of herbage grown per treatment per year. In the first year, S24 yielded 13% more dry matter than RvP; by the second year the advantage had increased to 48%.

Table 3 Grass grown (t DM/ha)

	L (14)	M (17)	H (20)	Mean	
		1973			
RvP	13.5	14.6	11.2	13.1)	
S24	13.7	16.2	14.4	14.8)	+ 13%
		1974			
RvP	7.4	8.0	7.1	7.5)	
S24	10.3	12.3	10.8	11.1)	48%

RvP did not withstand the intensive grazing pressure as well as S24, and by the second year there were less than 25% of the original plants of RvP left at the high stocking rate. In fact the intakes of grass by the ewes were higher on the RvP than on S24 but this was never translated into faster lamb growth rates. If grazing pressure is going to be high with frequent severe defoliation, the grass must be suited to intensive conditions.

TYPE OF GRAZING MANAGEMENT

Paddock grazing with forward creep for the lambs was found to increase the amount of meat produced at the three stocking rates (Table 4). Forward creep for the lambs is not likely to show any advantage at low or high stocking rates, but has been consistently better at middle range stocking rates, particularly where there is a high proportion of twins and triplets, which are much more dependent on grass than singles. Forward creep grazing enables the lambs to have the pick of the grass and the paddock system helps eliminate waste by encouraging even defoliation. It also promotes regular application of nitrogen just after defoliation.

Table 4 Carcass output kg meat/ha

	L (14)	M (17)	H (20)
RG	427	451	453
SS	380	379	419

STOCKING RATE

There are two pertinent questions; what is the optimal stocking rate and is there any value in an intensive situation in being slightly overstocked? The stocking rate is dependent on the amount of grass grown, the size of the ewe and the number of lambs reared at pasture. The generalised relationship between optimal stocking rate for Mashams (65kg ewes) with 1.8 lambs and amount of grass grown is shown in Figure 1, as worked out by a computer model.

Whether a sheep farmer aims at self-sufficiency depends on his attitude to buying in hay and barley concentrates. We have had experience of high stocking rates where the ewes and lambs have had to be supplemented during the grazing season and where not enough conservation has been made for the winter months. These have been more profitable than lower stocking rates but the financial gain depends on the price structure in the individual year, particularly for hay, barley and store lambs. If however the high stocking rate, and this can be very high locally with the 8 paddock system (160 ewes/ha), affects the grass yield adversely, subsequent sheep production will be impaired.

EFFECT OF TIME OF LAMBING

Time of lambing has a major effect on a sheep grazing system and is also likely to affect reproduction. Table 5 shows the impact of time of mating on reproductive performance and also the effect of body condition at mating on lambing percentage.

Figure 1

FIG. 1 OPTIMAL STOCKING RATE v AMOUNT OF GRASS GROWN

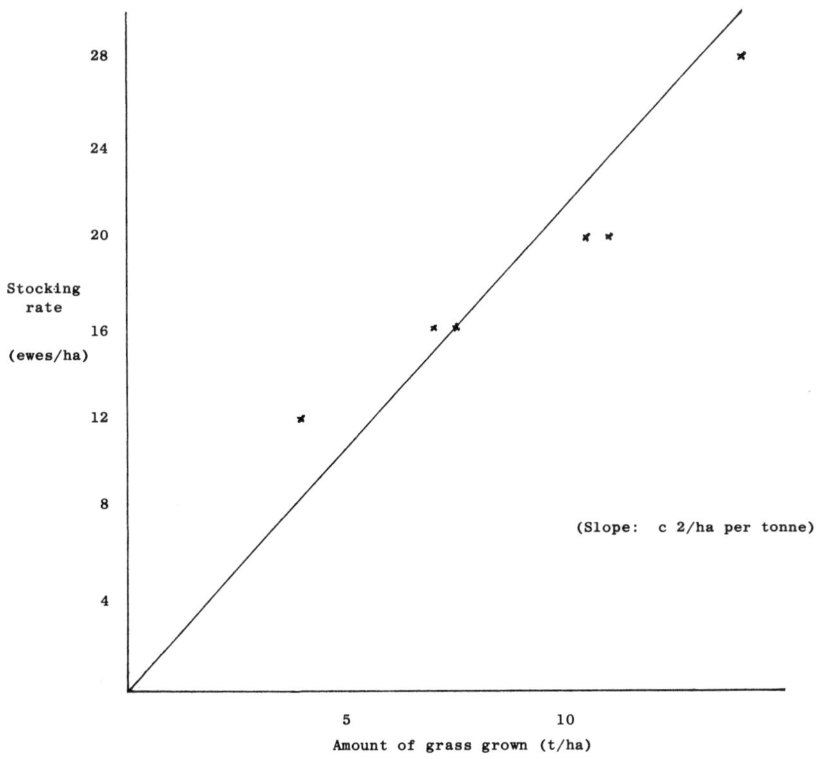

Table 5 The effect of time of mating and ewe body condition on reproductive performance

	Mean Litter Size		
	Early (Sept)	Mid (Nov)	Late (Dec)
		1976	
Fat (BC 4)	2.1	2.0	1.9
Thin (BC 2)	1.6	1.5	1.8
		1977	
Fat (BC 4)	2.2	2.3	1.8
Thin (BC 2)	1.9	1.8	1.6

Mean litter size = $\frac{\text{number of lambs born}}{\text{number of ewes lambing}}$

For ewes in good body condition the late mating (December) appeared to be the least fertile. The effect of improving body conditioin from a score of 2 to 4 (about 12kg liveweight) was most marked at mid mating (October-November) and least effective at the late mating. Because no measurements were taken of ovulation rate at the time of mating it is not clear whether ovulation rate was similar in October and December matings and embryonic loss was much higher at the later mating, or whether the ovulation rate was dissimilar. The same problem of interpretation exists for the difference between the fat and thin ewes at the two times of mating. Why is this difference much larger for the mid mating? Is it ovulation rate or embryonic loss which is causing the difference?

Time of lambing also affected conservation policy, the amount of concentrates fed, lamb growth rate and lamb price; we do not know yet whether it has affected grass growth and therefore yield differentially. The early lambing system requires more silage because the ewes have to be fed for a cnsiderable time at pasture after lambing. The grazing pressure (as opposed to the stocking rate) is much higher early on and consequently silage is made in July-August when the lambs have gone, rather than in May. The benefits of the early system are a high and predictable lamb growth rate (while the grass is highly digestible), high lamb prices and dry ewes with no pressing nutritional requirement during the difficult months of July and August. The mid-lambing time has less extreme grass surpluses and shortages, because the grazing pressure is more closely related to grass growth, but does have the problem of finishing lambs in July-August at a time when lamb price is usually at its lowest. The late lambing system has no concentrate input at all, because the winter months coincide with the first 3 months of pregnancy. Grass growth is maximal and digestibility is high during the more critical last 2 months of pregnancy. The birth weights of twin lambs were as high for this late, all grass system, as for the more conventional nutritional regimes of silage and concentrates of the two earlier times of lambing. More than enough silage is made for the winter because at a given stocking rate (17 ewes/ha) the nutritional demand and therefore the grazing pressure is lowest in April and May on this late lambing treatment. The crunch comes in August and September when animal nutritional demand is high and grass growth is not. Few lambs are sent fat off grass at this late lambing time but store prices have been high recently and these lambs are well suited to graze forage crops such as stubble turnips and rape during the autumn. A forage crop is easily slotted into a cereal rotation containing winter and spring barley. Gross margin returns for the 3 systems are shown in Table 6.

Table 6 The effect of time of lambing on production (1977)

Time of lambing	Early	Mid	Late
Carcass output (kg/ha)	490	493	430
Gross margin (£/ha)	322	336	370

Output from medium stocking rate (16 ewes/ha)

The amount of carcass produced per ha depends on the amount of grass grown and its utilization. Replication either in space or time overcomes some of the problems of variation in land potential, and a measurement of the amount of grass grown helps the interpretation of stocking rate treatments (Figure 2). More important, from the point of view of grazing, is whether severe defoliation at different times of the year (March/April in the early, August/September in the late) affects the grass sward adversely and thus depresses subsequent yield of grass. This would be more important for a long term ley than for a short one.

The importance of the ewe replacement cost in the gross margin highlights the need to keep ewes for as long as possible, providing physical performance does not decline too severely with age. The effect of age on reproduction, body size, lamb birth weight and lamb growth rate, which is an indirect measure of ewe milk yield, is shown in Figure 3. Reproduction and lamb birth weight have increased with age (up to 7 so far), ewe weight and milk yield appear to have declined slightly. There was a significant increase in perinatal lamb mortality (Table 7) between ewes aged 3 to 7, and overall lamb mortality has increased slightly with ewe age (3 to 6). Thus although mean litter size has increased with age, selling percentage has remained constant. There was an insufficient number of seven year old ewes to constitute a trend.

Table 7 The effect of age on mean litter size and lamb mortality

Ewe age (years)	Mean litter number	% Perinatal mortality	% lamb mortality (7 days - sale)	Selling %
2	1.90	5	5	171
3	2.12	2	3	201
4	2.12	3	3	199
5	2.22	4	8	195
6	2.25	5	6	200
7	2.37	6	2	218

Figure 2

FIG. 2 GRASS GROWTH AND GRAZING PRESSURE

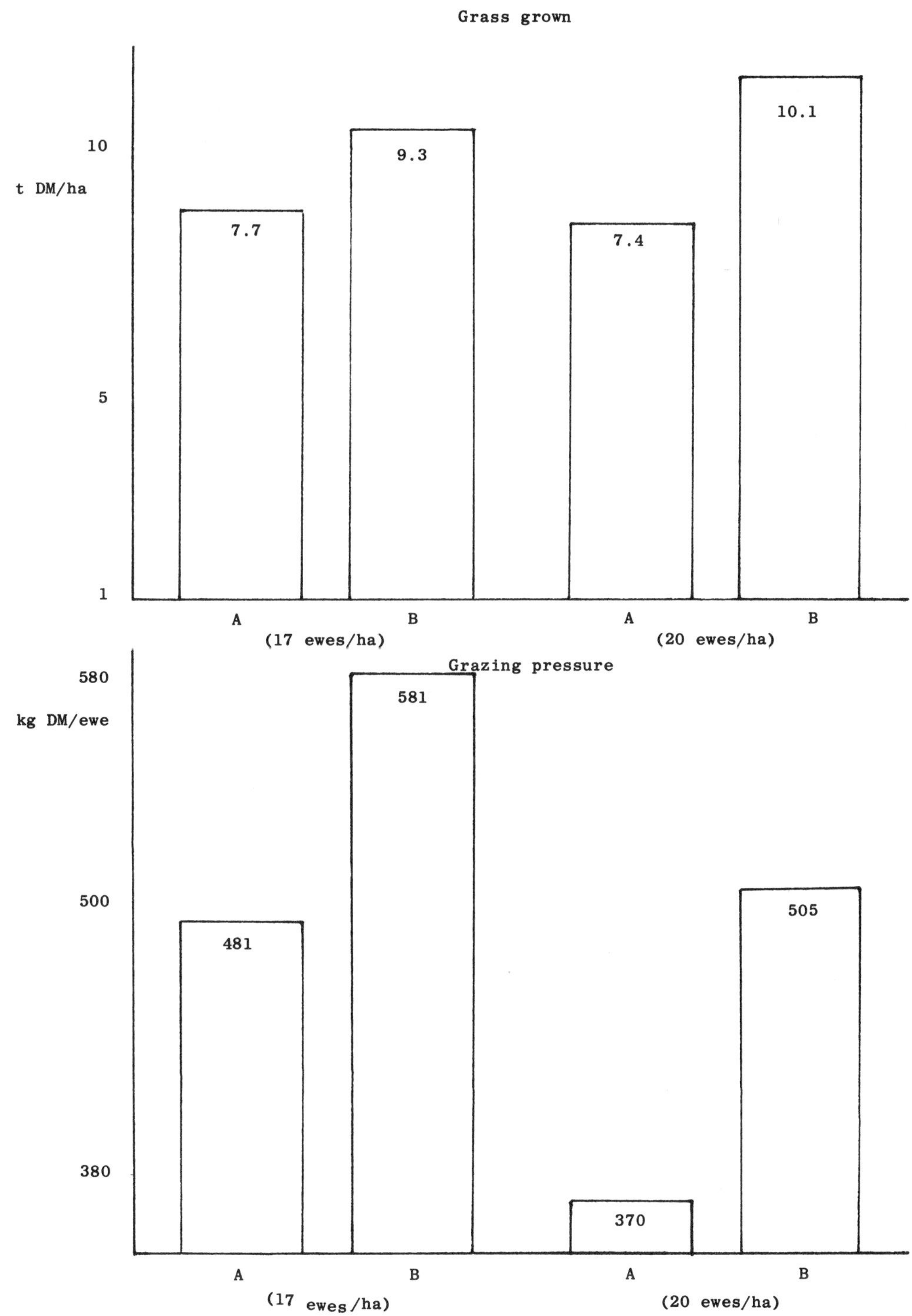

Figure 3

FIG. 3 EFFECT OF AGE ON EWE PERFORMANCE

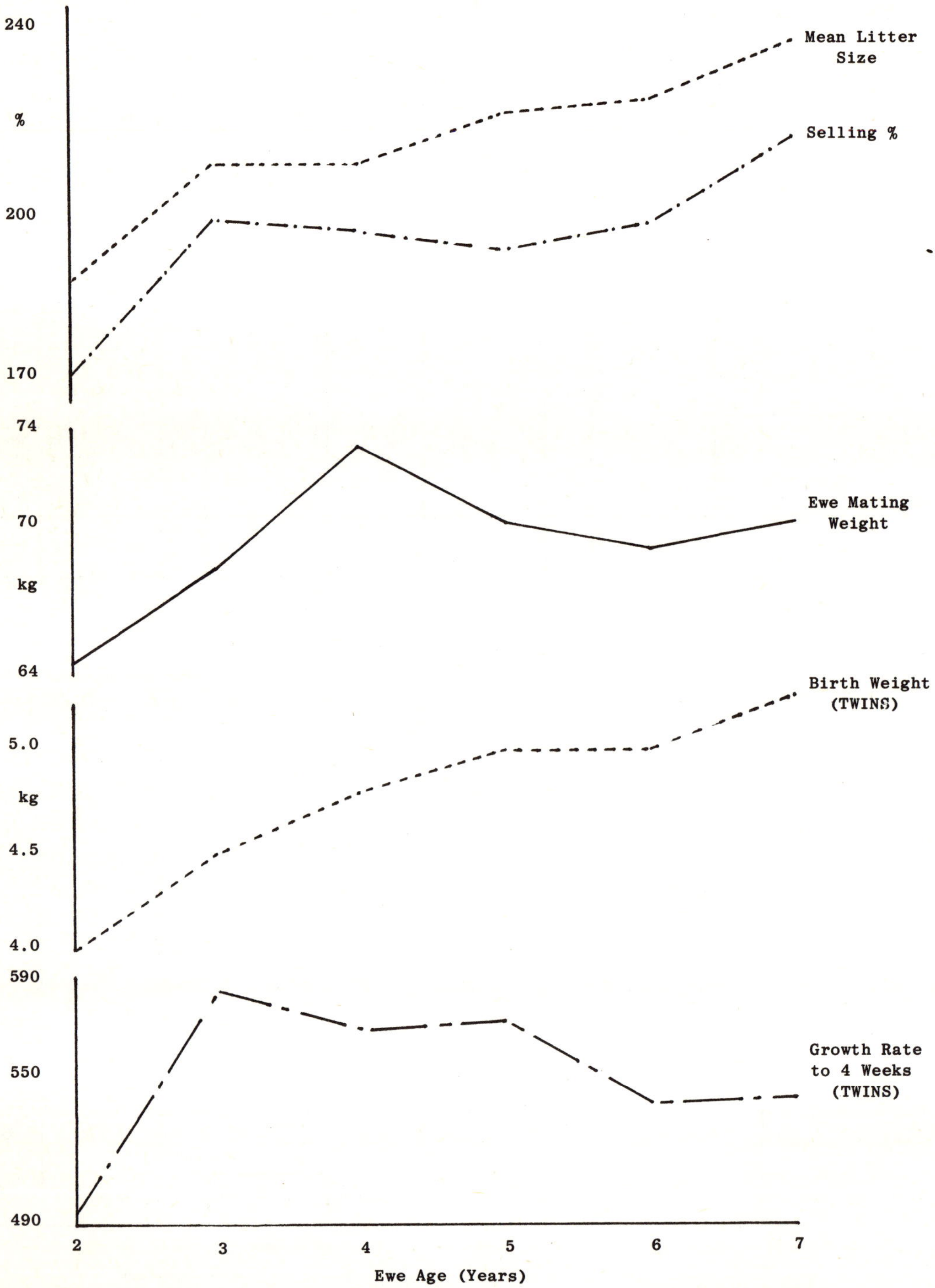

REFERENCES

Arnold, G.W. (1968) Pasture Management. Proceedings of the Australian Grasslands Conference, Perth Vol.2, 189-211.

Edelsten, P.R.;Newton, J.E. (1975) A simulation model of intensive lamb production from grass. Technical Report, Grassland Research Institute No.17.

Newton, J.E.;Young, N.E.;Betts, J.E. (1974) Lamb production from temporary grassland. Technical Report, Grassland Research Institute No.15.

Newton, J.E.;Young, N.E.;Betts, J.E. (1977) Sheep from lowland grass. Technical Report, Grassland Research Institute No.20.

Newton, J.E.;Young, N.E.;Orr, R.J. (1976) A comparison of the sown swards of perennial ryegrass S24 and Italian ryegrass RvP in a two year lamb production system, with different breeds of sheep. Journal of Agricultural Science, Cambridge 87,605-616.

Young, N.E.;Newton, J.E. (1975) A comparison between rotational grazing and set stocking with ewes and lambs at three stocking rates. Animal Production 21,303-311.

INTENSIVE SYSTEMS

DR. J.J. ROBINSON

Rowett Research Institute, Bucksburn, Aberdeen

INTRODUCTION

One of the primary aims of an intensive system of animal production is an increase in individual animal performance. For sheep the main variables contributing to individual animal performance are age at puberty, prolificacy, frequency of breeding, ewe longevity and growth rate and food conversion efficiency of the lamb. My aim is to draw together recent findings on the manipulation of the first four variables and to suggest how these findings may be suitably applied to the industry to enhance productivity.

AGE AT PUBERTY

While there is a distinct breeding season attributable to latitude, it is generally accepted that the older the ewe lamb at the beginning of the normal breeding season the greater her chance of attaining puberty in the first year (Dyrmundsson 1973).

While the stimulatory effects of the ram and exogenous hormones (Keane 1975) can be used to modify this age relationship, the extent to which these procedures can be successfully applied to ewe lambs destined for subsequent frequent breeding systems largely depends on the ewe lamb having attained at least 60% of its mature body size at mating (see review by Robinson et al 1978). The importance of body size at first mating cannot be overemphasized for, unlike the ewe bred once per year, the frequently bred ewe has little opportunity to increase her body size except during pregnancy, and attempts to achieve a high maternal growth rate during early pregnancy can result in small lambs with a reduced viability (Robinson et al 1971, Quirke et al 1976). These problems are more likely to be encountered in the more precocious breeds such as the Finnish Landrace and its crosses, in which a fertile oestrus can occur as early as three months of age (Robinson & Ørskov 1975). Indeed, in comparison with traditional breeds such as the Clun Forest with less than 60% lambing after mating at 6 to 9 months of age (Bichard et al 1974), Finn crosses excel in both conception and litter size. Recent embryo transfer studies (Quirke 1977) provide convincing evidence that the much higher incidence of barrenness in the traditional breeds of ewe lamb is due to the inherently low potential for development of their embryos and not to an abnormal uterine environment. Thus any breed which emulates the Finn crosses in this characteristic must be an obvious choice when attempting to improve early reproductive performance.

PROLIFICACY

There are numerous reports drawing attention to the theoretical 300% improvement in efficiency of food utilization that could result from increasing the annual production per ewe from 1 to 10 lambs. What many of these fail to point out however is that an increase to 4, which could be considered a realistic practical target, is likely to account for over a 240% improvement in efficiency (Robinson & Ørskov 1975).

Practical methods for improving ovulation rate and litter size

These have been reviewed by Robinson et al (1978) and the improvement of ovulation rate by selection, nutrition and the artifical manipulation of the ewe's endocrine status has been well documented elsewhere in this publication. Those of us concerned with increasing ewe productivity await with interest the practical developments which may arise from the recent observation that active immunisation against the ovarian steroid androstenedione is effective in increasing ovulation rate in the ewe (Scaramuzzi et al 1977). Excessive stimulation of ovulation rate may however become counterproductive in practice for there is increasing evidence (Rhind et al 1976, Hanrahan 1976, Land & Wilmut 1977) to suggest an intermediate optimum ovulation rate for maximum litter size. The optimum value is likely to depend on genotype. For the Finn x Dorset Horn, ovulation rates in excess of 5 appear to be counterproductive to litter size (Rhind et al 1976).

A large litter size leads to considerable physical stress in late pregnancy. The total lamb birth weight produced by the prolific ewe is considerably greater in relation to her body size than that of any other domestic livestock species (Figure 1). This has important implications for nutrition and management in late pregnancy, and this topic will be dealt with later. Published observations show that increasing litter size invariably results in an increase in perinatal lamb mortality (see review by Robinson et al 1978). This need not be so; observations on the anatomical development of the foetus (McDonald et al 1977) have clearly demonstrated that the small lamb from a large litter is more developed than its weight would suggest. Thus its reduced viability cannot be attributed to under-development *per se*. For this reason we routinely administer, by stomach catheter and syringe and as soon after birth as possible, approximately 60 ml of either ewe or cow colostrum to each lamb from a large litter. This easy, quick procedure has reduced perinatal lamb mortality in large litters to under 10% (Robinson 1974).

FREQUENCY OF BREEDING

From a review of the reproductive performance of ewes in frequent breeding programmes (Robinson et al 1978), it is clear that the high lamb production of the Finn and Finn crosses not only arises from their larger litter size but also from their reduced lambing interval. It is becoming clear that the Finn has a relatively long breeding season under a wide range of environmental and management conditions (Maijala & Osterberg 1976, Wheeler & Land 1977); this characteristic is also apparent in the Romanov (Veress et al 1976, Ricordeau et al 1976).

A practical difficulty encountered in most frequent breeding systems based solely on natural oestrus activity is the year to year variation in the onset of the breeding season in what appear to be fairly standard

Figure 1

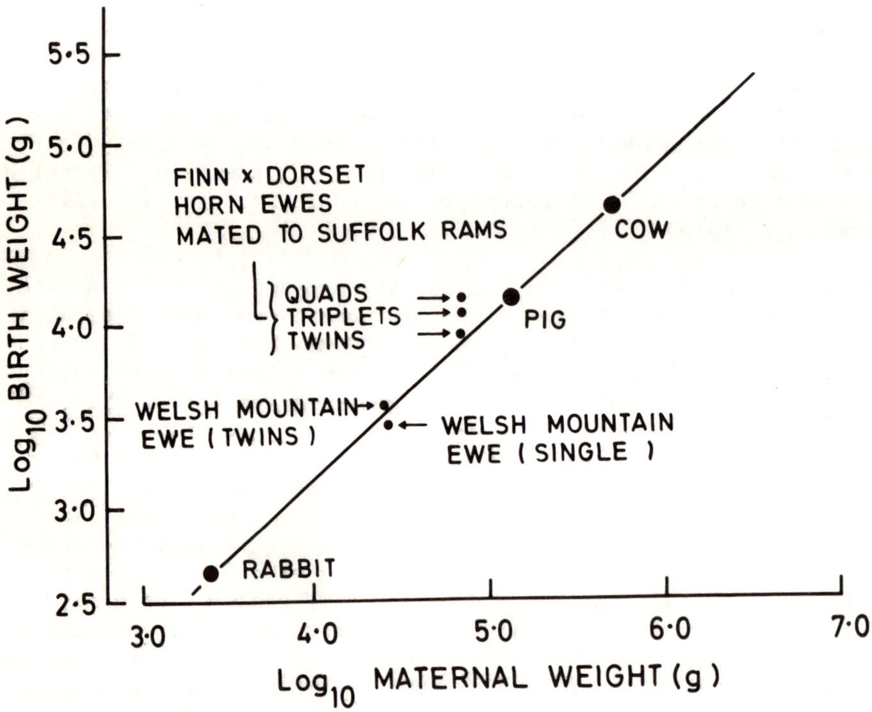

Fig.1. Total weight of newborn in relation to weight of dam. Data for the cow, pig, Welsh Mountain ewe and rabbit taken from Leitch et al (1957) and for the Finn x Dorset Horn ewes from Robinson et al (1977).

environmental conditions. This, coupled with a dearth of information on the duration of the breeding season, has undoubtedly led to interest in the control of oestrus activity by other methods, eg photoperiodism and exogenous hormones.

Photoperiodism

Altering the daylength pattern can increase breeding frequency but there is little evidence in the literature to indicate that it results in a substantial improvement in annual lamb production above that for frequent breeding systems under natural daylength conditions (Robinson et al 1978). Undoubtedly the main problem is the large variation between ewes in oestrus onset. This makes difficult the precise nutrition and management of the flock, which is essential to maximum productivity.

Exogenous hormones

The development of highly potent progesterone analogues, notably FGA (9α-fluoro-11-hydroxy-17α-acetoxy progesterone) and MAP (6α-methyl-17α-acetoxy progesterone), and the demonstration that, when applied on polyurethane intra-vaginal pessaries they could effectively emulate the action of the corpus luteum (Robinson 1967), inspired hope that, in conjunction with exogenous gonadotrophin, they could be used to bring precision to the timing of events in intensive systems. Unfortunately problems associated with the method of impregnating the pessaries, and the hostile uterine environment following their withdrawal, led in many instances to unacceptable results during their development phase. A number of workers persevered with the technique (Gordon 1971) and showed that commercially acceptable results could be obtained. It is largely through Professor Gordon's efforts that hormone treatments in conjunction with set time artificial insemination, as described by Colas et al (1968), are widely used on commercial farms in Ireland for advancing the breeding season. Attempts to use the technique for rebreeding in the spring the Galway ewe, the most widely used breed in Irish fat lamb production systems, have met with only limited success (Table 1), particularly when the ewes are lactating.

Table 1 Lamb production in relation to season of hormone treatment (Data from Gordon 1975)

	Season of treatment		
	Spring	Summer	Autumn
Number of clocks	83	594	37
Number of ewes	2508	21545	1453
% ewes lambing to induced oestrus	34.7	64.0	75.3
Litter size	1.58	1.62	1.72

There are however a number of farms in Ireland where hormone treatment in conjunction with artificial insemination (Cooper et al 1971) is being used to advance the breeding season of Suffolk, Dorset Horn and Finn cross ewes to give a December lambing. Lactation is terminated at one month and the ewes are again treated with hormones for the induction of oestrus in February/March (Northern Ireland Agricultural Trust 1974).

EWE LONGEVITY

It has been shown (Robinson et al 1975) that Finn x Dorset Horn ewes can sustain an average litter size of 2.0 for five consecutive frequent breeding cycles, each at seven month intervals. This is of considerable practical significance since the improvement in biological efficiency (carcass per unit of food energy) in going from a mean flock life of five to seven pregnancies in a frequent breeding system only increases the biological efficiency of ewes with this level of lamb production by 5%.

DEVELOPMENT OF INTENSIVE SYSTEMS BASED ON FREQUENT BREEDING

The successful development of intensive lamb production systems largely depends on recognising the maximum extent to which each of the preceding components can be manipulated before becoming counterproductive. I have already drawn attention to how excessive ovulation rates and litter sizes can become counterproductive. Similarly, striving to achieve too short a lambing interval of for example 6 months can also have a counterproductive effect; for example, compare the results of Speedy et al (1976) with those of Speedy & Fitzsimons (1977). Although individual ewes can conceive within 6 months of their previous conception, the observations of Van Wyk et al (1972), coupled with those of Choffardet (1971), suggest that time taken for uterine involution, particularly for ewes lambing in the normal period of anoestrus, is too long (45 to 60 days) to permit a continuous 6 month lambing interval on a flock basis. The likely minimum interval is therefore probably about 7 months.

The 'Nouzilly' frequent breeding system

The most 'streamlined' frequent breeding system seems to be the one developed by the French researchers at Nouzilly (Thimonier et al 1975), in which Romanov and Prealpe crossbred ewes produce on average 3 lambs each per year. The timing of events in the system (Table 2) is such that 'week-end' work is reduced to a minimum. The overall system involves a series of seven flocks each separated in physiological state by 7 weeks. Ewes which do not conceive to the hormonally-induced oestrus in one flock are transferred to the subsequent flock for remating. Recent results for an essentially similar system, under test in Northern Ireland and based on Border Leicester x Scottish Blackface ewes, have yielded only modest improvements in ewe productivity (Foster et al 1977). While the poorer performance in this study compared with the 'Nouzilly' system may be partly due to latitude, it is tempting to conclude that the prolificacy and duration of the breeding season of the Romanov and Prealpe make them more suitable candidates for intensive breeding systems.

Table 2 Details of the timing of the management procedures in the frequent breeding programme at the Station de Physiologie de la Reproduction, INRA, Nouzilly, France (After Thimonier et al 1975)

Day		Procedure
-16 -14	Wed or Fri	Progestagen impregnated intravaginal pessary inserted for either 12 or 14 days.
-2	Wed	Sponges removed and PMS injected (a.m.)
0	Fri	Artificial insemination using fresh diluted semen (a.m. and p.m.)
18	Tue	Pregnancy test using plasma progesterone concentration. Ewes diagnosed as barren 'slip' to subsequent flock.
144	Tue	Induction of parturition using dexamethazone (Bosc 1970).
146	Thu	Lambing.
147	Fri	Artificial rearing of lambs when lambing coincides with period of natural anoestrum.

The development of frequent breeding studies at the Rowett Research Institute

This involved, in the first instance, subjecting Finn x Dorset Horn ewes to optimum conditions for frequent breeding. Artificial daylength regimes and exogenous hormones were used for the induction and synchronisation respectively of oestrus; breeding frequency was every 7 months and lactation was terminated at one month with the lambs being weaned abruptly on to cereal based diets. Details of all the procedures used and the nutrition and management of the ewes have been given by Robinson & Ørskov (1975). The procedures were highly effective in removing the seasonal effect on fertility and there was little evidence for a decrease in performance as the study progressed (Tables 3 and 4). The mean annual production of 3.5 lambs per ewe over the five reproductive cycles demonstrates that, given the correct nutrition and environment, there is no biological limitation on the part of the ewe to sustaining a lambing interval of 7 months on a flock basis.

Table 3 Per cent conception and litter size over five consecutive reproductive cycles in two flocks of Finn x Dorset Horn ewes, each bred at seven month intervals using photostimulation and progestagen treatment for the induction and synchronisation of oestrus respectively (From Robinson et al 1975)

	Reproductive cycle number					Overall mean
	1	2	3	4	5	
Month of mating						
Flock 1	May	Dec	July	Feb	Sept	
Flock 2	Aug	Mar	Oct	May	Dec	
Mean % conception						
Flock 1	94	88	85	82	85	
Fock 2	98	92	85	81	83	88
Mean litter size						
Flock 1	2.3	1.9	2.1	2.0	2.3	
Flock 2	2.1	2.0	2.0	2.1	2.1	2.1

Table 4 Lamb production over a period of five breeding cycles, all at a reduced lambing interval of seven months for 96 Finn x Dorset Horn ewes (From Robinson & Ørskov 1975)

	Number of pregnancies out of a possible five					Overall means
	5	4	3	2	1	
Ewes pregnant (%)	49.0	31.2	12.5	5.2	2.1	-
Mean litter size	2.1	2.1	2.3	2.0	2.5	-
Lambs per ewe per year (Assuming all the original 96 ewes were available for breeding in all five cycles)	3.9	3.1	2.6	1.5	0.9	3.3
Lambs per ewe per year (based on only those ewes which were given the opportunity to mate)	3.9	3.3	2.9	2.0	2.2	3.5

Lamb production under natural daylength conditions

Having established a high level of reproductive performance under ideal conditions, we considered if this production could be obtained under natural daylength conditions. The results obtained under field conditions are given in Table 5.

Of the total conceptions to the induced and first repeat oestrus cycles, approximately 85% were to the induced oestrus. In order to minimise the adverse effects of the progestagen treatment on sperm transport, the ewes were inseminated 12 to 18 hours after the first signs of behavioural oestrus in the flock. For the ewes which lambed in February (Table 5(c)) their high conception rate in May, their normal period of anoestrus, is of particular interest in view of the large number of lambs that they reared to 6 weeks, and suggests that this particular cross can be induced to breed successfully at 7 month intervals under natural daylength conditions. Their mean annual lamb production of 3.45 lambs each, after adjusting for lamb mortality, was similar to that obtained under the artificial daylength regime and suggests that the latter procedure can be replaced effectively by a small amount (400 i.u.) of exogenous gonadotrophin (PMSG) at progestagen withdrawal.

FEEDING THE 'FREQUENTLY-BRED' EWE

In our experience there is little problem in maintaining the body weight of highly productive ewes in a frequent breeding programme (Figure 2), provided they are weaned at one month and well fed in the three weeks before mating to ensure that the tissue loss in late pregnancy and early lactation is replaced before re-mating.

One of the most important factors in the nutrition of the frequently bred ewe is the pattern of food intake. If the high body condition at mating, essential for maximum ovulation rate, is maintained until late pregnancy it leads to inappetance and hypoglycaemia, particularly in polytocous ewes. The key to preventing this is to restrict food intake in mid-pregnancy thus ensuring a carefully controlled mobilization of body fat at a time that will have no detrimental effect on either the ewe or her foestuses. During this period the frequently bred ewe should be either kept at high stocking rates at pasture or transferred to diets with a low digestibility.

An example of the levels of energy intake, at different stages of the reproductive cycle to give the body weight change illustrated in Figure 2, is given in Table 6. These recommendations raise the question of whether or not production can be maintained at a high level without prior knowledge of foetal numbers. Our experience in field trials is that production can be maintained provided ewes are grouped and feeding levels adjusted according to body condition and body condition changes, particularly during late pregnancy and lactation and prior to re-breeding. The results presented in Table 5 were achieved using this management procedure.

Figure 2

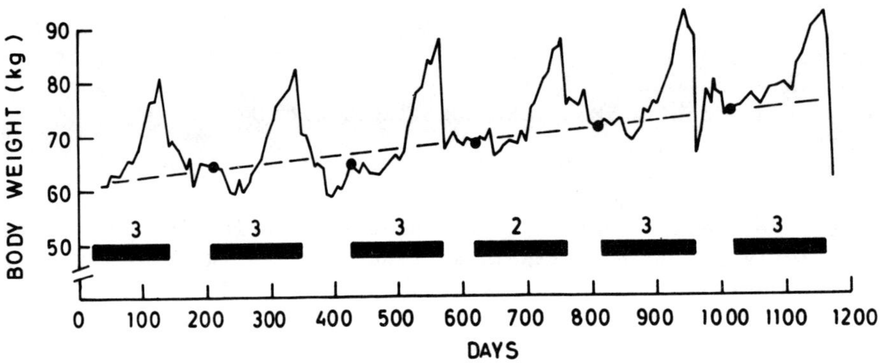

The pattern of body weight change in ewes offered feed according to the recommendations given in Table 6. The numbers refer to the lambs carried and reared respectively and the shaded areas indicate gestation periods.
● Mating weights (Source of data, Robinson & Ørskov 1975)

Table 6 Mean daily metabolizable energy intake (MJ) in relation to physiological state required to give the pattern of weight change shown in Figure 2 in highly productive ewes maintained on a seven month lambing interval

Weaning to end of first month of gestation		15.0		
30 to 90 days of gestation		9.4		
Number of lambs	1	2	3	4
Mean for 90 days gestation to paturition	12.4	14.5	15.8	16.6
At 90 days	9.4	9.4	9.4	9.4
At parturition	15.4	19.6	22.2	23.8
During lactation	18.8	24.5	26.2	-

Nutrition during lactation

Normally ewes rearing two or more lambs are in negative energy balance during early lactation. For such ewes we have shown that, if their diet does not provide sufficient amino acids to meet the requirements of their potential milk production, an increase in the dietary concentration of crude protein will stimulate milk production at the further expense of body fat reserves, provided the dietary protein is not completely degraded in the rumen. Table 7 shows the response obtained in ewes suckling twins and given a diet supplying about 80% of estimated energy requirements for maintenance and potential milk production.

Table 7 The effect of dietary crude protein concentration on ewe milk yield and ewe body weight loss in early lactation (From Robinson et al 1974)

Crude protein in diet (%)	10.3	13.6	16.9
CP (g) to ME (MJ) ratio	11.1	13.8	16.0
Milk production (kg/day)	2.42	2.93	3.10
Ewe live weight loss in early lactation (g/day)	118	170	265

The supplementary source of protein in this experiment was soya bean meal. The results of a recent experiment (Calderon Cortes et al 1977) show that, within 3 days of withdrawing a protein supplement (fish meal) with a low digestibility in the rumen, daily milk yield was reduced from 3.2 to 2.6 kg with a consequent decrease in body fat mobilization. Thus the removal of dietary protein supplements after 3 weeks of lactation is a useful practical procedure to prevent excessive mobilization of body reserves just before re-breeding in the frequently bred ewe; at the same time it encourages a greater intake of solid feed by the lamb in preparation for its abrupt weaning (Robinson et al 1974).

Concurrent pregnancy and lactation

With careful control of nutrition and management a mean annual lamb production of 3 to 3.5 can be achieved in a frequent breeding programme involving a seven month lambing interval and a shortened lactation of 4 to 6 weeks. This prompted us to ask whether or not conception rates could be maintained if lactation was extended. Initial investigations (Rhind et al 1977) have given conceptions of 100, 80, 70 and 58% to a hormonally induced oestrus in December for ewes mated two months after lambing and weaned after 30, 50, 70 or 100 days of lactation.

Corresponding values for a second flock of ewes mated in March were 92, 83, 36 and 33%. About 50% of those ewes not conceiving to the induced oestrus did so at the first repeat oestrus. It may be that extension of the lambing interval to 8 months will enable lactation to be extended to 10 weeks without impairing fertility. Also, the recent evidence that the seasonal increase in plasma prolactin concentrations and their further elevation during suckling may be luteolytic (Rhind et al 1978) suggests that the artificial suppression of prolactin concentrations may be of benefit, not only in improving the chances of achieving successful concurrent pregnancy and lactation.

Integrating the food requirements of a frequent breeding system with available resources

Knowing the reproductive potential of our ewes, the nutrient requirements of both the ewe and lamb for different levels of production and the availability of grass has enabled us, in conjunction with the three Scottish Colleges of Agriculture, to draw up a specification for the integration of frequent breeding systems on farms in Britain (Scottish Agricultural Colleges 1977). In doing so it was necessary to apportion the total food energy requirements between the three main food resources, i.e. grass, conserved forage and concentrates. The results of these calculations are illustrated in Figure 3. We adopted a 'three times in two years' programme of lambing and a 'two flock' system with matings in October, February and July. All lambs from the July mating are weaned solely on to a concentrate diet at one month and the ewes rebred in February. The ewes are then maintained at a high stocking rate at pasture during the spring and early summer when their energy requirements are minimal. During late pregnancy, stocking density is reduced in preparation for lambing. The resulting lambs remain on the ewes for 2 months and are finished on concentrates in the autumn. The ewes are 'flushed' in

Figure 3

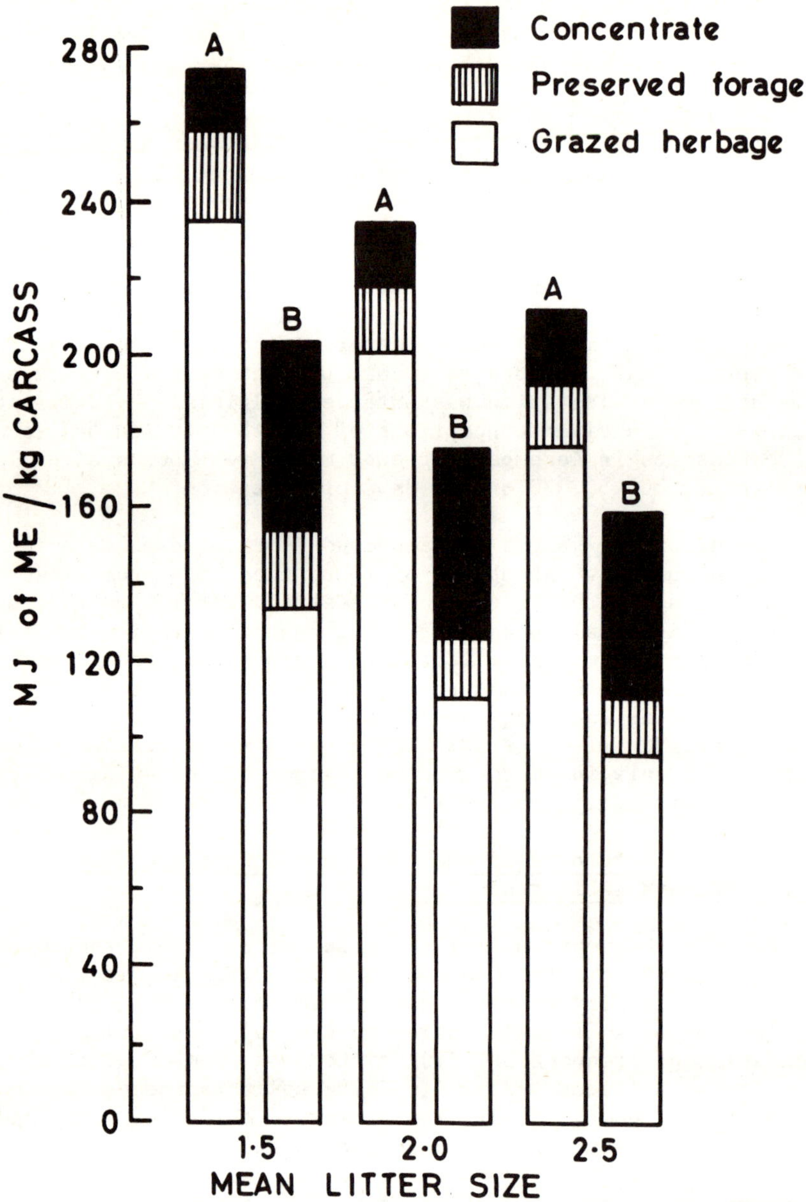

Fig.3. The effect of mean litter size and breeding frequency (A = breeding once per year; B = breeding three times in two years) on the total food energy (MJ of ME) and of the individual components of the diet required to produce 1 kg of carcass.

preparation for remating in October and for the production of 'normal season' lambs.

The data in Figure 3 not only demonstrate the improvements in biological efficiency of increasing both litter size and frequency of breeding of the ewe but, in conjunction with costs per unit of metabolizable energy for the different food sources, allow us to calculate how the food costs of lamb production are influenced by changes in ewe productivity. It is the extension of this approach to other environmental situations that is required in order to optimise productivity.

REFERENCES

Bichard, M.;Younis, A.A.;Forrest, P.A.;Cumberland, P.H. (1974) Animal Production 19,177.

Bosc, M.J. (1970) Compte Rendu de l'Academie des Sciences 270,3127.

Calderon Cortes, J.F.;Robinson, J.J.;McHattie, I.;Fraser, C. (1977) Animal Production 24,135.

Choffardet, R. (1971) Cited by Cognie, Y.;Hernandez-Barreto, M.; Saumande, J. (1975) Annales de Biologie Animale Biochemie, Biophysique 15,329.

Colas, G.;Dauzier, L.;Courot, M.;Ortavant, R.;Signoret, J.P. (1968) Annales Zootechnie 17,47.

Cooper, R.J.;Wallace, D.N.;Wishart, D.F.;Hoskin, B.D. (1971) Veterinary Record 88,381.

Dyrmundsson, O.R. (1973) Animal Breeding Abstracts 41,273-289.

Foster, W.H.;McCaughey, W.J.;Logan, E.F.;Irwin, D. (1977) Fiftieth Annual Report of the Agricultural Research Institute of Northern Ireland, p19.

Gordon, I. (1971) Journal of Agricultural Science, Cambridge 76,337.

Gordon, I. (1975) Proceedings of the British Society of Animal Production 4,79.

Hanrahan, J.P. (1976) Paper presented at the 27th Annual Meeting of the European Association for Animal Production, Zurich.

Honmode, D. (1977) Animal Breeding Abstracts 45,384.

Keane, M.G. (1975) Irish Journal of Agricultural Research 14,81.

Land, R.B.;Wilmut, I. (1977) Animal Production 24,183.

Leitch, I.;Hytten, F.E.;Billewicz, W.Z. (1959) Proceedings Zoological Society, London 133,11.

Maijala, K.;Osterberg, S. (1977) Livestock Production Science 4,355.

McDonald, I.;Wenham, G.;Robinson, J.J. (1977) Journal of Agricultural Science, Cambridge 89,373.

Northern Ireland Agricultural Trust (1974) Lamb Production. The development of a new system.

Quirke, J.F.;Sheenan, W.;Lawlor, M.J. (1976) Animal Production 22,163.

Quirke, J.F. (1977) Paper presented at the 28th Meeting of the European Association for Animal Production, Brussels.

Rhind, S.M.;Chesworth, J.M.;Robinson, J.J. (1978) Journal of Reproduction and Fertility. (In Press)

Rhind, S.M.;Robinson, J.J.;Fraser, C.;Phillippo, M. (1977) Animal Production 24,128.

Rhind, S.M.;Robinson, J.J.;Fraser, C.;McHattie, I. (1976) Animal Production 22,137.

Ricordeau, G.;Tchamitchian, L.;Thimonier, J.;Flamant, J.C.;Theriez, M. (1976) Paper presented at the 27th meeting of the European Association for Animal Production, Zurich.

Robinson, J.J. (1974) Proceedings of the British Society of Animal Production 3,31.

Robinson, J.J.;Fraser, C.;Corse, E.L.;Gill, J.C. (1971) Animal Production 13,653.

Robinson, J.J.;Fraser, C.;Gill, J.C.;McHattie, I. (1974) Animal Production 19,331.

Robinson, J.J.;Fraser, C.;McHattie, I. (1975) Annales de Biologie Animale Biochemie Biophysique 15,345.

Robinson, J.J.;Fraser, C.;McHattie, I. (1978) A Technical Publication of the US Feed Grains Council. (In Press)

Robinson, J.J.;McDonald, I.;Fraser, C.;Crofts, R.M.J. (1977) Journal of Agricultural Science, Cambridge 88,539.

Robinson, J.J.;Ørskov, E.R. (1975) World Review of Animal Production 11,63.

Robinson, T.J. (1967) The control of the ovarian cycle in sheep. Sydney University Press.

Scaramuzzi, R.J.;Davidson, W.G.;Van Look, P.F.A. (1977) Nature 269,817.

Scottish Agricultural Colleges, Technical Note No.16. (1977)

Speedy, A.W.;Black, W.J.M.;Fitzsimons, J. (1976) Animal Production 22,138.

Speedy, A.W.;Fitzsimons, J. (1977) Animal Production 24,189.

Thimonier, J.;Cognie, Y.;Cornu, C.;Schneberger, J.;Vernusse, G. (1975) Annales de Biologie Animale Biochemie Biophysique 15,365.

Van Wyk, L.C.;Van Niekerk, C.H.;Belonje, P.C. (1972) Journal of South African Veterinary Association 43,19.

Veress, L.;Stosz, J.;Lovas, L. (1976) Paper presented at the 27th Meeting of the European Association for Animal Production, Zurich.

Walton, J.S.;McNeilly, J.R.;McNeilly, A.S.;Cunningham, F.J. (1977) Journal of Endocrinology 75,127.

Wheeler, A.G.;Land, R.B. (1977) Animal Production 24,363.

THE QUESTION OF PURPOSE IN SHEEP PRODUCTION

PROFESSOR H.P. DONALD

School of Agriculture, University of Edinburgh

Most research on sheep is based on the assumption that the purpose is to improve the output of meat, wool or milk. For those countries where the research is being done this assumption may be justified, but it should be examined before being accepted elsewhere. This is because it refers to well-established sheep industries and markets and, perhaps unfortunately, to habits of thinking.

Sheep production, like other branches of agriculture, faces some serious problems of world significance. These are well known.

(i) Growing human populations and food requirements.

(ii) Limited resources especially of land, water and fertilizer.

(iii) Social aspects involving land ownership, labour, conservation of soil, and food habits of people.

These problems have economic, religious and political complications which affect research and development. Some analysis of this matter might begin by enquiring who is asking the question? And who is answering it? There are a number of interested groups:

(i) farmers and stockmen
(ii) consumers
(iii) processors
(iv) conservationists
(v) R and D workers
(vi) economists
(vii) politicians

Most if not all these groups tend to ask and to answer the question in accordance with their vested interests. We all know what these vested interests are and we all discount the corresponding opinions accordingly. There is no quantitative way of evaluating and balancing the views of these groups on the amount of land to be used for sheep, the kind of land, the kind of sheep or the kind of product. There is no way of determining objectively what level of intensity of capital and labour is best. We remain torn between the force of history which tells how agriculture has met the demand for food so far by intensifying production, and the force of the impending 'Tragedy of the Commons'. In his article with this title, Gilbert Hardin likened the demands on the world's resources and its consequences to the repeated overstocking of common grazings by unregulated flocks.

Those interested in research and extension have an obligation of which they are being frequently reminded; that they are expected to help in the solution of the problems and not to hide in their laboratories. Of course they will have to develop points of view about subjects on which they are not expert. That means behaving like everyone else. It means remembering that the growth of knowledge entails greater ignorance; and it means that the myth of scientific objectivity is not now believed.

By virtue of his training however a research and development worker is skilled in the handling and analysis of problems. He is used to variation and above all he has access to the work of his contemporaries and his predecessors. In the complex real world of sheep production, his value judgements should be much superior to those of people without his advantages. Without the art of persuasion however his contributions will count for little. For good reasons, the ancient Greeks rated Rhetoric very highly. It is not so elusive as the Truth no-one commands.

To be persuasive about his proposals for research and development or about his results, a worker will have to pay attention to all the aspects of sheep production. It is impossible to change one thing at a time. The principal aspects are:

(i) Economic. No firm objective conclusions can be reached in this aspect and therefore persuasion is all important. Pay attention to the conflicting interests, the short and the long term and alternative uses for the land and capital. Farmers cling to liberty and consumers to equality.

(ii) Social. The weight given to these aspects arises from power, whether political, military or rhetorical, and points to consider include conservation and amenity, ownership of land, employment, way-of-life, pollution and animal welfare.

(iii) Technological. Product changes - easy, quick, cheap.
Method " - less easy, quick, cheap.
System " - difficult, slow, expensive.
At a certain degree of quantitative change, a qualitative change becomes necessary, e.g. manual labour to mechanized.

(iv) Political. Government policies on price control, imports, taxes, regulations and land use change.

Two examples, the green revolution and factory methods of poultry and pig production, illustrate all these aspects. What can the sheep industry do towards solving the big problems?

Although exploited by man for 12000 years or more, the sheep still occupies a humble place in the evolution of food production methods. The basic reason for this is that man has not adapted it to the production of concentrated food, and it remains therefore an animal primarily used in areas unsuitable for intensive production, areas where the climate is very dry, very wet, very cold or very steep and producing food too fibrous for other stock. There are several consequences:

(i) It has been adapted to many different environments.

(ii) It is subject to more soil and climate variation and thus more managerial variation than other domestic animals.

(iii) Efforts to improve it are spread over a wide range of characters.

(iv) It attracts relatively little attention from the governments which support R & D most heavily.

Suggestions for R & D policy in the light of the foregoing considerations can be summarised as follows:

(i) Concentrate R & D on one or two general objectives.

(ii) Make changes profitable to farmers.

(iii) Work out ways of measuring achievement.

(iv) Organise production systems to attract practical and labour investment, e.g. carpet manufacture, processed meat, or cheese.

(v) Preserve grazing land by control of numbers of animals.

(vi) Control of one disease of high significance at a time.

(vii) Make sure Research and Development are integrated from the start of a project.

WOOL PRODUCTION IN SEMI-ARID AREAS WITH SPECIAL REFERENCE TO AFRICAN CONDITIONS

DR. F. DE K KOTZE

Department of Animal Production, University of Pretoria, South Africa

Australia and New Zealand produced 41.2% of the world's output of greasy wool in 1976 (Table 1). Russia and the South American states also made a significant contribution, whereas Africa only contributed 193 million kg, or approximately 7.4% to world production. The most important reason for this is probably the tropical climate in the central parts of Africa. It is interesting to note that a large proportion of the total world wool production comes from extensive areas (Table 2). Apart from Australia and Russia, both the Argentine and South Africa are countries with relatively low rainfalls where wool production is largely carried out under semi-arid conditions.

Table 1 Wool Production Statistics (1976)

	Greasy wool (kg x 10^6)	% of total
World	2590	100
Oceania (Aust. and NZ)	1066	41.2
USSR	431	16.6
South America	294	11.4
Asia	285	11.0
Europe	261	10.1
Africa	193	7.4
North America	61	2.3

Table 2 Major Wool Producing Countries (1976)

	Greasy wool (kg x 10^6)	% of total
World	2590	100
Australia	754.3	29.1
USSR	430.8	16.6
New Zealand	311.8	12.0
Argentina	167.8	6.5
South Africa	102.8	4.0
China	61.0	2.4
Uruguay	56.0	2.2
Turkey	54.0	2.1
USA	51.6	2.0
UK	47.0	1.8

The dry climate in the extreme north and the extreme south of Africa is admirably suited to sheep farming, especially wool production (Table 3). South Africa produces 53% of the total amount of wool in Africa, while South West Africa/Namibia in the south, and Morocco, Algeria, Tunisia, Sudan and Ethiopia in the north are the other important wool-producing countries. As the subject of this lecture is wool production under extensive conditions in Africa, I would like to discuss the climate of Africa briefly.

Table 3 Major Wool Producing Countries in Africa (1976)

	Sheep ($x 10^6$)	Greasy wool produced ($kg \times 10^6$)	% of total
Africa	159.5	192.7	100
South Africa	31.0	102.8	53.5
Morocco	16.8	21.0	10.9
Algeria	8.8	16.7	8.7
Sudan	15.3	14.8	7.7
Ethiopia	23.1	12.0	6.2
Tunisia	3.5	6.4	3.3
Namibia	5.0	4.6	2.4

CLIMATE AND VEGETATION

Africa extends from 37°N to 34°S and is therefore more or less bisected by the equator. This results in 90% of the surface being considered tropical, where sicknesses of man and animal are not conducive to economic activities. Except for the higher areas in the north and south, large temperature variations and frost are rare.

Rainfall is the most important single element in the climate and water is a limiting factor in 50% of the continent. Large areas of the continent actually receive sufficient total rain annually, but the seasonal character and unreliability in its distribution create considerable problems. According to the classification of one prominent climatologist, more than 40% of Africa can be considered dry.

The Republic of South Africa, where wool is produced on a large scale, has a dry climate. Only 20% of the country receives more than 625 mm rain annually, while one third of the country receives less than 250 mm annually. Seasonal droughts and periodic droughts of longer duration are part and parcel of the farming pattern, and, of course, at times lead to considerable losses of animals.

The most important wool-producing area in South Africa is the Karoo, which has an annual rainfall of between 125 and 450 mm. The vegation consists largely of Karoo-bushes, while a mixed Karoo-bush/grass vegetation is found further north. The nutritional value of the "veld" remains reasonably good throughout the year, despite a severe winter. Wool is also produced in the north-eastern parts of the country, where the predominantly grass vegetation has an extremely low nutritional value in the winter.

The natural vegetation of the Karoo has a carrying capacity that ranges from five ha per sheep throughout the year to one ha per sheep. Because every individual farm has a maximum carrying capacity, it is important that the farmer should maintain a suitable flock composition to ensure maximum profit per unit area in the long term. In the following section, the different breeds suitable for these climatic conditions are discussed.

BREEDS AND THEIR DISTRIBUTION

Of the 31 million sheep in South Africa, approximately 24 million are woolled sheep. Of the latter 20 million are Merinos. The Merino is also the breed that occurs in the greatest numbers in the world as a whole, some 930 million.

The Merino is a white-woolled sheep that carries a large amount of good quality fine wool while also having a reasonable carcass for meat purposes. In full fleece the sheep appear large, with wool reaching to the lower extremities and with the belly entirely covered with wool.

Mature rams reach a live mass of about 85 kg, while ewes have a mass of approximately 50 kg. A lambing percentage of 80 is generally considered acceptable, although this can vary immensely. In some cases lambing percentages of 120 to 140 are found. The wool is of good quality and the fibre diameter ranges from 16 to 27 microns. Mature ewes produce about 5 kg wool annually, while rams produce 10 to 15 kg and sometimes more.

The Merino is found in large numbers in the Karoo area, largely because of the adaptation of the breed to semi-arid conditions. In some parts of the country further north, the Merino also does well, although wintering problems do occur through the low nutritional quality of the natural grass vegetation.

The growth rate of Merino lambs differs considerably from the other breeds in South Africa. Lambs are usually weaned at an age of 120 days at a body mass of approximately 21 kg. This compares extremely adversely with breeds like the South African Mutton Merino which has an average mass of 35 kg at the same age. Because of the slow rate of growth of the Merino, it is uneconomical to market surplus animals (culls) earlier than at an age of 14 to 16 months. This naturally affects the flock composition adversely.

Because the farming system is extensive, the farmer is limited to a definite number of animals, determined by the carrying capacity of the farm. The flock composition must therefore be such as to ensure maximum return in the long term, with the percentage of breeding ewes as high as possible. Having to keep replacements and surplus animals until 16 months results in a diminished number of breeding ewes. This also applies to the young replacement ewes where they are mated for the first time at an age of two years or later. It often happens that young ewes lamb for the first time at 2½ years, although it is generally recommended that they be mated at 16 to 18 months, after final selection has been carried out.

Old ewes are culled at an age of about 7 years, largely because their teeth are by that stage so worn down by the hard "veld" conditions that their condition deteriorates rapidly. This results in an age-structure for the ewe flock that is reasonably favourable when viewed in relation to genetic progress. An average replacement percentage of 20 is the result.

In the case of rams the situation differs radically. Because of the high prices of rams, less than 2% rams (that is 2 rams per 100 ewes) are used in the mating season. Some of these rams are furthermore reasonably old. In practice, rams are not easily replaced, largely because their teeth remain in good condition for a longer period than those of ewes. This results in considerably higher average age for rams. To an increasing extent commercial farmers are breeding their own rams from their best flock ewes. The application and acceptance of group breeding schemes is a normal development in this respect.

Because of the slow growth rate of the Merino and its poorer carcass in comparison with other mutton breeds in the country, the Döhne Merino has been developed locally, specifically for the grassveld regions. The Döhne is the product of a cross between the Merino and the original German Merino in South Africa. The aim was to produce typical Merino wool on a carcass like that of the German Merino. The Döhne is today gaining rapid headway because of its production standards. This is primarily the result of the completely scientific grounds on which its breeding programme is based, something unique with sheep breeds in South Africa. Performance testing is compulsory and forms the basis of selection and registration.

The Döhne Merino is a well-balanced polled sheep with good conformation and good quality wool. Fewer skinfolds occur than in the Merino. Rams have a mass of approximately 100 kg and mature ewes approximately 70 kg. It has also been suggested that the Döhne grazes less selectively than does the Merino.

The Döhne maintains a lambing percentage of 100 to 150 and the lambs grow considerably faster than Merino lambs. Under good nutritional conditions they can achieve a live mass of 30 kg or more within four months. Mature rams and ewes produce approximately 11 and 4.5 kg of wool annually.

The Döhne is already widely dispersed in South Africa but actually requires nutritional conditions that can do justice to a dual purpose sheep breed.

Other woolled-mutton sheep in South Africa are the Letelle and Walrich Merino types. These have been evolved by strict selection from the Merino or the introduction of foreign genetic material over a period of 50 years. These breeds are polled, exhibit good mutton qualities and typically fine Merino wool. They are widely distributed in South Africa but in relatively small numbers. There are for instance only 13 Letelle stud breeders probably because of the extremely rigid control that the Breeders' Society exercises on members. In conformation, body mass and lambing percentage, these types are very similar to the Döhne Merino. The differences in approach between the respective Breeders' Societies is the most important reason for the difference in success under South African conditions. As a result of the similarity in breeding aims, Letelle and Walrich sheep are at present accepted for inclusion in the addendum to the Pedigree Register of the Döhne Breeders Society, an extremely positive step on the part of the latter.

As a matter of interest I would like to mention some of the other sheep breeds that are found in reasonable numbers in South Africa. The South African Mutton Merino is a development, over a period of more than 40 years, from the well-known German Merino. The breed has a high growth rate and a reasonable amount and quality of wool but performs better under intensive or semi-intensive conditions. The breed has an average lambing percentage of 150 and flocks are known where more than 200% has been obtained. Lambs reach 35 to 40 kg at four months and, in the absence of any fat localisation, can be marketed as super grade at an age of five to seven months. Various other wool breeds are found in South Africa, but in such small numbers that they do not warrant further discussion here.

The Dorper, developed in South Africa for meat production under semi-arid conditions, is becoming very popular. This sheep was developed from the Dorset Horn and the Black Head Persian. The latter is renowned for high fertility (fecundity) as well as a long breeding season. These properties were combined with the good growth and carcass properties of the Dorset Horn. Prominent features of the Dorper are a high milk production and growth rate and especially adaptability to widely varying climatic conditions. These sheep are kept in the arid parts of the country and are also excellent producers under intensive conditions. Because of the relatively long breeding season, ewes lamb every eight months without much difficulty. Lambs, with ideal fat distribution, reach 32-35 kg at an age of 80-90 days. The breed developed along two distinctive lines; the white Dorper and the black-headed Dorper. Apart from the colour distribution, there are no real differences. At present there are approximately 700 registered breeders of Dorpers as opposed to 900 registered Merino stud breeders in the country. The Dorper is widely used for cross-breeding purposes, although it can lead to hair-pollution of the wool of woolled sheep ewes.

Other indigenous breeds like the Namaqua and the Ronderib Afrikaner are extremely hardy and well adapted to the really dry parts of the country. The fat distribution on the carcass is poor but these breeds are sexually active throughout the year and the mating season can be selected and arranged according to prevailing climatic and nutrient conditions.

The Karakul is found in the dry western part of the country and in South West Africa/Nambibia and is kept for pelt production. Lambs are slaughtered 24 hours after birth and the pelts marketed. Mature Karakul sheep produce a relatively coarse wool that is fetching reasonable prices at present.

MANAGEMENT

The Merino is a breed of sheep that can be classified as having an intermediate breeding season. Sexual activity shows a reasonable decline in the spring with a minimum between October and December, the southern summer. Because of this it is traditional that ewes are mated in the autumn (April) so that the lambs are born in spring. The price of commercial Merino rams (flock rams) has increased to such an extent over the last four years that many farmers scarcely use 2% rams, as mentioned before. Ewes that are not fertilized during the first mating season are mated again during an additional mating season six months later. The argument is that should only 30 or 40% of the ewes be fertilized during the spring mating season, it is at least considerably fewer for the main mating season in the autumn.

Both mating seasons have pros and cons. Apart from the important problem of conception in the spring, nutritional conditions on the natural "veld" are of paramount importance. The dry winter usually persists for a long period before the first spring rains, which are usually reasonably late. Another important consideration is age at first mating. The spring lamb (the result of autumn mating) attains 18 months during the autumn, when sexual activity is high, and higher conception figures are obtained at first mating. The position of the autumn lamb is less favourable, as it is likely that she will only in the case of the Merino be fertilized for the first time at the age of two years.

The lambing season for the majority of ewes is therefore between August and October. This sometimes presents problems, as it coincides with the ideal shearing time. As the Merino has a heavy fleece it is advisable to shear the sheep after the cold winter and before the exhausting summer heat. Traditional shearing time is therefore between August and November. The shearing of ewes with lambs is an impossible task, due to the bleating in the shearing shed. Pregnant ewes should preferably not be handled in this manner during the last four weeks of pregnancy. Lambs of long-woolled ewes sometimes have problems in locating the teats and crutching is sometimes carried out before lambing. The best solution however is to shear the ewes in early August, with lambing commencing in mid-September.

Special attention must of course be given to newly shorn sheep as a little rain, combined with wind, can cause sudden mortality from cold.

SELECTION IN PRACTICE

The importance of scientific selection methods has been emphasised in recent years, especially for the stud industry. The well developed Merino in South Africa today is however largely the product of visual judging techniques. The success obtained with this is without doubt the most important stumbling block in the acceptance of scientific breeding methods. Some Breeders' Societies, such as that of the Döhne Merino, do however accept scientific methods as the basis for formulation of a breeding policy. Merino breeders unfortunately tend to regard science as a threat to the entrenched stud industry. The pressures for scientific breeding are however so great that reorientation, especially on the part of Merino breeders, will hopefully be obtained in the not too distant future.

The success achieved by the older school of breeders, using solely visual evaluation, is based on the fact that the most important properties of the Merino, namely body mass and fleece mass, are both highly hereditable. The reasonable accuracy with which selection could be carried out, due largely to years of experience, coupled with the high heredity, ensured this success. The Department of Agriculture in South Africa has however instituted a performance and progeny testing scheme for both meat and wool sheep, thus enabling breeders to obtain the necessary information and advice.

In selection of Merino sheep for extensive purposes, the type of sheep is of paramount importance. By this is meant the degree of development of skin folds, and the consequent greater skin area and density of fleece. This is however associated with a shorter staple length. Skin folding is important for high wool production but skin folds are genetically negatively correlated with fertility. An overdeveloped Merino also creates problems during shearing and infestation of external parasites remains a very real problem. The nutritional requirements of such animals are also such that the typical "veld" pastures of the extensive areas are inadequate.

For the commercial producer the most important consideration is the yield obtained per unit area. To effect this, the Merino is evaluated and selected on the basis of excellence in respect of:

(i) Amount of wool
(ii) Coverage on the belly and lower parts of the body
(iii) Presence of wool-oil (yolk)
(iv) Quality of wool.

Furthermore, the size of the animal is also important, as it is positively correlated with wool production and fertility. In judging wool quality, the prominence of crimps is considered important because of its connection with handling. The relationship between the number of crimps per unit length and fibre thickness is also of prime importance with respect to substance and handling of wool.

DISEASES UNDER EXTENSIVE CONDITIONS

Because of the low rainfall and the extent of the natural "veld", sheep kept under extensive conditions are far less prone to the animal diseases often experienced under more intensive conditions. During the past five years the drier areas of South Africa have experienced a higher rainfall than normal. This has had the advantage of allowing the natural "veld" to recover very well but, at the same time, animal diseases have increased.

It is recommended that rams and ewes be dosed three times and lambs four times a year, depending on circumstances. Roundworm in young sheep and tapeworm in lambs are the greatest problems but use of a broad spectrum remedy keeps stock handling to a minimum.

Enzootic abortion (Chlamydiosis) is a disease that suddenly struck in South Africa in 1972, claiming a large portion of the lamb crop. Ewes abort at different stages of pregnancy and lambs that are born alive never develop properly. This disease is most probably the most important economic disease of sheep at the present moment.

Blue tongue is a well-known viral disease that is carried by gnats, mosquitoes and other insects and particular care should be taken during wet seasons. Blue tongue is seasonal and will not necessarily occur every year. Because outbreaks are unpredictable it is absolutely essential that sheep be inoculated annually, even though the vaccine is supposed to give lifelong immunity. The disease attacks the mucous membrane of the nose, mouth, trachea and intestine and causes infection of the hooves.

Other diseases of importance are pulpy kidney, a bacterial disease caused by *Clostridium ovitoxicum*, and Rift Valley Fever, a viral disease that is also carried by mosquitoes and could also affect man.

All the above-mentioned diseases are well under control and, with the necessary caution and prevention, seldom cause problems of any severity.

CONCLUSION

The sheep and wool industry in South Africa is well-established on a sound footing. Scientists and leading farmers alike have the knowledge and proficiency to ensure maximum production. Nevertheless a lot can still be done to increase the average production figures for wool for the country as a whole, especially through the application of scientific principles. Extensive sheep farming is well established, but the incorporation of intensive sheep production systems as a supplementary branch of the mixed-farming and high rainfall areas requires urgent attention.

THE BREEDING SEASON OF ICELAND SHEEP

DR Ó.R. DÝRMUNDSSON

Agricultural Society of Iceland, Baendahollin, Box 7080, Reykjavik, Iceland

INTRODUCTION

There is only one breed of sheep in Iceland, the native Iceland breed, which belongs to the North European short-tailed race of sheep. This breed has existed in the country since the settlement some 1100 years ago and is noted for hardiness, longevity, prolificacy and good milking ability (Pálsson 1965). The national flock now totals some 890,000 winterfed sheep, housed in winter and grazed on extensive natural pastures in summer. Tupping begins by mid-December and is completed by mid-January. The mean length of gestation is 143 days and lambing takes place from early May to early June, when grass starts growing. While the main emphasis is placed on fat lamb production, primarily by increasing lambing rate as well as by improving body growth rate and conformation, the production of wool and pelts is of growing importance.

While in the past a number of studies have been made of various aspects of reproductive efficiency of Icelandic sheep (Dýrmundsson & Hallgrímsson 1978), detailed information has been lacking on the onset and duration of breeding activity in this breed (Hafez 1952, Dyrmundsson 1973a and b). This paper reports briefly on a study at Hvanneyri Experimental Station (latitude 64° 34 'N, longitude 21° 40 'W) from 1972 to 1977, and on information obtained from a nationwide survey of sporadic out-of-season breeding activity in ewes.

ATTAINMENT OF PUBERTY

Icelandic lambs have relatively fast growth rates and the majority are slaughtered immediately after weaning in September at 4 months of age, when an average lamb weighs some 36 kg and has a dressed carcass of 14.5 kg. Studies on sexual development indicate that ewe lambs selected for replacement experience their first oestrus at a mean age of 213 days and a mean body weight of 39 kg. Only a few of the lambs included in the studies failed to attain puberty. Normally some 60 to 80% of all ewe lambs exposed to rams will conceive and breeding from these young females for lambing at 12 months of age has become a common practice in Iceland (Dýrmundsson 1976). Mated ewe lambs require a high level of feeding and special attention is paid to proper mating management (ewe lambs penned separately).

The results of studies into penial, scrotal and testicular development of Icelandic ram lambs are indicative of relatively early attainment of physiological puberty. It appears that this stage of sexual development is generally reached by 4 months of age but with marked variation in both puberal age and body weight (Dýrmundsson 1978a). Ram lambs kept for breeding are normally used successfully at 7 months of age for mating with both adult ewes and ewe lambs.

ONSET OF THE BREEDING SEASON

In a preliminary study in 1972-1973 it was found that the mean date of onset of the breeding season was the 30th of November for 16 adult ewes and the 10th of December for 15 ewe lambs. Pooled data for three subsequent breeding seasons are shown in Table 1. It shows that the ewe lambs had a slightly later onset of breeding activity (12 December) than adult ewes (8 December) but on the whole there did not appear to be a relationship between adult ewe body weight and age, and the onset of breeding. Efforts to establish a relationship between the onset of breeding in successive years proved inconclusive.

Table 1 Onset of breeding

Age	n	Body wt. kg	Date of onset	
			Mean	Range
7 mo	137	39.0	12 Dec.	18 Nov. - 11 Jan.
19 mo	179	54.3	6 Dec.	20 Nov. - 6 Jan.
2 yr	124	54.6	10 Dec.	25 Nov. - 29 Dec.
3 yr	78	59.9	10 Dec.	30 Nov. - 5 Jan.
4 yr	56	59.1	10 Dec.	30 Nov. - 30 Dec.
5 yr	35	61.5	7 Dec.	19 Nov. - 29 Dec.
6 yr	40	59.5	7 Dec.	21 Nov. - 28 Dec.
7 yr	7	54.1	5 Dec.	29 Nov. - 17 Dec.
8 yr	27	59.3	3 Dec.	24 Nov. - 20 Dec.
9 yr	13	60.5	2 Dec.	12 Nov. - 13 Dec.

DURATION OF THE BREEDING SEASON

As shown in Table 2 adult ewes were found to have a much longer breeding season than ewe lambs as would be expected. Furthermore the results indicate that the mid-breeding season of Icelandic ewes and ewe lambs occurs some 4 to 7 weeks after the shortest day (21 Dec.), i.e. considerably later than what has generally been found in a number of breeds with seasonal breeding patterns (Hammond Jr 1944, Hafez 1952). Whether such differences in the distribution of sexual activity over the season are related to an inherent variation in the response to photoperiodic stimuli remains open to speculation. Moreover, it would seem reasonable to suggest that such a difference was probably due to natural selection throughout the evolution of the breed in the past but further studies are required to substantiate such a hypothesis (Dýrmundsson 1978b).

Table 2 Duration of breeding activity

	First oestrus	Last oestrus	Duration of breeding season (days)	Number of heats	Mid-breeding season
Ewes	30 Nov.	25 Apr.	145. 9	8. 9	11 Feb.
(n=16)	(22 Nov.- 14 Dec.)	(3 Apr.- 23 May)	(111-171)	(6-11)	
Ewe lambs	10 Dec.	23 Feb.	74. 5	4. 5	17 Jan.
(n=15)	(26 Nov.- 31 Dec.)	(25 Dec.- 1 Apr.)	(27-116)	(2-8)	

DURATION OF THE OESTRUS CYCLE

The ewe lambs were found to have a slightly shorter mean cycle length in the normal range (14-19 days) than the ewes, and the ewe lambs experienced a relatively larger number of long cycles ($>$19 days) indicating a higher incidence of silent heats. Pooled data in Table 3 show a slight increase in cycle length (in the normal range) with higher ewe age.

Table 3 Length of oestrus cycle

Age	No. of observations	Oestrus cycle
Ewe lambs	73	16.0
19 mo	198	16.2
2 yr	103	16.2
3 yr	53	16.4
4 yr	37	16.3
5 yr	32	16.4
6 yr	30	16.6
8 yr	28	16.5
9 yr	20	16.9

DURATION OF OESTRUS

While ewe lambs were found to have considerably shorter oestrus than ewes there were marked individual differences in the duration of oestrus as shown in Table 4.

Table 4 Duration of oestrus

	n	Estimated duration of oestrus	
		Mean (hours)	Range
Ewes (19mo-9yr)	160	48, 0	24-80
Ewe lambs	12	28, 3	12-40

SPORADIC BREEDING OUT-OF-SEASON

While it is clear that Icelandic ewes have a pronounced breeding season, there is ample evidence of limited sporadic out-of-season oestrus activity resulting in odd ewes conceiving and lambing in virtually any month of the year. According to data obtained from a number of farmers throughout the country such cases appear to be relatively rare particularly during the period of presumed anoestrum, i.e. June-July with maximum daylight.

Thus lambing in early winter (October-December) occurs most rarely. Some ewes appear to breed and lamb out-of-season two or more years in succession and occasionally ewes may lamb twice within the same year. The relative frequency of such sporadic breeding activity within individual flocks is unknown and is generally not considered to be of any practical importance.

HORMONAL TREATMENT OF NONCYCLIC EWES

Both PMSG for superovulation and progestagen sponges for oestrus synchronization have been used on a number of farms within the breeding season in December (Pálsson 1962, Dýrmundsson 1977). In a small trial 6 barren noncyclic ewes, which were treated with sheep sponges for 13 days and injected with 750 i.u. PMSG at sponge withdrawal, exhibited behavioural oestrus and were mated on 12 June. Three of the ewes conceived and lambed on 28 Oct, 30 Oct and 3 Nov respectively. The ewes were mated some 7 weeks later (December) while still suckling their lambs.

SEXUAL ACTIVITY OF RAMS

Although detailed information is lacking on the reproductive efficiency of Icelandic rams the above studies indicate that they may be sexually active throughout the year. At least some of them appear to achieve fertile matings in summer. In the out-of-season breeding trial reported above all the 8 rams (1-4yr) introduced individually to oestrus ewes in June exhibited normal mating ability; in fact their libido was similar to that observed in winter. Semen characteristics were not studied.

CONCLUSIONS

Traditionally, lambing in Iceland takes place during a relatively short period of some 4 weeks in spring. While the results reported in this short paper indicate that normal cyclic activity is more extended than generally assumed, it would seem possible to obtain lambings as early in the year as April and as late as September without manipulating sexual activity with hormonal treatments or photoperiodic control.

Moreover, in summer and autumn, out-of-season breeding using a combined sponge + PMSG treatment may be feasible. Further studies would however be required to test the practical significance of the findings in relation to the present system of sheep production in the country.

REFERENCES

Dýrmundsson, O.R. (1973a) Puberty and early reproductive performance in sheep. I. Ewe lambs. Animal Breeding Abstracts 41(6),273-289.

Dýrmundsson, O.R. (1973b) Puberty and early reproductive performance in sheep. II. Ram lambs. Animal Breeding Abstracts 41(9),419-430.

Dýrmundsson, O.R. (1976) Breeding from ewe lambs: a common practice in Iceland. Res.Inst. Neori As, Hver. Icel. Bull. No.24, 12 pp.

Dýrmundsson, O.R. (1977) Synchronization of oestrus in Iceland ewes with special reference to fixed-time artificial insemination. Acta Agriculturae Scandinavica 27(3),250-252.

Dýrmundsson, O.R. (1978a) A note on sexual development of Icelandic rams. Animal Production. (In press.)

Dýrmundsson, O.R. (1978b) Studies on the breeding season of Icelandic ewes and ewe lambs. Journal of Agricultural Science, Cambridge 90,275-281.

Dýrmundsson, O.R.;Hallgrímsson, S. (1978) Reproductive efficiency of Iceland sheep. Livestock Production Science 5,231-234.

Hafez, E.S.E. (1952) Studies on the breeding season and reproduction of the ewe. Journal of Agricultural Science, Cambridge 42,189-265.

Hammond, J.,Jr. (1944) On the breeding season in the sheep. Journal of Agricultural Science, Cambridge 34,97-105.

Pálsson, H. (1962) Augmentation of fertility of Iceland ewes with pregnant-mare serum in successive years. Journal of Reproduction and Fertility 3,55-63.

Pálsson, H. (1965) Methods of increasing meat production per sheep in Iceland. World Review of Animal Production 1,(1),139-145.

GESTATIONAL FAILURE: A MAJOR PROBLEM IN REPRODUCTIVE EFFICIENCY OF THE ANGORA GOAT BREED IN SOUTH AFRICA

P.S. PRETORIUS

University of the Orange Free State, Bloemfontein, South Africa

SOUTH AFRICAN MOHAIR INDUSTRY AND ITS ECONOMIC IMPORTANCE

The Angora goat is a small slender animal weighing from 27 to 40 kg. It originated in Central Asia as a descendant of the wild goat of Persia and was first introduced to South Africa in 1838. Its most distinctive feature is the coat of white hair, 13 to 15 cm in length with unique textile properties of lustre, resilience and warmth. Shearing is done twice a year with an annual yield of 2.5 to 3.0 kg mohair per animal. Unlike most other small stock breeds which also produce meat and/or wool, mohair is the only economically important product from the Angora goat.

The total annual world production of mohair amounts to about 12.5 million kg of which South Africa produces some 3.9 million kg high quality mohair, worth some R15 million (£9.3 million) (Mohair Board 1977). South Africa is the world's largest mohair exporting country; only small quantities of the clip of other major mohair producing countries like Turkey, USA (Texas) and Russia reach the world market.

Due to several limiting factors, mainly droughts and low mohair prices, production figures varied over the years and the South African goat population decreased to less than 0.5 million by the late 1940s. Increased demands for mohair since 1950 resulted in spectacular rises in the price and attempts to expand rapidly the goat numbers met considerable difficulties. At that time some farmers reported losing up to 80% of their kid crop through abortions (Van Heerden 1963). As such losses barely provided offspring for normal flock replacement, it was realised that abortion was a severe economic problem and a serious handicap to the local mohair industry.

There is no evidence of a similar problem in other mohair producing countries, but the majority of those goats are of the larger robust type which produce smaller quantities of fairly coarse hair and rarely abort. Reliable information indicates however that in flocks consisting predominantly of the smaller South African type, selected for quality hair, the abortion rate may be as high as 40% (Van Rensburg 1971).

EXTENT OF THE ABORTION PROBLEM

Although the Angora goat has proved to be fairly fertile with a mean ovulation rate of 1.32 (Pretorius 1970), pregnancy percentages up to 90% and post-natal kid losses as low as 4% (Van der Westhuysen 1975), the occurrence of habitual abortion is one of the main problems contributing to low reproductive efficiency.

In a field survey including some 114,000 Angora goats on 168 farms, Van Heerden (1963) clearly demonstrated that gestational failure through abortions was the major factor responsible for the low reproductive rate experienced in this goat breed. The incidence of abortion varied from 5 to well over 50%. A further 9.3% of the mated females failed to kid eventually but displayed no signs of abortion (Lamont 1964). Pretorius (1970) recorded a difference of 24% in the number of foetuses found in the uteri of pregnant does and the number of corpora lutea present in the ovaries, indicating a relatively high rate of unsuccessful fertilization or embryonic mortality.

The stage of pregnancy at which abortion occurred varied from two weeks after conception to full term with a peak around 100 to 110 days of gestation (Van Rensburg 1971). Under farming conditions peaks in the occurrence of abortion are experienced following cold and wet windy weather and after shearing or dipping just prior to the actual kidding season (Van Heerden 1963).

POSSIBLE CAUSES FOR GESTATIONAL FAILURE

In a systematic approach to this problem, Van Heerden (1963) excluded infectious diseases and mineral and vitamin deficiencies as possible causes of abortion. He concluded that abortion was due to an hereditary defect of the maternal pituitary gland resulting in a lack of maintenance of the corpus luteum through luteotrophic hormone secretion. The involvement of the pituitary in this syndrome was subsequently confirmed by Pretorius (1972). A strict selection programme against abortions was adopted by Van Heerden (1964) reducing the occurrence of abortion to first time aborters only but failing to eliminate the defect over a period of nine years (Van der Westhuysen & Roelofse 1971).

In studies on the genetic background of this phenomenon, Van der Westhuysen & Wentzel (1971) concluded that the progeny of does which were known to be aborters does not display a higher abortion rate than "normal" does. This observation stressed the complicated nature of this problem.

Investigations into the pattern of gestational failure indicated that the incidence of abortion increased rapidly with age and abortions seldom occurred during the first or second gestation period. In young does the abortion rate was only 5% while it could be as high as 47% in older does. Another feature of the abortion phenomenon is that it displayed a high degree of repeatability - up to 50%. Consequently if a doe had aborted once she was prone to abort again and become an habitual aborter.

Variation in the incidence of abortion from year to year indicates that environmental factors must also be encountered in this syndrome. Van Heerden (1963) concluded that adverse weather conditions precipitates abortions in susceptible does and also probably causes the expulsion of a large number of already dead foetuses, rather than acting as the primary cause of abortion. Although the contribution of drought conditions (feeding level) to the abortion problem was at first disputed (Van Heerden 1963), subsequent research clearly demonstrated that the overall nutritional level is the largest single factor affecting abortion rate

(Van der Westhuysen & Roelofse 1971). It has also been proved that the energy content of the ration plays a much greater role than the protein content (Wentzel 1973). In addition it was observed that underfed does were also more sensitive to the stress of cold environmental conditions.

The nature of the abortion problem has also been the subject of several detrailed physiological investigations. Van Rensburg (1963, 1964, 1965) attributed gestational failure to maternal adrenal hypoplasia, an hypothesis supported by Morgenthal (1966). In a later study Van Rensburg (1971) found a relationship between the potential of mohair production and the incidence of abortion, the former being linked to adrenal function. He also reported defective steriodogenesis preceded by high maternal plasma cortisol levels. It was postulated that physiological adaptation involves adrenal hyperplasia in order to assist the transfer of maternal nutrients to the foetus and that abortion is a consequence of failure of this mechanism.

A common feature observed in all cases of abortion was that advanced luteal regression preceded expulsion of the foetus (Van Heerden 1963, Pretorius 1970, Van Rensburg 1971, Wentzel et al 1975a). These studies proved through histological examinations and plasma progesterone levels that luteal function was impaired prior to abortion. Moreover, it is known that a functionally active corpus luteum is a prerequisite for maintenance of pregnancy in the goat (Meites et al 1951, Van Rensburg 1971). It is evident however that luteal failure is not the primary cause of abortion but rather a consequence of the malfunction of other endocrine mechanisms.

ENDOCRINE BASIS OF THE ABORTION PHENOMENON

Research proved that this phenomenon has an extremely complex endocrine basis which is often complicated by interactions of the various physiological mechanisms involved during gestation.

Observations on the condition of the aborted foetuses indicated a considerable time lapse between death and the actual expulsion of the foetus, while other abortuses appeared to be "fresh" and normal (Van Heerden 1963, Van Rensburg 1971, Wentzel 1973). Wentzel (1973) held the view that the exact time of foetal death in utero must be detected to determine the actual cause of foetal death. With the aid of an ultrasonic foetal pulse detector and regular detection of foetal heart rae, Wentzel (1973) identified two completely different types of abortion, which proved subsequently to have totally different endocrine bases.

In the first type of abortion foetal heart rate remains normal virtually up to expulsion of the foetus. In some cases the hearts of some foetuses were still beating after expulsion. These abortuses had a normal "fresh" appearance. In the second type of abortion the foetus suffered variable degrees of bradycardia before it died in utero. These abortuses were expelled from the uterus after variable periods of time - up to as much as four weeks following death. An outstanding feature of this type of abortion is the advanced degree of oedema found in the foetuses. This indicates that the foetus suffers prolonged unfavourable conditions in the uterus prior to death.

Further investigations into the pathology of these two types of abortion revealed that it originates from two completely different sets of factors.

"Fresh" abortions with a normal appearance

Wentzel et al (1974) demonstrated that abortions following stress conditions, such as exerted by submaintenance energy diets, were of the type where the foetus is expelled in the "fresh" and normal state. Foetal heart rate remained normal virtually up to time of expulsion. This finding substantiated Van Heerden's (1963) observation of "abortion storms" following unfavourable environmental conditions exerting stress on pregnant does.

Investigations into the possible endocrine pathways through which these stress factors mediated their effect revealed a remarkable increase in maternal plasma oestrogen level soon after reducing the plane of energy nutrition (Wentzel et al 1975), the primary trigger being a drop in maternal blood glucose concentration (Wentzel et al 1976), probably mediated through the foetal - hypothalamic - pituitary-adrenal-axis (Mellor & Slater 1972). This observation indicated that sub-energy nutrition and probably also other stress factors affects placental steroidogenesis and the elevated maternal steroid levels are intimately related to the abortion phenomenon. The previously mentioned impaired luteal function is probably mediated through the excessive maternal oestrogen levels resulting in premature regression of the corpus luteum (Akbar et al 1971). This eventually led to expulsion of the foetus.

Wentzel & Viljoen (1975) subsequently simulated the "fresh" normal type of abortion through the administration of exogenous oestrogens to pregnant does. Advanced luteal regression was recorded in all does which had aborted. They concluded that excessive oestrogens probably stimulate the mechanism responsible for luteolysis rather than acting directly on the corpus luteum itself. The luteolytic action of excessive circulating oestrogens is most probably mediated via the anterior pituitary resulting in lowered luteinising hormone activity through its negative-feedback action (Pretorius 1972). Furthermore, LH is proved to be luteotrophic in the goat (Cowie & Tindall 1960) and interference with this luteotrophic action results in luteal regression. Uterine prostaglandins may also be involved in this luteolytic action but this remains to be proved in the goat.

It was also reported by Wentzel et al (1975b) that stress conditions which initiated the "fresh" normal abortions did not affect maternal corticoid levels. This observation suggested that maternal adrenal function played a minor role, if any, in this type of abortion.

Oedematous abortions

In the second type of abortion, where the foetus had died in an oedematous state in utero, Van Rensburg (1971) and Wentzel et al (1975b) recorded abnormally high maternal blood corticosteroid levels and extensive adrenal hyperplasia. This results in an unfavourable environment for the foetus in utero and eventually foetal death. A direct relationship between adrenal corticosteroid level and degree of oedema of the abortus was observed which indicates that this condition is mediated through the electrolyte and water balance of the body.

The viability of the new-born kid seems to be also related to the condition of hyperfunction of the maternal adrenals. Wentzel et al (1975b) demonstrated that does whose kids had died perinatally also displayed higher corticosteroid levels during the last third of pregnancy. This substantiated the views of Van Rensburg (1971) that perinatal mortality is intimately related to the abortion syndrome. The period of hyperadrenocorticocism was probably too short to cause foetal death.

Wentzel & Roelofse (1975) simulated this type of abortion through the administration of excessive cortisone to pregnant does resulting in abortions in 78% of the treated does. The abortuses displayed without exception variable degrees of oedema corresponding to those found under practical conditions.

Sub-maintenance nutrition during gestation failed to effect notable alterations in the plasma corticoid levels, indicating that stress factors are most likely not the trigger mechanism for this type of abortion (Wentzel & Roelofse 1975).

APPLICATION OF RESEARCH FINDINGS

In retrospect, it is evident that the abortion syndrome in the Angora goat breed originates from two completely different pathological conditions. In the type where the adrenal is involved the primary cause for gestational termination results from a condition of maternal hyperadrenocorticocism which may have an hereditary basis. Limiting this type of abortion under farming conditions will obviously be difficult. Selection against all aborters should be practised and the internal generation should be kept as low as possible. The older animals which are more inclined to suffer from this type of abortion are then eliminated from the flock.

The type of abortion which is induced by stress factors, such as nutritional conditions, can most probably be eliminated or at least limited by energy supplementation during periods of unfavourable weather conditions.

The importance of the implementation of sound breeding and managerial practices to attain the highest possible breeding percentage cannot be over emphasized.

REFERENCES

Akbar, A.M.;Rowe, K.E.;Stormshak, F. (1971) Estradiol induced luteal regression in unilaterally hysterectomized and luteinizing hormone-treated ewes, Journal of Animal Science 33,426.

Cowie, A.T.;Tindal, J.S. (1960) Effects of hypophysectomy of the pregnant and lactating goat. Acta endocrinologica Copenhagen 35,679.

Lamont, J.C. (1964) Influence of vasectomised bucks on the reproductive performance of Angora does. South African Journal of Agricultural Science 7,305.

Meites, J.;Webster, H.D.;Young, F.W.;Thorp, F. Jr.;Hatch, R.N. (1951) Effects of corpora lutea removal and replacement with progesterone on pregnancy in goats. Journal of Animal Science 10,411.

Mellor, D.J.;Slater, J.S. (1972) Daily changes in foetal urine and relationships with amniotic and allantoic fluid and maternal plasma during the last two months of pregnancy in conscious, unstressed ewes with chronically implanted catheters. Journal of Physiology 227,503.

Mohair Board (1977) Mohair board's budget 1977/78. Angora goat & Mohair Journal 19,43.

Morgenthal, J.C. (1966) The haematology of the Angora goat with special reference to the habitual aborter. Onderstepoort Journal of Veterinary Research 33,363-377.

Pretorius, P.S. (1970) Gonadotrophic hormone activity of the anterior pituitary and related phenomena in the female Angora goat during various reproductive phases. D.Sc.Agric.thesis, Univ.Pretoria.

Pretorius, P.S. (1972) Gonadotrophic hormone activity in female Angora goats exhibiting normal and aberrant reproductive activity. Journal of the South African Veterinary Medical Association 43,35.

Van der Westhuysen, J.M. (1975) Present-day knowledge of the abortion problem. Angora goat & Mohair Journal 17,29.

Van der Westhuysen, J.M.;Roelofse, C.S. (1971) Effect of shelter and different levels of dietary energy and protein on reproductive performance in Angora goats with special reference to the habitual aborter. Agroanimalia 3,129.

Van der Westhuysen, J.M.;Wentzel, D. (1971) Progress through selection against the aborting Angora goat. South African Journal of Animal Science 1,101-102.

Van Heerden, K.M. (1963) Investigations into the cause of abortions in Angora goats in South Africa. Onderstepoort Journal of Veterinary Research 30,23-83.

Van Heerden, K.M. (1964) The effect of culling aborting ewes on the abortion rate in Angora ewes. Journal of the South African Veterinary Medical Association 35,19-20.

Van Rensburg, S.J. (1963) Endocrinological aspects of habitually aborting Angora goat ewes. South African Medical Journal 37,1114.

Van Rensburg, S.J. (1964) Possible endocrine mechanisms in gestational failure. Proceedings, Vth International Congress on Animal Reproduction Trento, 3,375-378.

Van Rensburg, S.J. (1965) Adrenal function and fertility. Journal of the South African Veterinary Medical Association 36,491-500.

Van Rensburg, S.J. (1971) Reproductive physiology and endocrinology of normal and habitually aborting Angora goats. Onderstepoort Journal of Veterinary Research 38,1-62.

Wentzel, D. (1973) The habitual aborting Angora doe: Recognition and endocrinology of two distinct types of abortion. PhD. (Agric.) thesis, Univ. Stellenbosch.

Wentzel, D.;Morgenthal, J.C.;Van Niekerk, C.H.;Roelofse, C.S. (1974) The habitually aborting Angora doe: II. The effect of an energy deficiency of the incidence of abortion. Agroanimalia 6,129-131.

Wentzel, D.;Morgenthal, J.C.;Van Niekerk, C.H. (1975(a)) The habitually aborting Angora doe: III Luteal function in normal and aborter does. Agroanimalia 7,15-22.

Wentzel, D.;Morgenthal, J.C.;Van Niekerk, C.H. (1975(b)) The habitually aborting Angora doe. IV Adrenal function in normal and aborter does. Agroanimalia 7,27-34.

Wentzel, D.;Morgenthal, J.C.;Van Niekerk, C.H. (1975(c)) The habitually aborting Angora doe. V Plasma oestrogen concentration in normal and aborter does with special reference to the effect of an energy deficiency Agroanimalia 7,35-40.

Wentzel, D.;Viljoen, K.S. (1975) The habitually aborting Angora doe. VI Induction of abortion by administration of exogenous oestrogens. Agroanimalia 7,41-43.

Wentzel, D.;Roelofse, C.S. (1975) The habitually aborting Angora doe. VII Induction of abortion by administration of cortisone acetate. Agroanimalia 7,45-47.

Wentzel, D.;Le Roux, Marita, M.;Botha, L.J.J. (1976) Effect of the level of nutrition on blood glucose concentration and reproductive performance of pregnant Angora goats. Agroanimalia 8,59-61.